Handbook of Atmospheric Science

Handbook of Atmospheric Science

Editor: Ronin Massey

R CALLISTO
REFERENCE
www.callistoreference.com

Callisto Reference,
118-35 Queens Blvd., Suite 400,
Forest Hills, NY 11375, USA

Visit us on the World Wide Web at:
www.callistoreference.com

ISBN: 978-1-64116-174-9 (Hardback)

Trademark Notice: Registered trademark of products or corporate names are used only for explanation and identification without intent to infringe.

Cataloging-in-Publication Data

Handbook of atmospheric science / edited by Ronin Massey.
    p. cm.
Includes bibliographical references and index.
ISBN 978-1-64116-174-9
1. Atmospheric physics. 2. Atmosphere. I. Massey, Ronin.
QC861.3 .H36 2019
551.5--dc23

Table of Contents

Preface

The study of the Earth's atmosphere, its processes and the effects of other systems on its atmosphere and vice versa is known as atmospheric science. This field has three significant domains- meteorology, aeronomy and climatology. Atmospheric science uses lasers, radiosondes, rocketsondes, satellites and weather balloons for different studies. The discipline of atmospheric science can be divided into three categories- atmospheric chemistry, atmospheric dynamics and atmospheric physics. Atmospheric chemistry studies the chemistry of the Earth's atmosphere. Some of the issues dealt in this domain include acid rain, global warming and photochemical smog. Atmospheric dynamics studies diverse phenomena such as tornadoes, tropical cyclones, jet streams, thunderstorms, etc. Atmospheric physics strives to model the atmosphere using wave propagation models, statistical mechanics, cloud physics, etc. This book provides significant information of this discipline to help develop a good understanding of atmospheric science and related fields. It aims to shed light on some of the unexplored aspects and the recent researches in this field. Scientists and students actively engaged in this field will find atmospheric science full of crucial and unexplored concepts.

All of the data presented henceforth, was collaborated in the wake of recent advancements in the field. The aim of this book is to present the diversified developments from across the globe in a comprehensible manner. The opinions expressed in each chapter belong solely to the contributing authors. Their interpretations of the topics are the integral part of this book, which I have carefully compiled for a better understanding of the readers.

At the end, I would like to thank all those who dedicated their time and efforts for the successful completion of this book. I also wish to convey my gratitude towards my friends and family who supported me at every step.

Editor

Assimilating GPM hydrometeor retrievals in HWRF: choice of observation operators

Ting-Chi Wu* and Milija Zupanski

Cooperative Institute for Research in the Atmosphere, Colorado State University, Fort Collins, CO, USA

*Correspondence to:
T.-C. Wu, Cooperative Institute for Research in the Atmosphere, Colorado State University, 1375 Campus Delivery, Fort Collins, CO 80523, USA.
E-mail: ting-chi.wu@colostate.edu

Abstract

A recent study investigated the capability to assimilate hydrometeor retrievals in the National Oceanic and Atmospheric Administration (NOAA) Hurricane Weather Research and Forecasting (HWRF) system. Hydrometeor retrievals were obtained from a hurricane-specific retrieval that utilizes data from the Global Measurement Mission Microwave Imager. Data assimilation system used in HWRF is the hybrid ensemble-variational Gridpoint Statistical Interpolation (GSI). As a first attempt to assimilate hydrometeor retrievals in HWRF, observation operators for solid and liquid integrated water content were developed using the GSI standard control variables, such as temperature, pressure, and specific humidity, under the assumption that all water vapor in excess of saturation is condensed out. To improve the usefulness of assimilating hydrometeor retrievals in HWRF, this study extends this previous work by introducing new observation operators that (1) directly use hydrometeor species estimated from HWRF microphysics and (2) include total cloud condensate as a control variable. Hurricane Gonzalo (2014) was used to examine the performance of the old and new observation operators on HWRF analysis and subsequent forecasts with three pairs of experiments. Results suggest that when new observation operators are used the analysis is an improved fit to observations and realistic adjustments to control variables are evident as seen in analysis increment. In addition, the forecasts also indicate some improvements in hurricane intensity.

Keywords: data assimilation; hydrometeor retrievals; HWRF; observation operators; cloud microphysics

I. Introduction

Hurricane intensity and structure are fundamentally related to clouds. Among all the available observations, the use of all-sky satellite radiances and satellite-retrieved hydrometeors (e.g. Boukabara *et al.*, 2013; Kummerow *et al.*, 2015) are especially important for improving hurricane forecasting, because they contain valuable information about cloud and precipitation that often occur in sensitive regions in terms of forecast impact (Bauer *et al.*, 2011). While techniques to assimilate these observations (Bauer *et al.*, 2010; Zhu *et al.*, 2016) are currently used on a global scale (Global Forecast System; GFS), implementation on a regional scale is yet to be done. In this study, the National Oceanic and Atmospheric Administration (NOAA) operational Hurricane Weather Research and Forecasting (HWRF) system (Tallapragada *et al.*, 2015) is used as a representation of such regional scale model that currently assimilates neither all-sky satellite radiances nor hydrometeor retrievals.

As a preliminary attempt toward assimilating hydrometeor retrievals in HWRF, techniques were developed. In Wu *et al.* (2016), two observation operators to assimilate integrated solid–water content (SWC) and liquid–water content (LWC) are implemented in the hybrid Gridpoint Statistical Interpolation (GSI; Wu

et al., 2002; Kleist *et al.*, 2009; Wang, 2010). The integrated SWC and integrated LWC were obtained from a hurricane-specific microwave retrievals that utilizes data from the Global Measurement Mission (GPM) Microwave Imager (GMI), referred to as Hurricane Goddard PROFiling algorithm (GPROF) (Brown *et al.*, 2016). These two observation operators were built based on the assumption that super-saturated water vapor will condense out; the one for integrated SWC, referred to as $h_{s\_noHydro}$, is,

$$h_{s\_noHydro} = \sum_{k=k_0}^{k_{max}} \left[\left(\frac{q^k}{1-q^k} \right) - 0.622 \frac{e_s(T^k)}{P^k - e_s(T^k)} \right] \cdot \frac{\Delta P^k}{g} \quad (1)$$

and the one for integrated LWC, referred to $h_{l\_noHydro}$, is,

$$h_{l\_noHydro} = \sum_{k=1}^{k_{mix}} \left[\left(\frac{q^k}{1-q^k} \right) - 0.622 \frac{e_s(T^k)}{P^k - e_s(T^k)} \right] \cdot \frac{\Delta P^k}{g} \quad (2)$$

where T is temperature, P is pressure, q is specific humidity, e_s is saturation vapor pressure, the superscript k denotes the model level index, k_0 is the vertical level where temperature is $T_0 = 273.15$ K, k_{mix} is the vertical level where temperature is $T_{mix} = 253.15$ K, k_{max} is the index for the top model level, ΔP^k is pressure difference between two vertical levels k and $k+1$ (i.e.

$\Delta P^k = P(k) - P(k+1))$, and g is the acceleration due to gravity.

Due to the super-saturation assumption that only depends on temperature, specific humidity, and temperature, there was no need to include cloud condensate as an additional control variable. However, as pointed out by Wu et al. (2016), using $h_{s\_noHydro}$ and $h_{l\_noHydro}$ may potentially create a negative bias of the guess due to the use of already saturated-then-condensed water vapor. Because of the potential negative bias and the lack of cloud condensate updates, new techniques have been developed that include new observation operators, which directly use cloud microphysical variables. This is done by including total cloud condensate mass (CWM) as control variable and adding individual hydrometeor species (cloud water, rain, ice, snow, graupel, and hail) as state variables; a technique that follows the GSI all-sky implementation developed by Zhu et al. (2016).

2. Methodology

2.1. Preparation of background hydrometeor species

The HWRF v3.7a release (Tallapragada et al., 2015) is employed in this study. Among the various HWRF physics packages, cloud microphysics scheme is most directly related to the assimilation of hydrometeor retrievals because microphysics explicitly handles the behavior of hydrometeor species. The Ferrier–Aligo microphysics scheme (Aligo et al., 2014) employed by HWRF predicts changes in water vapor mixing ratio (q_v) and CWM, which is the combined sum of individual hydrometeor species that include cloud water (q_l), rain (q_r), cloud ice (q_i), snow (q_s), graupel (q_g), and hail (q_h). Since individual hydrometeor species are not prognostic variables, they are diagnosed from CWM with the use of partition parameters that include fraction of ice (F_ICE), fraction of rain (F_RAIN), and values of riming rate (F_RIMEF).

2.2. CWM partition

There exist one formula in GSI that can be used to partition CWM into individual hydrometeor species (q_l, q_i, q_r, q_s, q_g, q_h) with the use of F_ICE, F_RAIN, and F_RIMEF. This formula, referred to as P6, is part of a routine called *cloud_efr_mod.f90*. This routine takes CWM from Ferrier–Aligo microphysics and provides mass mixing ratios and particle sizes of individual hydrometeor species as input to the Community Radiative Transfer Model (CRTM; Han et al., 2006), which is used by GSI to compute top-of-the-atmosphere radiances for a given HWRF background.

The first step of P6 formula uses F_ICE to determine the solid and liquid phases of CWM, that is,

the solid phase of CWM = F_ICE · CWM   (3)

and

the liquid phase of CWM = (1 − F_ICE) · CWM   (4)

Then, F_RAIN is used to partition the liquid phase of CWM into q_l and q_r as

$$q_l = (1 - \text{F\_RAIN}) \cdot (1 - \text{F\_ICE}) \cdot \text{CWM} \quad (5)$$

and

$$q_r = \text{F\_RAIN} \cdot (1 - \text{F\_ICE}) \cdot \text{CWM} \quad (6)$$

Similarly, the solid phase of CWM is decomposed into q_i and *precip_ice* as

$$q_i = w \cdot \text{F\_ICE} \cdot \text{CWM} \quad (7)$$

and

$$precip\_ice = (1 - w) \cdot \text{F\_ICE} \cdot \text{CWM} \quad (8)$$

where *precip_ice* is precipitating ice as opposed to q_i being the nonprecipitating ice, using an empirical weighting coefficient w

$$w = \begin{cases} 0.05 \cdot \frac{T-T_2}{T_1-T_2} + 0.1 \cdot \frac{T-T_1}{T_2-T_1} & \text{if } T \leq T_1 \\ 0.05 & \text{if } T > T_1 \end{cases} \quad (9)$$

where T is temperature, $T_1 = 243.15$ K, and $T_2 = 233.15$ K.

Finally, depending on the values of F_RIMEF, *precip_ice* will be equal to either q_s, q_g, or q_h as follows

$$precip\_ice = \begin{cases} q_s & \text{if } 1 \leq \text{F\_RIMEF} \leq 5 \\ q_g & \text{if } 5 < \text{F\_RIMEF} \leq 20 \\ q_h & \text{if } \text{F\_RIMEF} > 20 \end{cases} \quad (10)$$

After the partition, individual hydrometeor species (q_l, q_i, q_r, q_s, q_g, and q_h) are used by the new observation operators, which are discussed next.

2.3. New observation operators

The new observation operator for integrated SWC, referred to as h_s, is defined as a vertical integration of the solid hydrometeor species including cloud ice, snow, graupel, and hail, that is,

$$h_s = \sum_{k=k_0}^{k_{max}} \left(q_i^k + q_s^k + q_g^k + q_h^k \right) \cdot \frac{\Delta P^k}{g} \quad (11)$$

Similarly, the new observation operator for integrated LWC, referred to as h_l, is defined as a vertical integration of the liquid hydrometeor species including cloud water and rain, that is,

$$h_l = \sum_{k=1}^{k_{mix}} \left(q_l^k + q_r^k \right) \cdot \frac{\Delta P^k}{g} \quad (12)$$

A horizontal smoothing procedure is carried out prior to the vertical integration in Equations (11) and (12). This horizontal smoothing is done by applying a formulation of recursive filter (Hayden and Purser, 1995) to the individual hydrometeor species. Such smoothing procedure was necessary because values of integrated SWC and integrated LWC computed by h_s and h_l from a given HWRF background were found much larger compared to observations during the preliminary development. As a result, large values of innovation were created and observations were rejected. The horizontal smoothing was used to reduce large innovations by smoothing individual hydrometeor fields. Reasons causing the large innovation values are left for future investigation, because they may be related to the cloud microphysics scheme and the P6 formula. In addition, the spatial resolution of Hurricane GPROF retrievals (5–10 km) is incomparable to the HWRF grid that has a grid spacing of 2 km. Due to the different resolutions, fine features that are associated with clouds and precipitations, which may be inferred from an HWRF background, may not be detected in the retrievals due to a coarse footprint size.

3. Experiments

Hurricane Gonzalo (2014) is used in this study. In order to have a large portion of Gonzalo covered by Hurricane GPROF data, only three analysis times are selected and they are (1) 0600 UTC 13 October, (2) 1200 UTC 16 October, and (3) 0600 UTC 17 October. Examples of integrated SWC and integrated LWC valid at these three times in the innermost domain of HWRF are displayed in Figure 1. In Figures 1(a) and (b), a less organized precipitation structure with no indication of the center of Gonzalo is seen in analysis time (1). In contrast, a well-defined spiral feature in Figures 1(c)–(f) is evident in both analysis times (2) and (3), which is indicative a mature hurricane.

3.1. Prepare for assimilation

Currently, HWRF only assimilates radiances under clear-sky condition. Therefore, there is no preparation of microphysical variables for a background field. Prior to data assimilation, vortex initialization is carried out to improve the HWRF hurricane vortex through a series of adjustments. After vortex initialization, a background field that contains an improved vortex is then provided and used by GSI. However, currently, adjustment of cloud microphysical variables was not considered by vortex initialization, hence, values of CWM and partition parameters are set to zero. In order to include cloud condensate variables in HWRF assimilation, techniques are developed in HWRF, which can be summarized in three procedures: (1) reinitialize CWM and F_ICE, F_RAIN, and F_RIMEF in the background prior to data assimilation, (2) add CWM as control variable and add individual hydrometeor species as state variables, and (3) include tangent linear and adjoint parts of P6 formula (Equations (3)–(10)) in the minimization of the cost function.

Specifically, in (1), a possible solution to avoid the zero-cloud-condensate situation was proposed. That is, values of CWM, F_ICE, F_RAIN, and F_RIMEF in the 6-h HWRF forecast from a previous HWRF cycle are mapped onto a background (same grid spacing) provided by vortex initialization. Since the center of a hurricane in the 6-h HWRF forecast may be different from the center of the observed one, three-dimensional fields of CWM and partition parameters from the 6-h HWRF forecast are horizontally shifted to the true center before mapping to the background. In (2), the cross-variable correlations between CWM and other variables are described by the hybrid background error covariance, in which 20% comes from a static component embedded in GSI (Wu et al., 2002) and 80% comes from an ensemble component. For the static component that is based on the National Meteorological Center (NMC), the former name of National Centers for Environmental Prediction (NCEP) method (Parrish and Derber, 1992), the same correlations for q_v are used to describe the correlations for CWM. For the ensemble component, the 80-member GFS ensemble forecasts are used. Since Ferrier–Aligo microphysics scheme is employed by both GFS and HWRF, CWM is also a prognostic variable to GFS. By adding CWM as an additional control variable, correlations between CWM and other control variables estimated by both static and ensemble components will be included in the hybrid background error covariance. Consequently, the resulting analysis will contain a consistent update of CWM and other control variables. This is important as pointed out by Huang (1996) that a consistent treatment of cloud and other control variables can avoid imbalanced initial condition. In (3), by including tangent linear and adjoint parts of Equations (3)–(6) in the cost function minimization, conversions between individual hydrometeor state variables and CWM control variable are carried out during each iteration.

As mentioned above, the 80-member GFS ensemble is used by HWRF. There exists a potential undesirable impact on the HWRF analysis and forecast when non-native ensemble is used (e.g. vortex spindown as discussed by Pu et al., 2016). Due to the different scales and architectures of the two models, background error covariance estimated by the GFS ensemble may be unable to accurately describe the cross-variable correlations of HWRF.

3.2. Experimental design

Two experiments are conducted that use two different pairs of observation operators and they are (1) CTL, which uses the 2015 HWRF operational configuration (Tallapragada et al., 2015) and assimilates integrated SWC and integrated LWC retrievals using $h_{s\_noHydro}$ and $h_{l\_noHydro}$ while conventional observations are also assimilated in the innermost domain of HWRF, and (2) USEP6, the same as CTL, except that h_s and h_l are used instead. Both CTL and USEP6 experiments are conducted for Hurricane Gonzalo (2014) at the three analysis times.

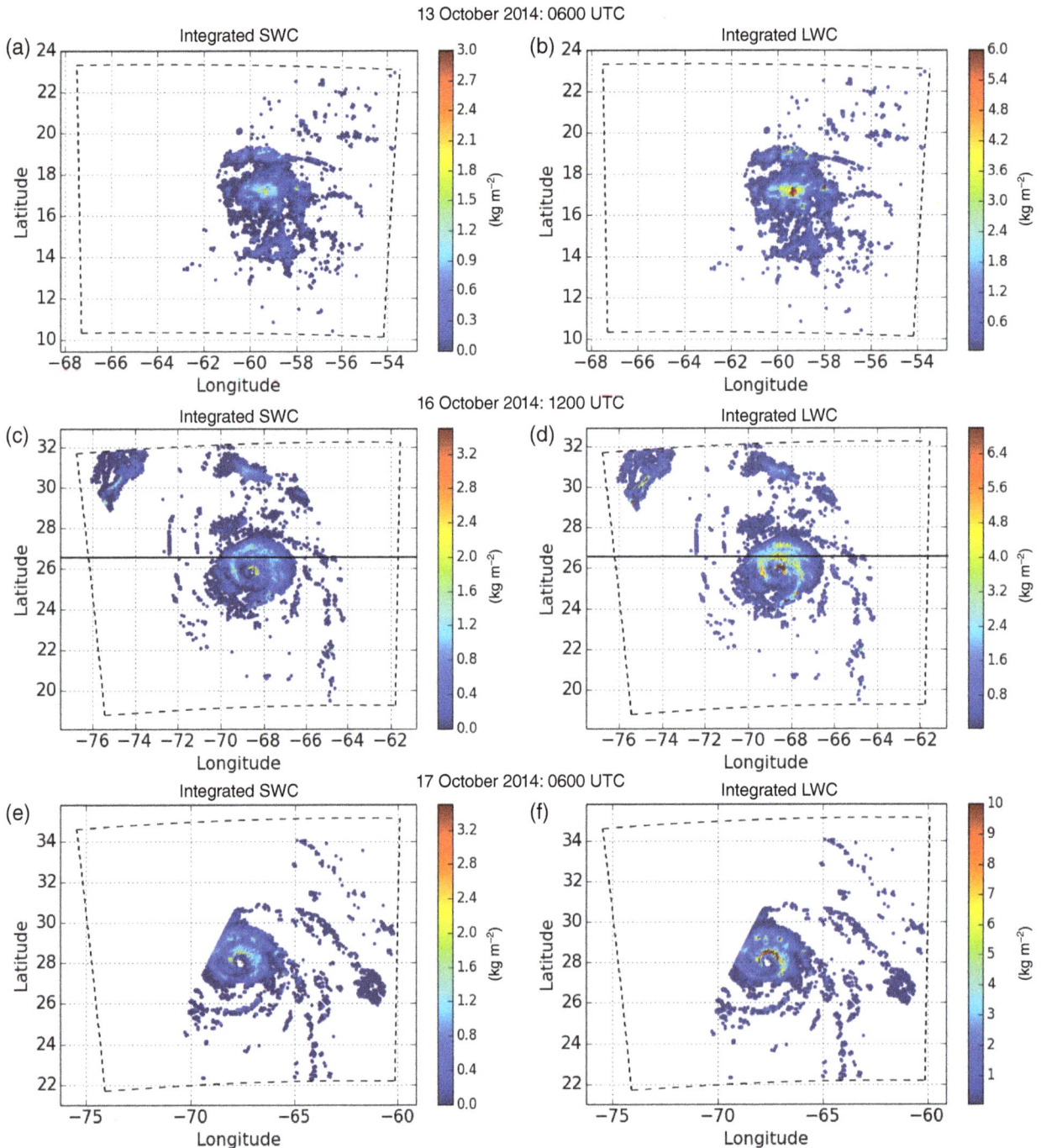

Figure 1. Hurricane GPROF retrieved (a) integrated SWC (kg m^{-2}) and (b) integrated LWC (kg m^{-2}) at 0600 UTC 13 October during Hurricane Gonzalo (2014). (c), (d) and (e), (f) are the same as (a) and (b), except that they are valid at 1200 UTC 16 October and 0600 UTC 17 October. A horizontal black line along 26.5°N in (c) and (d) indicates the latitude at which the cross-sections shown in Figure 3 were taken.

4. Results

4.1. Observed versus simulated

A scatterplot is produced to summarize the assimilation statistics from the three pairs of CTL (dots) and USEP6 (triangles) experiments. In Figure 2, statistics are presented in terms of the absolute values of background innovation (observation minus background), |O − B|, plotted on the y-axis and the absolute values of analysis innovation (observation minus analysis), |O − A|, plotted on the x-axis. A diagonal solid line will be referred to as the 1-1 line.

In general, there are more data points from USEP6 located above the 1-1 line. In Figure 2, 42% (45%) of the data points of integrated SWC (integrated LWC) from the CTL experiment (circles) are above the 1-1 line, while 55% (77%) of the data points of integrated SWC (integrated LWC) from the USEP6 experiment are above the 1-1 line. One may also notice that there exhibits a wider spread of USEP6 data points, which

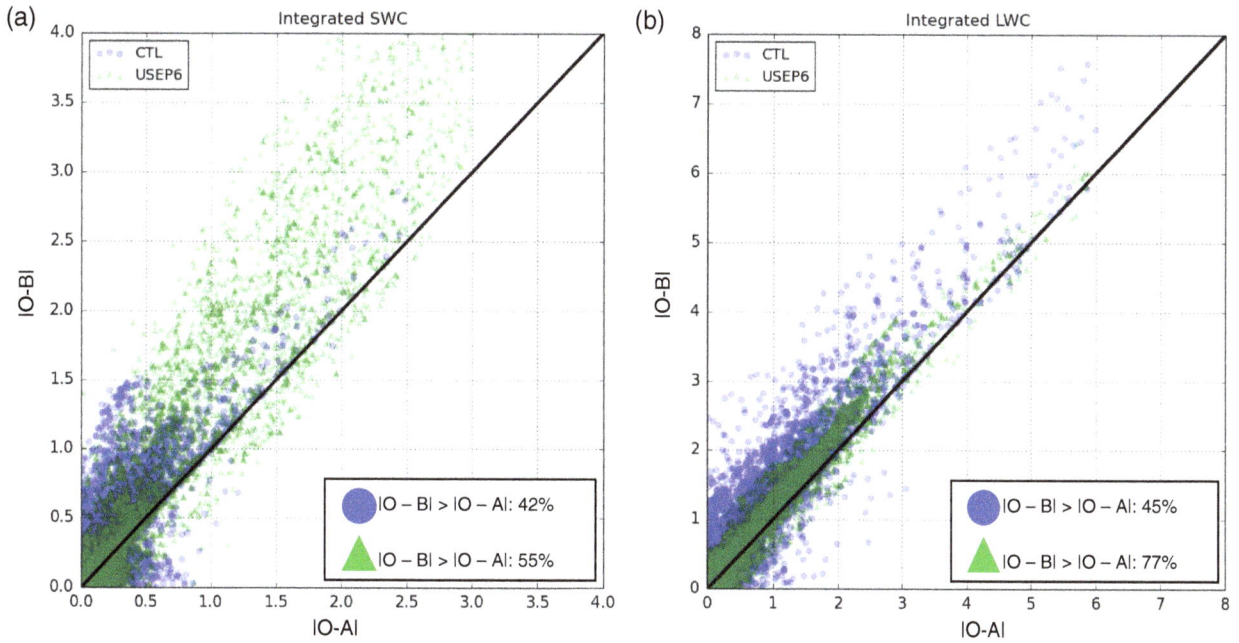

Figure 2. Scatterplot of absolute values of observation minus analysis ($|O-A|$) versus absolute values of observation minus background ($|O-B|$) from the three pairs of CTL (blue dots) and USEP6 (green triangle) experiments.

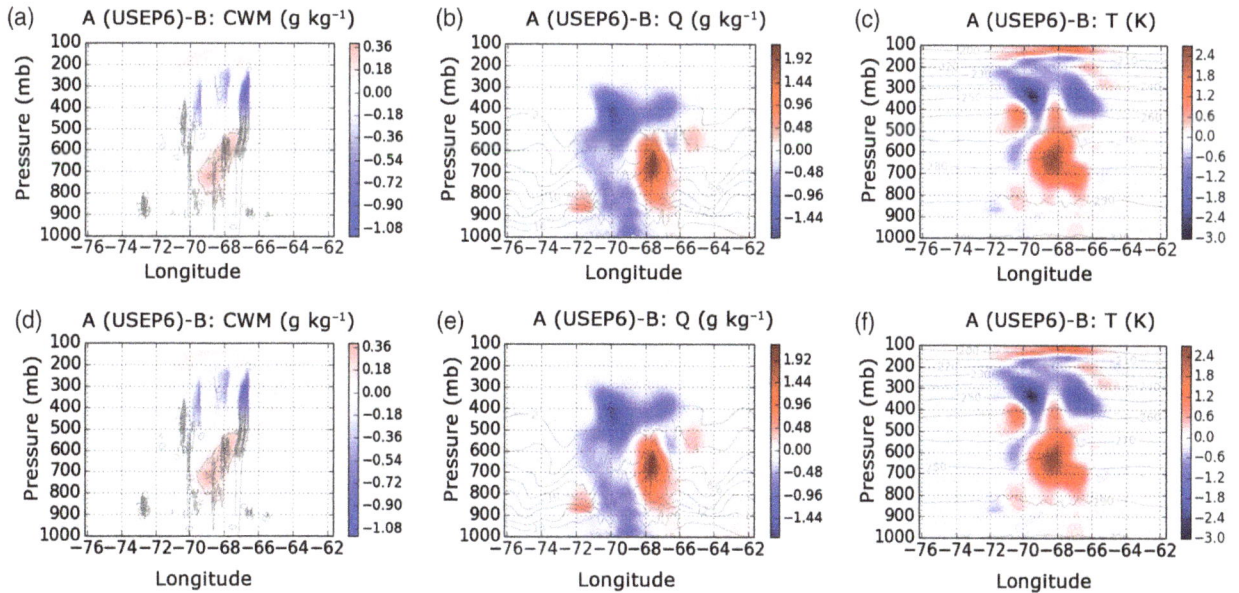

Figure 3. East–west cross-section of analysis increments (color) overlapped with background field (contour) from the USEP6 experiment: (a) CWM (g kg^{-1}), (b) specific humidity (g kg^{-1}), and (c) temperature (K). (d)–(f) are the same as (a)–(c), except for analysis increments from the CTL experiment.

suggest that there are many larger values of background innovation of integrated SWC produced when using h_s. In contrast, a narrower spread of USEP6 data points are seen in Figure 2(b) when using h_l is used. Nevertheless, USEP6 analysis was an improvement of fit to observations over the background (i.e. $|O-A| < |O-B|$) when compared to that of CTL, which is encouraging.

4.2. Analysis increments

A vertical cross-section of analysis increments (analysis minus background) for CWM, specific humidity, and

temperature is produced. In Figure 3, analysis increments from both CTL and USEP6 experiments valid at 1200 UTC 16 October is displayed along 26.5°N (Figures 1(c) and (d)).

Nonzero CWM increments in the USEP6 experiments are evident in the region where observations are located (Figure 3(a)), while no CWM increments are displayed in CTL experiment (Figure 3(d)), as expected. Focusing on the USEP6 experiment, the increase of condensate between 600 and 900 hPa may be related to the liquid component of CWM, while the reduction between 200 and 500 hPa is likely due to the solid

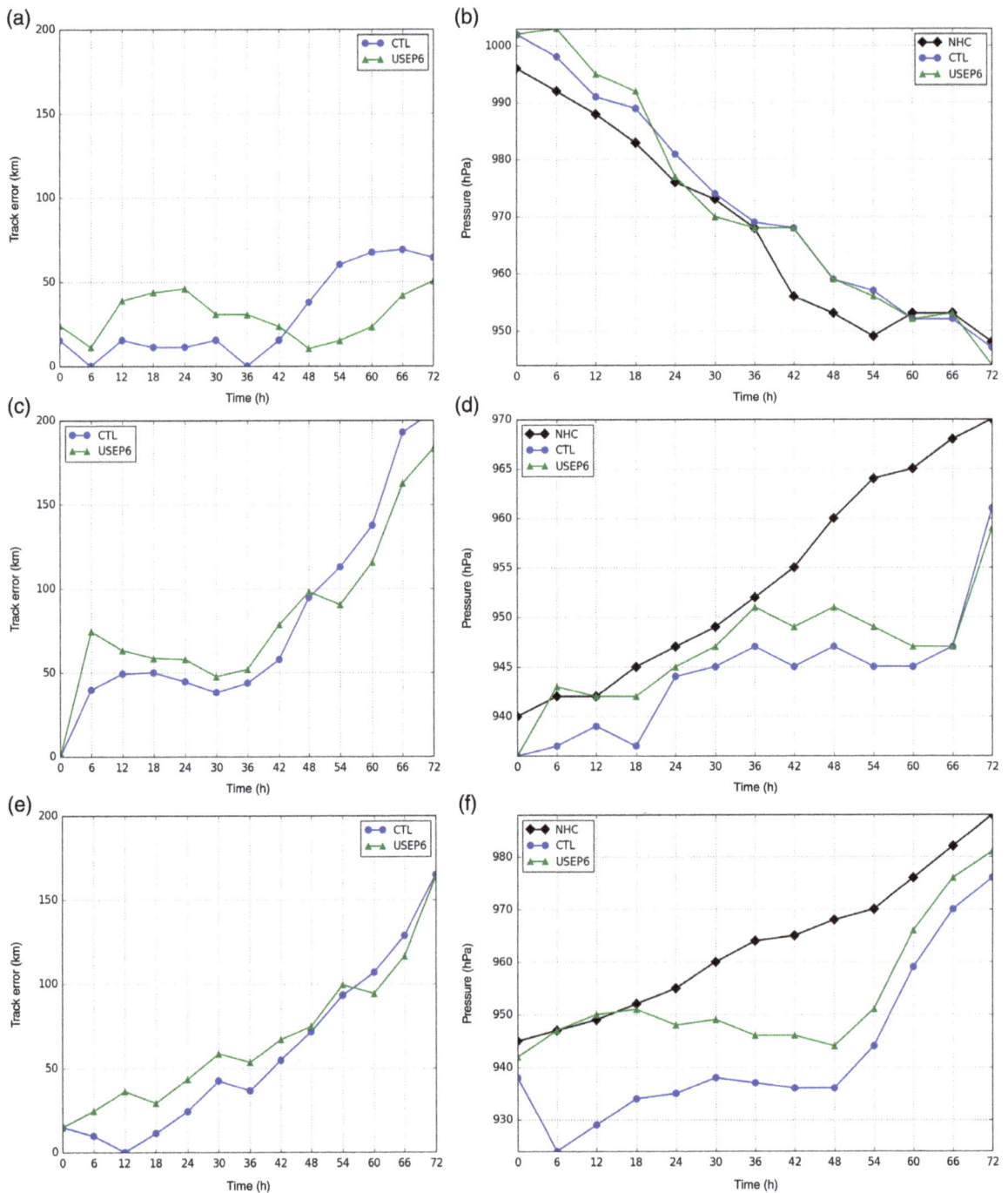

Figure 4. (a) Track error (km) and (b) MSLP (hPa) from the 72 h HWRF forecasts initialized using the CTL (blue) and USEP6 (green) analyses valid at 0600 UTC 13 October during Hurricane Gonzalo (2014). (c)–(d) and (e)–(f) are the same as (a)–(b), except for forecasts initialized at 1200 UTC 16 October and 0600 UTC 17 October, respectively. Black solid lines represent the corresponding estimates from NHC best-track data.

component (Figure 3(a)). These responses in CWM appear to coincide with the corresponding increments in specific humidity and temperature where warming and moistening is evident in lower troposphere and cooling and drying is more pronounced in upper troposphere (Figures 3(b) and (c)). On the other hand, the adjustments of specific humidity and temperature in the CTL experiment (Figures 3(e) and (f)) show a tendency to reach super-saturation by moistening and cooling the air column. Such adjustments are considerably different

from the USEP6 experiment with additional constraints from CWM.

4.3. Hurricane track and intensity forecasts

Three pairs of 72-h HWRF forecasts initialized with CTL and USEP6 analyses are conducted. Forecast track error and intensity (minimum sea-level pressure; MSLP) from CTL and USEP6 are compared with best-track data from National Hurricane Center (NHC). Results suggest that USEP6 forecasts have slightly larger track errors during the first 48 h, but the errors become smaller compared to CTL forecasts in the last 24 h (Figures 4(a), (c) and (e)). In general, the differences between the track errors from CTL and USEP6 forecasts are small. The intensity forecast as measured by MSLP from USEP6 is found closer to the NHC best-track values than CTL for the two forecasts initialized at 1200 UTC 16 October and 0600 UTC 17 October (Figures 4(d) and (f)). There is no obvious difference between the intensity forecasts initialized by CTL and USEP6 analyses valid at 0600 UTC 13 October when Gonzalo was a tropical storm (Figure 4(b)).

5. Summary and discussion

New observation operators are developed to extend upon the work by Wu *et al.* (2016). Due to the limitations that include a potential negative bias and the lack of cloud condensate updates from using observation operators based on super-saturation ($h_{s\_noHydro}$ and $h_{l\_noHydro}$), new observation operators are developed by directly using hydrometeor species estimated from cloud microphysical variables. Since the Ferrier–Aligo microphysics scheme predicts CWM rather than the individual hydrometeors, an empirical formula embedded in GSI is used to partition CWM into individual hydrometeor species. Then, the new observation operators, h_s and h_l, are defined as a vertical integration of solid and liquid hydrometeor species, respectively.

Hurricane Gonzalo (2014) is selected to perform three pairs of CTL and USEP6 experiments to examine the impact of assimilating retrieved integrated SWC and integrated LWC using $h_{s\_noHydro}$ and $h_{l\_noHydro}$ (CTL) and h_s and h_l (USEP6) on HWRF analysis and subsequent forecast. For USEP6 experiments, encouraging results are obtained, in which the analysis was an improvement of fit to observations over the background, when compared to CTL. Including CWM as a control variable is found to have a clear impact on other variables, such as specific humidity and temperature. In general, USEP6 experiment appears to result in similar forecast track to that of the CTL experiment. Nevertheless, the MSLP forecast from the USEP6 experiment shows some improvement over the CTL forecast in two of the three cases.

As stated earlier, cloud microphysics scheme is essential to the development of the new observation operators, h_s and h_l. As an operational model, HWRF is currently restricted to use a rather simple but computationally efficient scheme that predicts total cloud condensate instead of individual hydrometeor species. If switched to a more sophisticated microphysics scheme that predicts individual hydrometeor species, the partition formula may be avoided. As a result, individual hydrometeors are more accurately accounted for during data assimilation, thus, improved assimilation of the retrieved integrated SWC and integrated LWC is anticipated. In addition, the resulting analysis may be further improved by using the HWRF-based ensemble, which is a more optimal choice over the GFS ensemble.

Acknowledgements

This research is primarily funded by NOAA's Sandy Supplemental Award Number NA14OAR4830122. The assimilation experiments were performed using the Joint Center for Satellite Data Assimilation 'S4' supercomputer located at the University of Wisconsin-Madison. The authors would like to thank Drs. Krishna Kumar and Samuel Trahan for their help on porting the HWRF system onto S4. Encouraging discussions with Drs. Lewis Grasso and Paula Brown are greatly appreciated. The authors are also grateful to the two anonymous reviewers for their careful reviews and valuable comments.

References

Aligo E, Ferrier BS, Carley J, Rogers E, Pyle M, Weiss SJ, Jirak IL. 2014. Modified microphysics for use in high resolution NAM forecast. In 27th AMS Conference on Severe Local Storms, Madison, WI, 3–7 November 2014. https://ams.confex.com/ams/27SLS/webprogram/Paper255732.html (accessed 8 April 2017).

Bauer P, Geer AJ, Lopez P, Salmond D. 2010. Direct 4D-Var assimilation of all-sky radiances. Part I: Implementation. *Quarterly Journal of the Royal Meteorological Society* **136**: 1868–1885. https://doi.org/10.1002/qj.659.

Bauer P, Ohring G, Kummerow C, Auligne T. 2011. Assimilating satellite observations of clouds and precipitation into NWP models. *Bulletin of the American Meteorological Society* **92**: 25–28. https://doi.org/10.1175/2011BAMS3182.1.

Boukabara S-A, Garrett K, Grassotti C, Iturbide-Sanchez F, Chen W, Jiang Z, Clough SA, Zhan X, Liang P, Liu Q, Islam T, Zubko V, Mims A. 2013. A physical approach for a simultaneous retrieval of sounding, surface, hydrometeor, and cryospheric parameters from SNPP/ATMS. *Journal of Geophysical Research: Atmospheres* **118**: 12600–12619. https://doi.org/10.1002/2013JD020448.

Brown PJ, Kummerow CD, Randel DL. 2016. Hurricane GPROF: an optimized ocean microwave rainfall retrieval for tropical cyclones. *Journal of Atmospheric and Oceanic Technology* **33**: 1539–1556. https://doi.org/10.1175/JTECH-D-15-0234.1.

Han Y, van Delst P, Liu Q, Weng F, Yan B, Treadon R, Derber J. 2006. *JCSDA Community Radiative Transfer Model (CRTM) – version 1*. NOAA Technical Report NESDIS 122; 40 pp.

Hayden CM, Purser RJ. 1995. Recursive filter objective analysis of meteorological fields: applications to NESDIS operational processing. *Journal of Applied Meteorology* **34**: 3–15. https://doi.org/10.1175/1520-0450-34.1.3.

Huang XY. 1996. Initialization of cloud water content in a data assimilation system. *Monthly Weather Review* **124**: 478–486.

Kleist DT, Parrish DF, Derber JC, Treadon R, Wu W-S, Lord S. 2009. Introduction of the GSI into the NCEP global data assimilation system. *Weather and Forecasting* **24**: 1691–1705. https://doi.org/10.1175/2009WAF2222201.1.

Kummerow CD, Randel DL, Kulie M, Wang NY, Ferraro R, Joseph Munchak S, Petkovic V. 2015. The evolution of the Goddard profiling algorithm to a fully parametric scheme. *Journal of Atmospheric and Oceanic Technology* **32**: 2265–2280. https://doi.org/10.1175/JTECH-D-15-0039.1.

Parrish DF, Derber JC. 1992. The national meteorological center's spectral statistical-interpolation analysis system. *Monthly Weather Review* **120**: 1747–1763. https://doi.org/10.1175/1520-0493(1992)120<1747:TNMCSS>2.0.CO;2.

Pu Z, Zhang S, Tong M, Tallapragada V. 2016. Influence of the self-consistent regional ensemble background error covariance on hurricane inner-core data assimilation with the GSI-based hybrid system for HWRF. *Journal of the Atmospheric Sciences* **73**: 4911–4925. https://doi.org/10.1175/JAS-D-16-0017.1.

Tallapragada V, Bernadet L, Biswas MK, Ginis I, Kwon Y, Liu Q, Marchok T, Sheinin D, Thomas B, Tong M, Trahan S, Wang W, Yablonsky R, Zhang X. 2015. *Hurricane Weather Research and Forecasting (HWRF) Model: 2015 Scientific Documentation*. Developmental Testbed Center: Boulder, CO; 113.

Wang X. 2010. Incorporating ensemble covariance in the gridpoint statistical interpolation variational minimization: a mathematical framework. *Monthly Weather Review* **138**: 2990–2995. https://doi.org/10.1175/2010MWR3245.1.

Wu W-S, Purser RJ, Parrish DF. 2002. Three-dimensional variational analysis with spatially inhomogeneous covariances. *Monthly Weather Review* **130**: 2905–2916. https://doi.org/10.1175/1520-0493(2002)130<2905:TDVAWS>2.0.CO;2.

Wu T-C, Zupanski M, Grasso LD, Brown PJ, Kummerow CD, Knaff JA. 2016. The GSI capability to assimilate TRMM and GPM hydrometeor retrievals in HWRF. *Quarterly Journal of the Royal Meteorological Society* **142**: 2768–2787. https://doi.org/10.1002/qj.2867.

Zhu Y, Liu E, Mahajan R, Thomas C, Groff D, Van Delst P, Collard A, Kleist D, Treadon R, Derber JC. 2016. All-sky microwave radiance assimilation in NCEP's GSI analysis system. *Monthly Weather Review* **144**: 4709–4735. https://doi.org/10.1175/MWR-D-15-0445.1.

Binary mesovortex structure associated with southwest vortex

Kuo Zhou,[1] Haiwen Liu,[2,3]* Liang Zhao,[2] Yuxiang Zhu,[4] Yihua Lin,[2] Fuying Zhang[5] and Ning Fu[3]

[1]College of Atmospheric Sciences, Chengdu University of Information Technology, China
[2]LASG, Institute of Atmospheric Physics, Chinese Academy of Sciences, Beijing, China
[3]Department of Aviation Meteorology, Civil Aviation University of China, Tianjin, China
[4]CMA Training Center, China Meteorological Administration, Beijing, China
[5]College of Atmospheric Sciences, Nanjing University of Information Science and Technology, Nanjing, China

*Correspondence to:
H. Liu, Department of Aviation
Meteorology, Civil Aviation
University of China, 2898 Jinbei
Road, Dongli District, Tianjin,
300300, China.
E-mail: lhw499@126.com

Abstract

Previous work has concluded that the southwest vortex (SWV) is a single mesoscale vortex. Applying the National Centers for Environmental Prediction Final Operational Global Analysis, the Interim European Centre for Medium-Range Weather Forecasts Re-Analysis, and the non-hydrostatic mesoscale Advanced Research Weather Research and Forecasting (WRF) model to a case study, we discovered a new type of SWV associated with another coexisting mesoscale warm and moist vortex. In the case study, meso-β-scale vortex-A was generated at 1800 UTC 17 July, and dissipated around 0500 UTC 18 July 2013, with a lifespan of approximately 11 h. Vortex-B occurred at 0600 UTC 17 July and moved out of the Sichuan Basin at 0800 UTC 18 July 2013, remaining over the basin for approximately 26 h. Stronger atmospheric upward motion and two mesoscale rainbelts associated with each of the vortices further demonstrate the binary mesoscale vortex structure related to the SWV using the WRF model. The quasi-geostrophic balance of the two mesoscale cyclonic circulations is responsible for the generation and maintenance of the two closed mesoscale vortices.

Keywords: southwest vortex; WRF model; binary vortex structure; geostrophic adjustment

I. Introduction

The Tibetan Plateau (TP) is the largest plateau in the world, with the Sichuan Basin located on its eastern flank. A mesoscale vortex, known as a southwest vortex (SWV) because of its location over southwest China (Tao and Ding, 1981), is commonly found at the eastern and southeastern flanks of the TP (Wang et al., 1993). Unlike the mesoscale convective vortex that is most evident between 500 and 600 hPa over the central United States (Davis and Trier, 2007), the SWV is normally most visible in the lower troposphere centered at 700 or 850 hPa (Lu, 1986; Fu et al., 2013, 2014b). The SWV has been studied extensively. Lu (1986), e.g. showed that in a SWV the lower troposphere and the middle troposphere were dominated by convergence, while the upper troposphere is controlled by divergence (Zhao and Fu, 2007). Huang (1986) found that ascending motion existed in the center and periphery of the SWV. A strong and well-developed SWV can stretch up to 100 hPa, which is extremely deep (Chen et al., 1998). In general, the SWV has complex temperature and humidity characteristics during its formation and development stages. Kuo et al. (1988) found that a high equivalent potential temperature (θ_e) occurred in the center of the SWV. These characteristics have many similarities with the vortices found in the United States, as summarized by Menard and Fritsch (1989). Recently, Li et al. (2014) studied the SWV using ensemble-based analyses and forecasts methods, indicating that a more baroclinic environment facilitated the evolution of the

SWV. The mechanisms associated with the formation and development of the SWV are largely influenced by the unique thermo-dynamical and dynamical environments of the plateau, including topography (Wu and Chen, 1985; Jiang et al., 2012; Wang and Tan, 2014), latent heat release (Kuo et al., 1986; Wang and Orlanski, 1987; Kuo et al., 1988), and the interaction between SWV and other systems, such as the eastward movement of the plateau vortex (Wang, 1987; Yu et al., 2016).

As a mesoscale vortex, SWV can comprise small-scale weather systems. Tao (1980) found that three mesoscale vortices that were lined up with one shear line were associated with the SWV. However, due to the lack of high spatial and temporal resolution datasets, it is difficult to find multiple mesoscale systems associated with the SWV. This paper identifies and studies a new binary vortex structure related to the SWV. The article is structured as follows: data and methods are described in Section 2; Section 3 presents the results, including the observed and simulated results of the binary mesoscale vortices, and an analysis of the potential mechanism for the formation and development of the binary mesoscale vortices; and Discussion and conclusion are presented in Section 4.

2. Data and methods

Hourly gauged rainfall data, including the conventional meteorological observations and automated weather

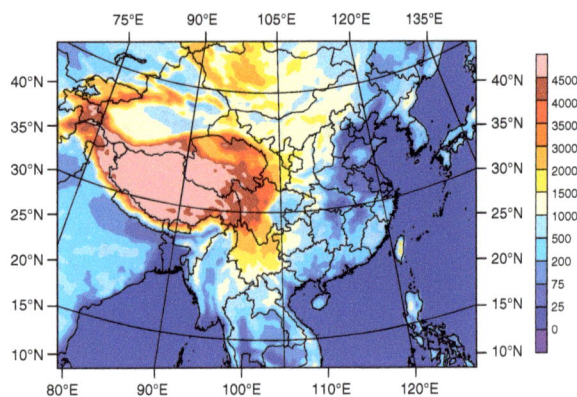

Figure 1. Model domains and topography.

stations from China, were provided by the Meteorological Information Center of China Meteorological Administration.

The large-scale circulations associated with the SWV were analyzed based on the National Centers for Environmental Prediction (NCEP) Final (FNL) Operational Global Analysis with $1° \times 1°$ latitude–longitude grids at 6-h intervals. The Interim European Centre for Medium-Range Weather Forecasts Re-Analysis (ERA-Interim) with a temporal resolution of 6 h and a spatial resolution of $0.5° \times 0.5°$ was also used to analyze the weather circulation.

To investigate the evolution of the SWV, version 3.4.1 of the non-hydrostatic mesoscale Advanced Research Weather Research and Forecasting (WRF) model was applied. A one-way nested run was performed in this simulation, and its initial and lateral conditions were derived from the NCEP FNL Operational Global Analysis. The WRF horizontal grid spacing is 30 km (200×150 grid points) in the domain, which has 28 vertical levels. The numerical simulation was initialized at 0000 UTC 17 July 2013 and the results were output hourly with a total simulation length of 36 h. The model parameterization schemes in the domain were as follows. The Rapid and Accurate Radiative Transfer Model scheme (Mlawer *et al.*, 1997) was used for longwave radiation and the Dudhia scheme (Dudhia, 1989) for shortwave radiation. We used the Kain–Fritsch convective scheme (Kain, 2004) for cumulus convection. The Monin–Obukhov scheme (Janjić, 2002) was used for the surface layer physics. The Noah land surface scheme (Ek *et al.*, 2003) was applied along with the Yonsei University planetary boundary layer (PBL) scheme (Hong *et al.*, 2006). The microphysics scheme was used by Thompson *et al.* (2008). The model grids are shown in Figure 1.

3. Results

3.1. Observations of the binary mesoscale vortices associated with the SWV

A mesoscale SWV was observed over Sichuan Province and an isoline of 3055 gpm occurred in the vortex

at 1200 UTC 17 July 2013 (Figure S1, Supporting information). The vortex had been generated in Sichuan Province at 0600 UTC 17 July 2013 (not shown). Following the criteria of Lu (1986), it was categorized as a typical SWV. In accordance with the SWV in the lower troposphere, a longwave trough was observed at the 500-hPa level behind the lower level SWV. This further demonstrates that the SWV is a shallow system and has baroclinic features.

To study the detailed structure of the mesoscale SWV, the 700-hPa FNL analyses over the region are given in Figure 2. At 1800 UTC 17 July, the shape of the SWV began to change. Mesoscale cyclonic vortices appeared in the northeast (labeled 'A') and south ('B') of Sichuan Province, although the 3080 gpm isoline was not closed in mesoscale cyclonic vortex-A. Based on the criteria of Orlanski (1975), vortex-A was a meso-β-scale vortex. Although vortex-A satisfied the criteria of meso-β-scale, the temporal scale of vortex-B was 2 h beyond the guidelines of Orlanski (1975) on the meso-β-scale; therefore, vortex-B was classified as a mesoscale vortex. The weather patterns presented in Figure 2(b) persisted until 0000 UTC 18 July 2013. The two mesoscale vortices were observed from 1800 UTC 17 July to 0600 UTC 18 July. Compared with the scope of SWV (Bao and Li, 1985), the location of vortex-A was only 1°N of SWV. Therefore, further study is required on whether vortex-A belonged to SWV. At 0600 UTC 18 July 2013, vortex-A began to move to the northeast, and vortex-B moved south. Similar weather patterns were also present in the ERA-Interim dataset.

3.2. Simulation of the binary mesoscale vortex structure

Many studies, including two major TP scientific field experiments in 1979 and 1998 (Xu and Chen, 2006), have examined the formation and development of SWVs (Li *et al.*, 2014). Figure 3 shows the two mesoscale vortices simulated by WRF at 0600, 1800, and 2100 UTC 17 July, and at 0300, 0500, and 0800 UTC 18 July 2013. The WRF model configuration, initial and lateral boundary conditions, and physical parameterization schemes were introduced as described previously. At 0600 UTC 17 July 2013, a parent mesoscale vortex was produced by WRF (Figure 3(a)). Twelve hours later at 1800 UTC 17 July, the parent mesoscale vortex split into two mesoscale vortices ('A' and 'B'). Meso-β-scale vortex-A was located over the northeast of Sichuan Province, and mesoscale vortex-B was located over the south of the province.

Vortex-A began to form from 1800 UTC 17 July 2013, and then merged to another low system (labeled 'L') at 0500 UTC 18 July 2013. The lifespan of vortex-A was about 11 h. Mesoscale vortex-B began to form at 0600 UTC 17 July 2013, then moved southward at 0800 UTC 18 July 2013, and lasted for 26 h. The two mesoscale vortices were most apparent at 2100 UTC 17 July 2013. As both mesoscale vortices had closed

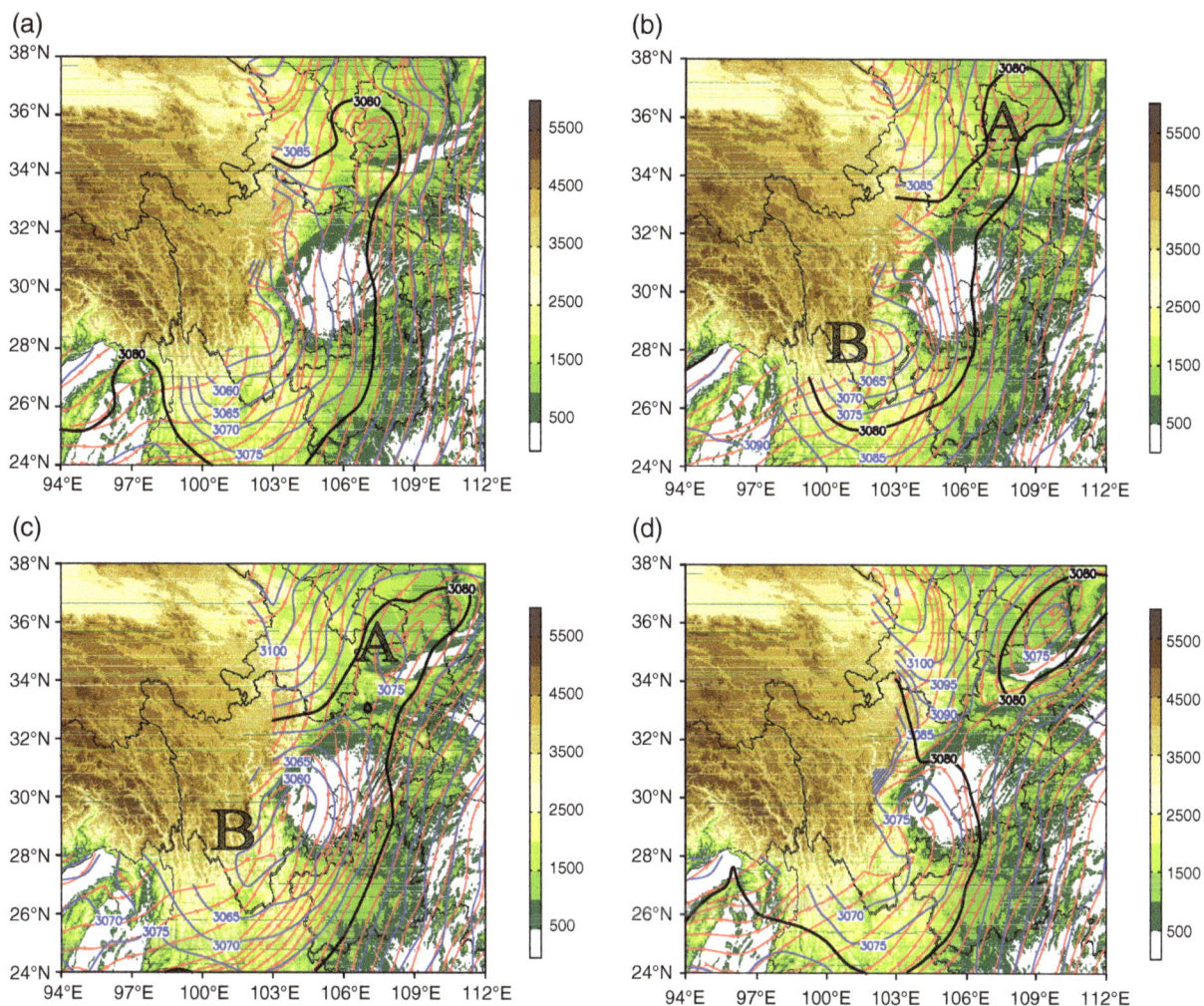

Figure 2. Synoptic conditions of 700-hPa geopotential heights (blue contours, units: gpm) and flow field analyzed from FNL data at (a) 1200 UTC 17 July, (b) 1800 UTC 17 July, (c) 0000 UTC 18 July, and (d) 0600 UTC 18 July 2013. Shading shows the terrain of Sichuan Basin (units: m) and the value of 3080 gpm in black contour is used to highlight the binary vortex structure. The symbols 'A' and 'B' in panels (b) and (c) denote centers of the binary vortices, respectively.

isolines, they should be considered as two mesoscale vortices associated with the developing SWV, as shown in Figure 3(c).

Fu *et al.* (2014a) proposed that 88.5% of the detected SWVs caused rainfall. In heavy rainfall, an oblique meso-β-scale vortex appears frequently in the interior of the SWV (Gu *et al.*, 2008). Figure 4 shows the observed and WRF-simulated hourly precipitation. The two rainbelts ('A' and 'B') for the observed and WRF-simulated precipitation are separated in space (Figure 4). The differences between them, including the location and the strength, were beyond the scope of this study. Rainbelt-A was located roughly to the south of vortex-A, and rainbelt-B to the south of vortex-B, which further demonstrates that the two mesoscale vortices (Figure 3) were two separate features.

To investigate the vertical structure of these binary vortices, Figure 5 shows the vertical cross-section of the WRF-simulated potential pseudo-equivalent temperature and vertical motion between the center points of A and B. At 0600 UTC 17 July 2013, there was a strong

upward motion center between the vortices; however, the two centers were observed in the typical time of the two mesoscale vortices. Furthermore, two warm and moist centers also existed above vortices A and B. These two warm and moisture centers lasted until 0800 UTC 18 July 2013, although vortices A and B had changed by this time.

3.3. Potential mechanism for the generation and maintenance of the binary mesoscale vortices

Based on the geostrophic adjustment theory (Blumen, 1972), mesoscale mass fields tend to adjust to wind fields. At 0600 UTC 17 July 2013 (Figure 3(a)), an extensive area of positive relative vorticity was produced by WRF over Sichuan Province. After a few hours, two closed mesoscale vortices began to emerge as a result of geostrophic adjustment. This situation continued until 0500 UTC 18 July 2013 (Figure 3(e)). Following the movements of the relative vorticity centers, vortex-A merged with another low system, L, in the northeast of Sichuan Province, and vortex-B moved to

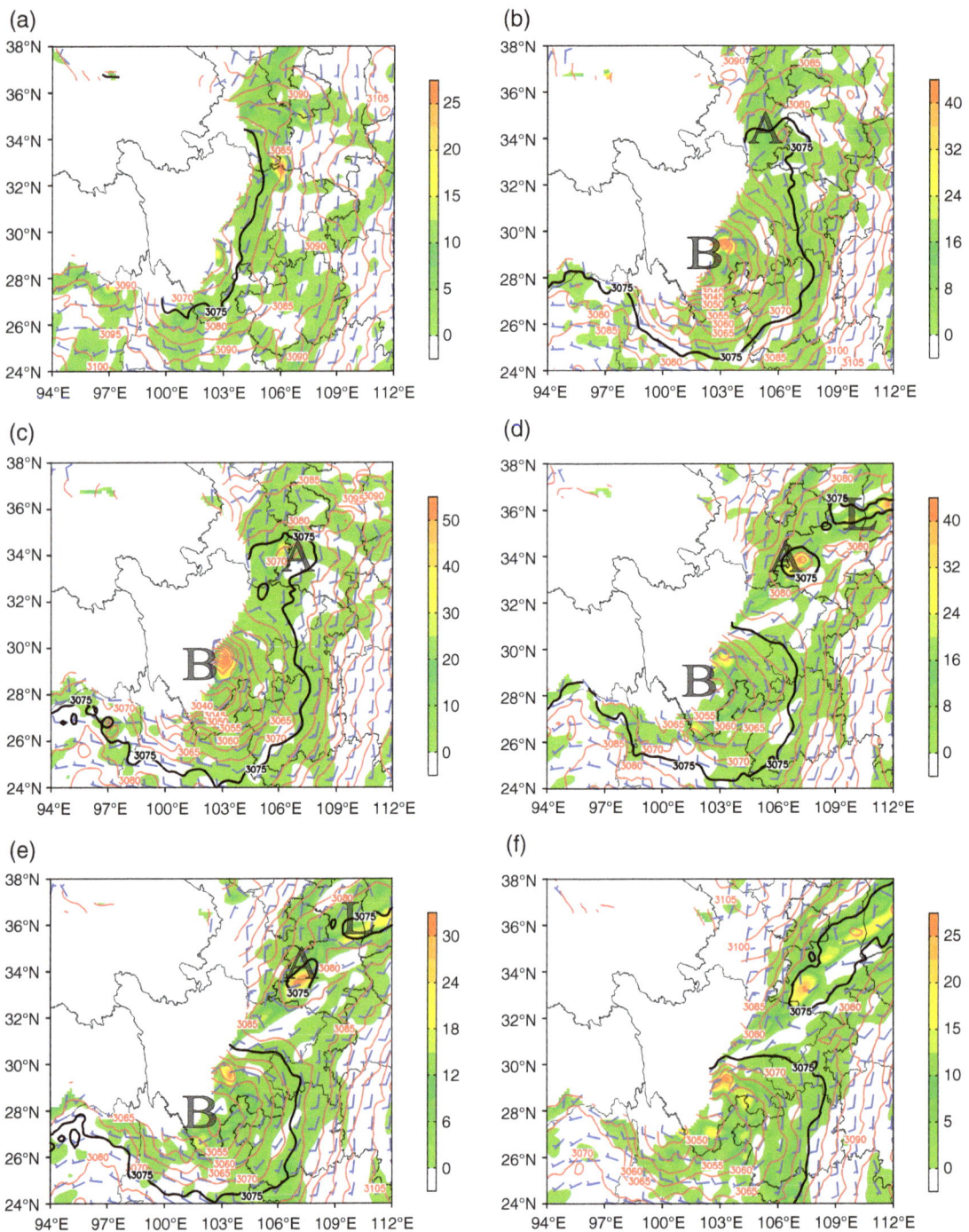

Figure 3. Composite simulated relative vorticity (shadings, units: 10^{-5} s^{-1}), 700-hPa geopotential heights (red contours, units: gpm), and wind field (blue barbs, units: m s^{-1}) at (a) 0600 UTC 17 July, (b) 1800 UTC 17 July, (c) 2100 UTC 17 July, (d) 0300 UTC 18 July, (e) 0500 UTC 18 July, and (f) 0800 UTC 18 July 2013. The value of 3075 gpm in black contour is used to highlight the binary vortex structure. The symbols 'A' and 'B' denote centers of the binary vortices, whereas the symbol 'L' in panels (d) and (e) denotes the center of another low system. Blank area in the upper left denotes the topography over 3000 m.

the south of the province. The mesoscale cyclonic flows played an important role in the formation of the binary vortices associated with SWV.

4. Discussion and conclusion

The SWV is often associated with extreme weather, especially with heavy rain. The severity of heavy precipitation caused by SWV is second only to that caused by tropical cyclones in China (Wang *et al.*, 1996). Due in part to the fact that their spatial and temporal scales are too small to be captured by the conventional observational network, the two mesoscale cyclonic vortices were difficult to observe in the relatively coarse data. However, the more distinguishable structures of the two mesoscale closed cyclonic vortices could be demonstrated with the

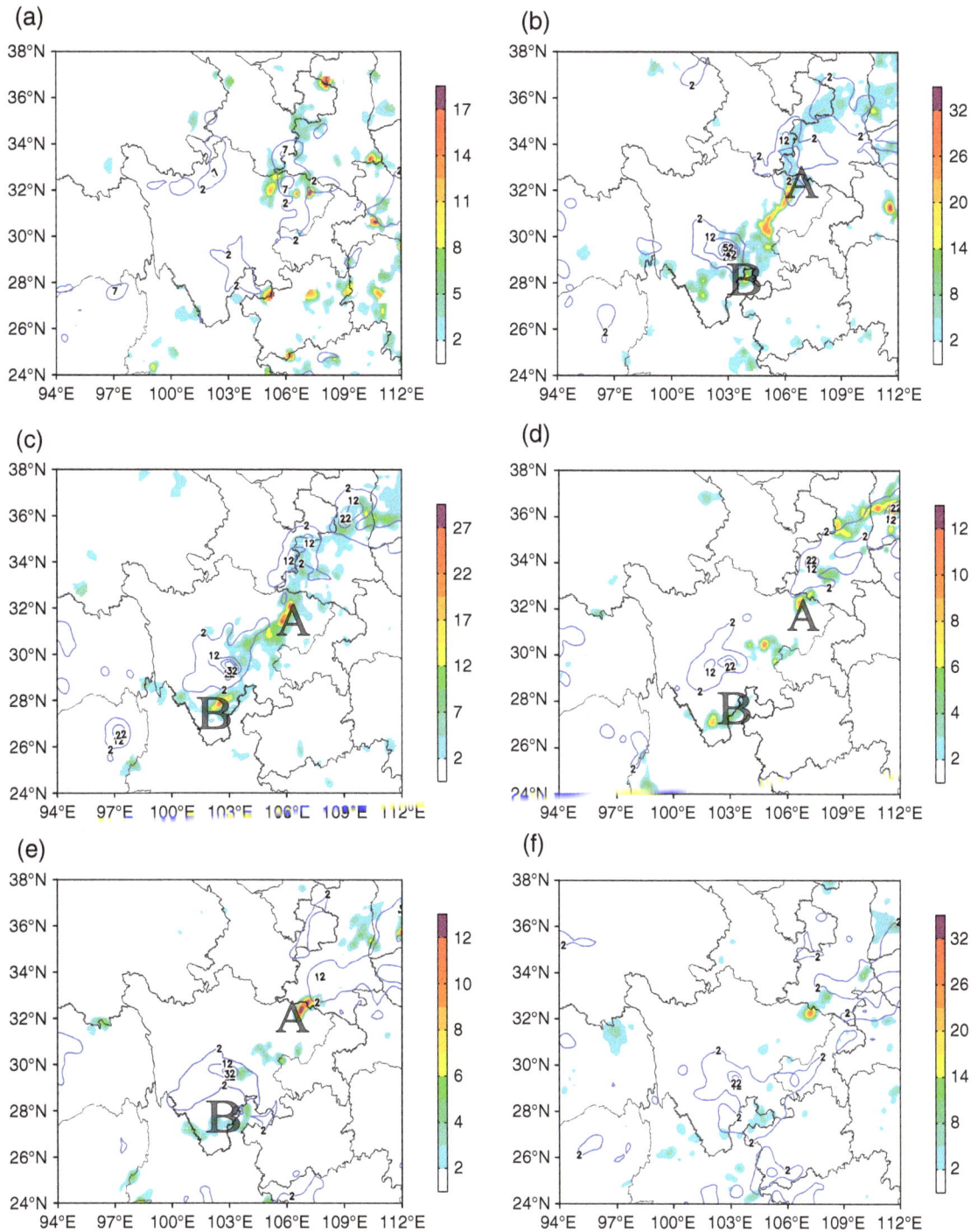

Figure 4. Composite observed precipitation (shadings, units: mm) and simulated precipitation (blue contours, units: mm) for 1 h ending at (a) 0600 UTC 17 July, (b) 1800 UTC 17 July, (c) 2100 UTC 17 July, (d) 0300 UTC 18 July, (e) 0500 UTC 18 July, and (f) 0800 UTC 18 July 2013. The symbols 'A' and 'B' in panels (b)–(e) denote centers of rainbelt.

WRF model. In this case, meso-β-scale vortex-A began to form at 1800 UTC 17 July 2013, and was maintained until 0500 UTC 18 July 2013. Vortex-B occurred at 0600 UTC 17 July 2013, and ended at 0800 UTC 18 July 2013. The spatial scale of vortex-B was larger than that of vortex-A, and the lifespan of vortex-A was shorter than that of vortex-B. From generation to dissipation, vortex-A lasted approximately 11 h, whereas vortex-B persisted for about 26 h. The mesoscale cyclonic circulation makes an important contribution to the formation of the binary vortices. To satisfy the quasi-geostrophic balance of the two mesoscale cyclonic circulations, the binary mesoscale vortex structure associated with SWV was generated and maintained.

Questions still remaining for future work include whether vortex-A belongs to the SWV, whether vortex-B is responsible for the generation and maintenance of vortex-A, and how vortex-B and vortex-A interact with each other.

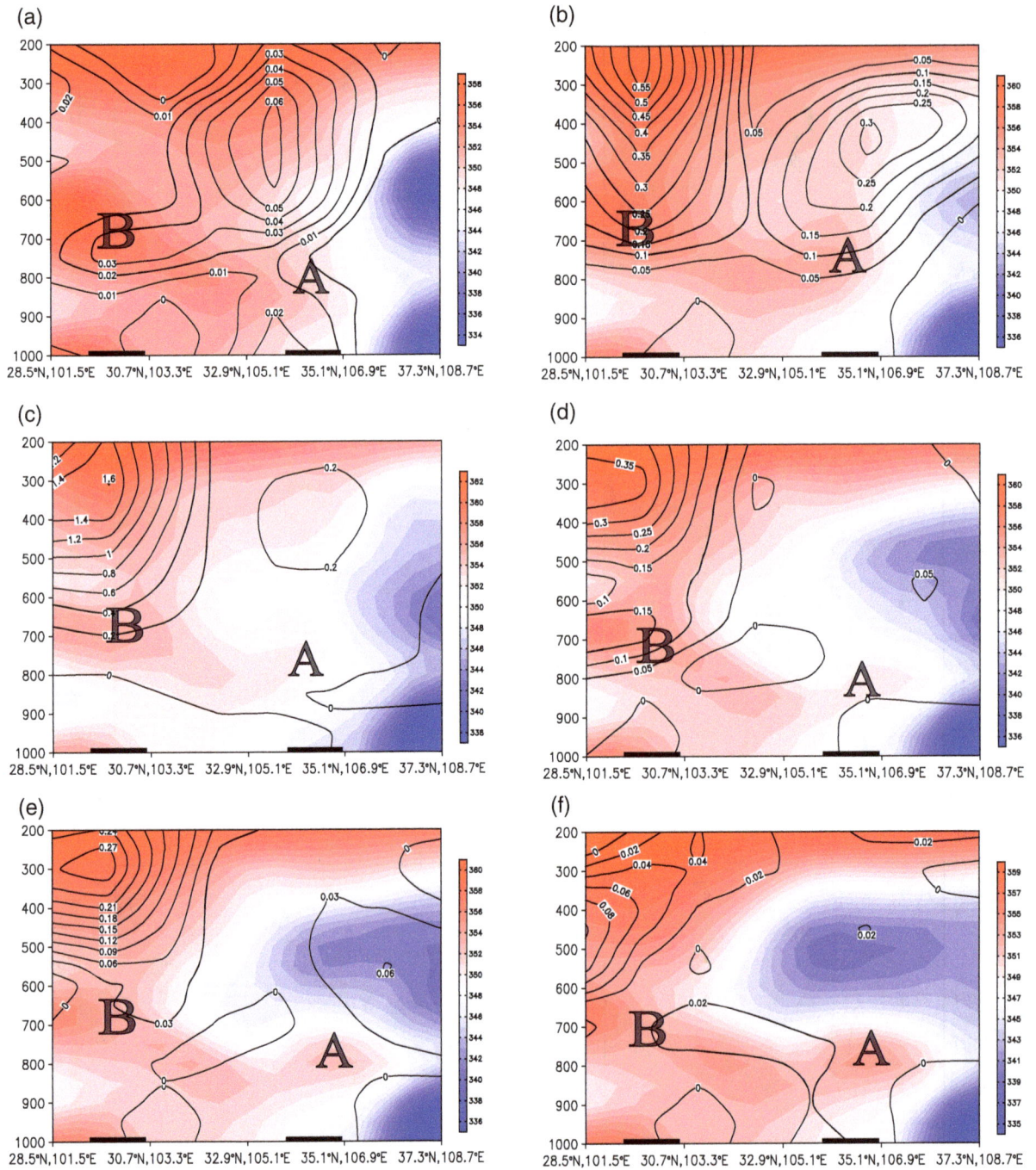

Figure 5. Vertical profile of simulated potential pseudo-equivalent temperature (shadings, units: K), and vertical velocity (black contours, units: m s^{-1}) along the connection line of A and B at (a) 0600 UTC 17 July, (b) 1800 UTC 17 July, (c) 2100 UTC 17 July, (d) 0300 UTC 18 July, (e) 0500 UTC 18 July, and (f) 0800 UTC 18 July 2013. The symbols 'A' and 'B' in panels (a)–(f) denote warm centers of the binary vortices. The location of the binary vortices is shown by the heavy lines at the bottom.

Acknowledgements

This work was jointly supported by the State Key Program of National Natural Science of China (91337215, 41575059), the Applied Basic Research Programs of the Sichuan Provincial Department of Science and Technology (2015JY0109) the Opening Foundation of Chongqing Meteorological Bureau Grants KFJJ-201102 and YU-XM-2011030, the China Special Fund for Meteorological Research in the Public Interest (GYHY201406020), Special Fund for Climate change (CCSF201706), and Special Fond for Development of weather forecasting key technologies (YBGJXM(2017)03-13).

Supporting information

The following supporting information is available:

Figure S1. Synoptic conditions of 700-hPa geopotential heights (black contours, units: gpm), wind field (vectors, units: m s^{-1}) and 500-hPa geopotential heights (green contours, units: gpm) analyzed from FNL data at (a) 1200 UTC 17 July, (b) 1800 UTC 17 July, (c) 0000 UTC 18 July, and (d) 0600 UTC 18 July 2013. The value of 3080 gpm in red contour at 700 hPa is used to highlight the binary vortex structure. The symbols 'A' and

'B' in panels (b) and (c) denote centers of the mesoscale vortices, respectively. Blue shading shows topography more than 3000 m.

References

Bao CL, Li SC. 1985. Preliminary study on the formation of southwest vortex (in Chinese). *Meteorological Monthly* **11**: 2–6.

Blumen W. 1972. Geostrophic adjustment. *Reviews of Geophysics and Space Physics* **10**: 485–528.

Chen ZM, Miu Q, Min WB. 1998. A case analysis on mesoscale structure of severe southwest vortex (in Chinese). *Journal of Applied Meteorological Science* **9**: 273–282.

Davis CA, Trier SB. 2007. Mesoscale convective vortices observed during BAMEX. Part I: kinematic and thermodynamic structure. *Monthly Weather Review* **135**: 2029–2049.

Dudhia J. 1989. Numerical study of convection observed during the winter monsoon experiment using a mesoscale two-dimensional model. *Journal of the Atmospheric Sciences* **46**: 3077–3107.

Ek MB, Mitchell KE, Lin Y, Rogers E, Grunmann P, Koren V, Gayno G, Tarpley JD. 2003. Implementation of Noah land surface model advances in the National Centers for Environmental Prediction operational mesoscale Eta model. *Journal of Geophysical Research: Atmospheres* **108**: 8851.

Fu SM, Yu F, Wang DH, Xia RD. 2013. A comparison of two kinds of eastward-moving mesoscale vortices during the mei-yu period of 2010. *Science China Earth Sciences* **56**: 282–300.

Fu SM, Zhang JP, Sun JH, Shen XY. 2014a. A fourteen-year climatology of the southwest vortex in summer. *Atmospheric and Oceanic Science Letters* **7**: 510–514.

Fu SM, Li WL, Sun JH, Zhang JP, Zhang YC. 2014b. Universal evolution mechanisms and energy conversion characteristics of long-lived mesoscale vortices over the Sichuan Basin. *Atmospheric Science Letters* **16**: 127–134.

Gu QY, Zhou CH, Qing Q, Zhang J. 2008. Mesoscale characteristics analysis of severe torrential rain caused by a southwestern low vortex process (in Chinese). *Meteorological Monthly* **34**: 39–47.

Hong SY, Noh Y, Dudhia J. 2006. A new vertical diffusion package with an explicit treatment of entrainment process. *Monthly Weather Review* **134**: 2318–2341.

Huang FJ. 1986. A composite analysis of the southwest vortex (in Chinese). *Chinese Journal of Atmospheric Sciences* **10**: 402–408.

Janjić ZI. 2002. Nonsingular implementation of the Mellor–Yamada level 2.5 scheme in the NCEP Meso model. NCEP Office Note No. 437; 61 pp.

Jiang XW, Li YQ, Zhao XB, Koike T. 2012. Characteristics of the summertime boundary layer and atmospheric vertical structure over the Sichuan Basin. *Journal of the Meteorological Society of Japan* **90**: 33–54.

Kain JS. 2004. The Kain–Fritsch convective parameterization: an update. *Journal of Applied Meteorology* **43**: 170–181.

Kuo YH, Cheng LS, Anthes RA. 1986. Mesoscale analyses of the Sichuan flood catastrophe, 11–15 July 1981. *Monthly Weather Review* **114**: 1984–2003.

Kuo YH, Cheng LS, Bao JW. 1988. Numerical simulation of the 1981 Sichuan flood. Part I: evolution of a mesoscale southwest vortex. *Monthly Weather Review* **116**: 2481–2504.

Li J, Du J, Zhang DL, Cui CG, Liao YS. 2014. Ensemble-based analysis and sensitivity of mesoscale forecasts of a vortex over southwest China. *Quarterly Journal of the Royal Meteorological Society* **140**: 766–782.

Lu JH. 1986. *Introduction to the Southwest Vortices (in Chinese)*. China Meteorological Press: Beijing; 276 pp.

Menard RD, Fritsch JM. 1989. A mesoscale convective complex-generated inertially stable warm core vortex. *Monthly Weather Review* **117**: 1237–1261.

Mlawer EJ, Taubman SJ, Brown PD, Iacono MJ, Clough SA. 1997. Radiative transfer for inhomogeneous atmospheres: RRTM, a validated correlated-k model for the longwave. *Journal of Geophysical Research: Atmospheres* **102**: 16663–16682.

Orlanski I. 1975. A rational subdivision of scales for atmospheric processes. *Bulletin of the American Meteorological Society* **56**: 527–530.

Tao SY. 1980. *Rainstorms in China (in Chinese)*. Science Press: Beijing; 225 pp.

Tao SY, Ding YH. 1981. Observational evidence of the influence of the Qinghai–Xizang (Tibet) Plateau on the occurrence of heavy rain and severe convective storms in China. *Bulletin of the American Meteorological Society* **62**: 23–30.

Thompson G, Field PR, Rasmussen RM, Hall WD. 2008. Explicit forecasts of winter precipitation using an improved bulk microphysics scheme. Part II: implementation of a new snow parameterization. *Monthly Weather Review* **136**: 5095–5115.

Wang B. 1987. The development mechanism for Tibetan Plateau warm vortices. *Journal of the Atmospheric Sciences* **44**: 2978–2994.

Wang B, Orlanski I. 1987. Study of a heavy rain vortex formed over the eastern flank of the Tibetan Plateau. *Monthly Weather Review* **115**: 1370–1393.

Wang QW, Tan ZM. 2014. A numerical simulation study of a heavy rain event induced by a southwest vortex. *Journal of Geophysical Research: Atmospheres* **119**: 11543–11561.

Wang W, Kuo YH, Warner TT. 1993. A diabatically driven mesoscale vortex in the lee of the Tibetan Plateau. *Monthly Weather Review* **121**: 2542–2561.

Wang ZS, Wang Y, Liang Y. 1996. A numerical simulation study of a heavy rain event induced by a southwest vortex (in Chinese). In *Experimental, Synoptical and Dynamical Studies of Heavy Rain*. China Meteorological Press: Beijing; 257–267.

Wu GX, Chen SJ. 1985. The effect of mechanical forcing on the formation of a mesoscale vortex. *Quarterly Journal of the Royal Meteorological Society* **111**: 1049–1070.

Xu XD, Chen LS. 2006. Advances of the study on Tibetan Plateau experiment of atmospheric sciences (in Chinese). *Journal of the Applied Meteorological Science* **17**: 754–772.

Yu SH, Gao WL, Xiao DX, Peng J. 2016. Observational facts regarding the joint activities of the southwest vortex and plateau vortex after its departure from the Tibetan Plateau. *Advances in Atmospheric Sciences* **33**: 34–46.

Zhao SX, Fu SM. 2007. An analysis on the southwest vortex and its environment fields during heavy rainfall in eastern Sichuan Province and Chongqing in September 2004 (in Chinese). *Chinese Journal of Atmospheric Sciences* **31**: 1059–1075.

3

Inflating transform matrices to mitigate assimilation errors with robust filtering based ensemble Kalman filters

Yulong Bai,[1,*] Zhuanhua Zhang,[1] Yanli Zhang[2] and Lili Wang[1]

[1]College of Physics and Electrical Engineering, Northwest Normal University, Lanzhou, China
[2]College of Geography and Environment Science, Northwest Normal University, Lanzhou, China

*Correspondence to:
Y. Bai, College of Physics and
Electrical Engineering,
Northwest Normal University,
Lanzhou, Gansu 730070, China.
E-mail: yulongbai@gmail.com

Abstract

An error covariance matrix plays an important role in maintaining the statistical property of the ensemble in an ensemble Kalman filter method. However, data assimilation filter divergence may occur from an inaccurate estimate of the covariance matrix. In this study, based on an ensemble time-local H-infinity filter, which inflates the eigenvalues of the analysis error covariance matrix, a new robust ensemble data assimilation method is proposed, referred to as an inflation transform matrix eigenvalues. By design, new filters may be preferred over other traditional ensemble filters, when model performances are not well known, or change unpredictably. The primary aim is to improve the performance of the assimilation system in the framework of the ensemble filtering, according to the minimum/maximum rule of robust filtering. The proposed estimation method is tested using the well-known Lorenz-96 model, in order to investigate how the ensemble time-local H-infinity filter method of the inflation transform matrix impacts the robustness of the assimilation system under selected special conditions, such as the assimilation steps, force parameters, ensemble sizes, and observation information. The experiments show that the proposed inflation transform matrix method displays good robustness to the changes in the system's parameters. Also, when compared with the traditional filtering methods, this robust filtering method is found to improve the assimilation performance.

Keywords: ensemble transfer Kalman filter; time-local H-infinity filter; Lorenz-96 chaos system; robust Kalman filter

1. Introduction

In data assimilation, the ensemble Kalman filter (EnKF) (Evensen, 2003) provides advantages over the traditional Kalman filter (KF) by overcoming the shortcomings with reduced computational costs. However, due to model errors, initial background conditions, and limitations of ensemble sizes, the EnKF generally leads to an inaccurate estimate of the error covariance matrix, and also usually reduces the filtering performance, which eventually results in filter divergence. To compensate for those defects, Anderson and Anderson (1999) proposed a method referred to as covariance inflation. Sakov and Bertino (2011) introduced localization to address the background error spurious correlations. Luo and Hoteit (2013, 2014) conducted residual nudging. However, in practical atmospheric problems, the model and observation errors are often non-Gaussian and/or biased, and the statistical properties of the errors are often unknown, or not fully known. Therefore, there is a need to study the assimilation method in a general sense.

Motivated by robust control concepts that have been developed in the control engineering field for numerous years, a robust filter applied the norm of performance index H_∞ to the filters, which is also used to

solve the uncertainty problems in the data assimilation system (Luo and Hoteit, 2011), and is referred to as an H-infinity filter (HF) (Simon, 2006). Wang and Cai (2008) compared the HF and KF, and showed that the HF utilized deterministic interference with unknown but finite energy, instead of the white noise process driven state space system. For convenience in applying the H_∞ filtering theories to sequential data assimilation, and in order to solve the problem of high dimensional data assimilation systems, Luo and Hoteit (2011) proposed a variant of the HF, known as the time-local HF (TLHF), and then applied the ensemble idea to the TLHF. Altaf et al. (2013) improved the accuracy of ensemble Kalman prediction using a robust adaptive inflation method.

For the purpose of constructing an improved error covariance matrix, Luo and Hoteit (2011) directly inflated the eigenvalues of the analysis error covariance matrix (hereafter referred to as IA). In this study, based on an ensemble time-local H-infinity filter (EnTLHF), a new robust data assimilation method is proposed in order to inflate the transform matrix eigenvalues (hereafter referred to as IT). The essence of the method is to conduct singular value decomposition (SVD), and to obtain the eigenvalues of the transform matrix. To apply the positive semi-definite condition of analysis error covariance matrix inverse, the eigenvalues of the transform matrix are inflated, which indirectly inflates the analysis error covariance matrix. The application of this method can potentially avoid the complex SVD of the analysis error covariance matrix.

In this article, a robust EnTLHF is developed and implemented in the Lorenz-96 model for state forecasting based on the ensemble transform Kalman filter (ETKF) (Bishop *et al.*, 2001). Similar experiments are reported using the IA method (Luo and Hoteit, 2011). Although the IA method was proved to improve the filter performance by significant amounts, it still lacks accuracy when the parameter changes, even filter divergence occurs, which motivates this study. New methods proposed in this study are compared with ETKF and IA for the special conditions. The results suggest that the IT implemented in the framework of the EnTLHF strongly mitigate assimilation errors as compared to the ETKF and IA.

2. Data assimilation methods

2.1. Time-local H-infinity filter

HF (Simon, 2006) is one of the robust filters involving the nature of the model and observation errors which the KF relies on. No matter what the size of the model and observation errors, if the norm of the HF has less than a preset positive value $1/\gamma$, then the performance of the HF estimator can be guaranteed. The minimum/maximum rule is utilized to compute the estimation values in the uncertainty background, model errors, and observation errors.

To satisfy the sequential data assimilation in certain circumstances, Luo and Hoteit (2011) proposed that a TLHF would be more flexible than the HF.

Define a local cost function of the TLHF:

$$J_{x,i}^{\text{HF}} = \frac{\left\| x_i - x_i^a \right\|_{S_i}^2}{\left\| x_i - x_i^b \right\|_{(\Delta_i^b)^{-1}}^2 + \left\| u_i \right\|_{Q_i^{-1}}^2 + \left\| v_i \right\|_{R_i^{-1}}^2} \qquad (1)$$

$$\left\| x_i - x_i^a \right\|_{S_i}^2 \leq \frac{1}{\gamma_i} \left(\left\| x_i - x_i^b \right\|_{(\Delta_i^b)^{-1}}^2 \right.$$

$$\left. + \left\| u_i \right\|_{Q_i^{-1}}^2 + \left\| v_i \right\|_{R_i^{-1}}^2 \right) \qquad (2)$$

where x_i and x_i^a are the system true value and the corresponding true value estimate, respectively. Δ_i^b, Q_i, and R_i are the uncertainty weight matrices of the background x_i^b, the model error u_i, and observation error v_i, respectively, and their inverses, $\left(\Delta_i^b\right)^{-1}$, Q_i^{-1}, and R_i^{-1} are the corresponding information matrices. The weight matrixes S_i, Q_i, and R_i are free choice by the designers, and γ_i is a suitable local performance level, which satisfies Equation (3).

$$\frac{1}{\gamma_i} > \frac{1}{\gamma_i^*} \equiv \inf_{x_i^a} \sup_{x_i, u_i, v_i} J_{x,i}^{\text{HF}} \qquad (3)$$

where $\sup_{x_i, u_i, v_i}$ calculates the supremum of the cost function $J_{x,i}^{\text{HF}}$ with respect to the variables x_i, u_i, v_i (which are functions of x_i^a); and $\inf_{x_i^a}$ evaluates the infimum with respect to x_i^a along the previously obtained supremum plane of x_i, u_i, v_i.

2.2. EnTLHF-IA (IA)

It is assumed that the system is nonlinear in assimilation, with $M_{i,i-1}$ possibly being nonlinear, and H_i is the linear operators. Concretely, let $x_i^b = \left[x_{i,j}^b : x_{i,j}^b = M_{i,i-1}\left(x_{i-1,j}^a\right), j = 1, \dots, n \right]$ be the n-member background ensemble at time instant i, which is the prediction of the analysis ensemble $x_{i-1}^a = \left(x_{i-1,j}^a, j = 1, \dots, n \right)$ at the previous cycle. The steps of the EnTLHF can be derived as follows:

Prediction step:

$$\bar{x}_i^b = \text{mean}\left(x_i^b\right) \qquad (4)$$

$$\Delta_i^b = \text{cov}\left(x_i^b\right) + Q_i \qquad (5)$$

Filtering step:

$$\bar{x}_i^a = \bar{x}_i^b + G_i \left[y_i - H_i(\bar{x}_i^b) \right] \qquad (6)$$

$$(\Delta_i^a)^{-1} = (\Delta_i^b)^{-1} + (H_i)^T (R_i)^{-1} H_i - \gamma_i S_i \qquad (7)$$

$$G_i = \Delta_i^a (H_i)^T (R_i)^{-1} \qquad (8)$$

where Δ_i denotes the uncertainty matrix; superscript 'a' and 'b' are the analysis and background, respectively; and G_i represents the gain matrix.

Subject to the constraints:

$$(\Delta_i^a)^{-1} = (\Delta_i^b)^{-1} + (H_i)^T (R_i)^{-1} H_i - \gamma_i S_i \geq 0 \qquad (9)$$

It should be noted that $(\Delta_i^b)^{-1} + (H_i)^T (R_i)^{-1} H_i$ in Equation (7) corresponds to the inverse of the analysis covariance matrix obtained in the KF (Luo and Hoteit, 2011). Therefore, the EnTLHF mainly differs from the KF in that it introduces an extra term $-\gamma_i S_i$ into the inverse covariance update formula Equation (7). Therefore, the KF can be considered as a special case of the EnTLHF with a performance level of $\gamma_i = 0$.

From the above it can be concluded that, when $\gamma_i = 0$ in Equation (2), the estimation error 'energy' $\left\| x_i - x_i^a \right\|_{S_i}^2$ is less than infinite. Therefore, that ensemble filtering algorithm does not guarantee that the estimated error 'energy' is bounded in the assimilation. In other words, this algorithm is not robust for the errors. However, the TLHF can guarantee the minimum energy of the estimation error when the system is disturbed even though the filtering accuracy is not high for highly nonlinear systems. Therefore, when the idea of an ensemble is applied to the TLHF method, a robust framework for ensemble filtering can be formed.

For Equation (7), if $\gamma_i > 0$, then $-\gamma_i S_i \leq 0$, and thereby the uncertainty matrix Δ_i^a is larger than that obtained in the EnKF ($\gamma_i = 0$). Therefore, the presence of the extra term $-\gamma_i S_i$ in Equation (7) introduces inflation into the covariance matrix obtained by the EnKF. The EnTLHF provides a general mathematical mechanism of the robust control theory for conducting covariance inflation.

2.3. EnTLHF-IT (IT)

Similarly, a more 'sophisticated' inflation scheme based on Equation (7) can also be derived. For instance, by choosing $S_i = I_m$, since $(\Delta_i^b)^{-1} + (H_i)^T (R_i)^{-1} H_i$ is equal to the analysis error covariance matrix inverse of ensemble idea in Equation (7). Then in the standard formulas of the ETKF: (1) $x_{i,j}^a = \bar{x}_i^a + \sqrt{n-1}\left(X_i^a\right)_j$ ($j = 1, \ldots, n$), where $(X_i^a)_j$ denotes the jth column of the square roots of analysis error covariance matrix; (2) $X_i^a = X_i^b T_i U_i$, where U is the centering matrix (Julier, 2003), and satisfies $UU^T = I$ and $U1_m^T = 0$ (Livings et al., 2007), with 1_m being the m-dimensional vector whose elements are all equal to 1; (3) the transform matrix $T_i = C_i(\Gamma_i + 1)^{-0.5}$, where C and Γ are the eigenvectors and eigenvalues by conducting SVD on $(HX^b)^T R^{-1} HX^b$, respectively; (4) $X_i^b = \frac{1}{\sqrt{n-1}}\left[x_{i,1}^b - \bar{x}_i^b, \ldots, x_{i,n}^b - \bar{x}_i^b\right]$, where X_i^b represents the square roots of the sample covariance matrices of the background ensemble; (5) the analysis error covariance matrix $P_i^a = X_i^a \left(X_i^a\right)^T = X_i^b T_i T_i^T \left(X_i^b\right)^T = X_i^b C_i \left(\Gamma_i + 1\right)^{-1} \left(X_i^b C_i\right)^T$; so one has

$$\left(\Delta_i^a\right)^{-1} = \left(X_i^b C_i\right)^{-T} \left(\Gamma_i + I\right) \left(X_i^b C_i\right)^{-1} - \gamma_i I_m \quad (10)$$

where $\Gamma = \text{diagonal}(\sigma_{i,1}, \sigma_{i,2}, \ldots, \sigma_{i,n-1})$ is a diagonal matrix containing the corresponding eigenvalues $\sigma_{i,j}$ ($\sigma_{i,j} \geq \sigma_{i,e}$ if $j < e$). To guarantee that $\left(\Delta_i^a\right)^{-1}$ is positive semi-definite, it is required that $\sigma_{i,j} - (\gamma_i - 1) \geq 0$. A convenient choice is to take $\gamma_i - 1 = \alpha \sigma_{i,n-1}$ with $\alpha \in [0, 1)$. So, the α is named as the performance level coefficient (PLC). By selecting this choice, the following is obtained:

$$\left(\Delta_i^a\right)^{-1} = \left(X_i^b C_i\right)^{-T} \left(\Gamma_i - \alpha \sigma_{i,n-1} I_m\right) \left(X_i^b C_i\right)^{-1}$$
$$= \left(X_i^b C_i\right)^{-T} \text{diagonal}\left[\sigma_{i,1} - \alpha \sigma_{i,n-1}, \ldots, \right.$$
$$\left. (1-\alpha)\sigma_{i,n-1}\right] \left(X_i^b C_i\right)^{-1} \quad (11)$$

In Equation (11), the eigenvalues Γ is modified using $\sigma_{i,j} - \alpha \sigma_{i,n-1}$ instead of $\sigma_{i,j}$ from the above. The following one can be realized: $T = C(\Gamma + I)^{-0.5}$, where $(\Gamma + I)^{-0.5}$ is the eigenvalues of the transform matrix T. That is to say, the transform matrix is inflated with the increase of α, and referred to as the inflation transform matrix basic of the inflation form of EnTLHF. The essence of this method is to inflate the analysis error covariance matrix by means of the inflation of the eigenvalues of the transform matrix.

This inflation mechanism is similar to that introduced by Luo and Hoteit (2011). They introduced the different inflation techniques from the essence, and the basic principles are as follows. Firstly, an SVD is conducted to the analysis error covariance matrix. Secondly, inflation to the eigenvalues $\eta_{i,j}(j = 1, \cdots, m)$ is proposed. In other words, $\frac{\eta_{i,j}}{1 - \alpha \eta_{i,j}/\eta_{i,1}}$ is used instead of $\eta_{i,j}$ as the eigenvalues of the analysis error covariance matrix, where $\alpha > 0$ is the PLC. The analysis error covariance matrix is inflated with the α growing and regarded as the EnTLHF-IA, which directly inflates the analysis error covariance matrix eigenvalues.

Although both IT and IA conduct SVD, the IA may attempt to directly modify the eigenvalues of the analysis covariance, which will be expensive for large-scale problems. Therefore, this is a selling point for the modifications of the eigenvalues of the transform matrix T, as it is more economical than the IA.

In the conventional update scheme, the inflation factor is invariant in time. In contrast, in the new inflation scheme, $\gamma_i = \alpha \sigma_{i,n-1} + 1$ is not only adaptive in time (with i), but it is also different for each mode of the system (column of $X_i^b C_i$), even for a fixed value of α. Also, alternative adaptive inflation schemes have already been proposed in the research conducted by Hoteit et al. (2002).

3. Numerical experiments

3.1. Lorenz-96 model

The Lorenz-96 model (Lorenz and Emanuel, 1998) is a strongly nonlinear dynamical system, with quadratic nonlinearity, and the governing equations are given by the following:

$$\frac{dX_k}{dt} = \left(X_{k+1} - X_{k-2}\right)X_{k-1} - X_k + F \quad (12)$$

where $k = 1, 2, \ldots, 40$, and for consistency, are defined as $X_{-1} = X_{39}, X_0 = X_{40}, X_1 = X_{41}$. It can be assumed that the true value of parameter F is eight. However, in the assimilation other values for F may be chosen, which represent different model error degrees.

In this study, a fourth-order Runge–Kutta method is used to integrate (and discretize) the system from time 0 to 125, including a constant integration step of 0.05 (overall 2501 integration steps). The first 500 steps are discarded to avoid transition effects, and the remaining 2000 steps were used for the data assimilation.

3.2. Observation system

$$y_i = H(x_i) + v_i \quad (13)$$

The above system is used to record the observation of the state vector x_i at the i time step, $x_i = (x_{i,1}, x_{i,2}, \ldots, x_{i,40})^T$, where v_i is the observation error that follows the Gaussian distribution

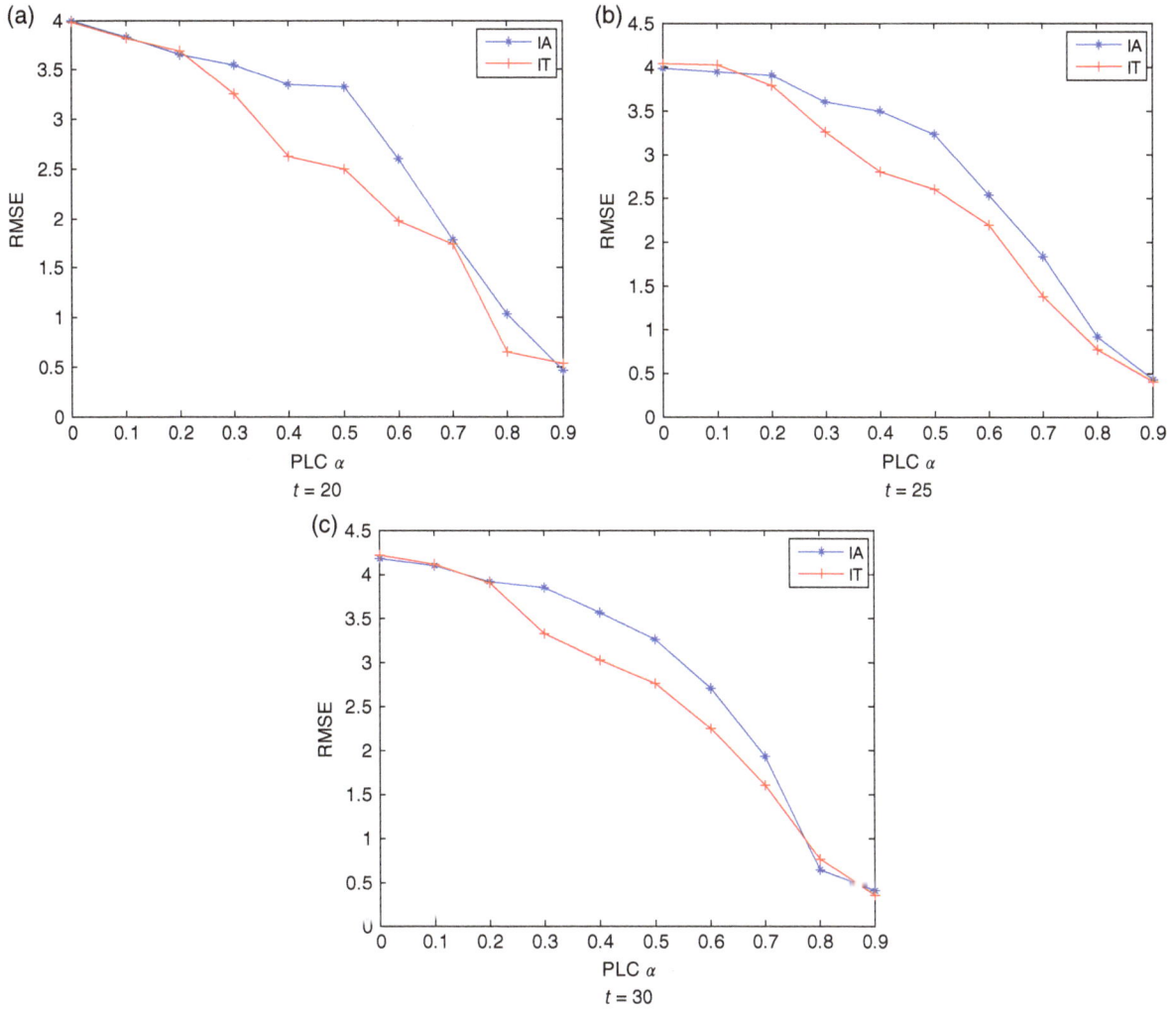

Figure 1. Time mean RMSE (using Equation (15), $L = 20$) of the IA and IT as a function of the PLC in assimilating the L96 model. The values of the parameter t are 20, 25, and 30, respectively. The assimilation results of the IA and IT are the same as the ETKF when the PLC $\alpha = 0$.

$N(v_i : 0, I_{40})$, with I_{40} being the 40-dimensional identity matrix. The observations are made for every four integration steps.

3.3. Validation statistics

Given a set of m-dimensional state vectors $\{x_i : x_i = (x_{i,1}, \ldots, x_{i,m})^T, i = 1, \ldots, i_{max}\}$, with i_{max} being the maximum time index, then the resulting root mean square error (function of PLC α) could be defined as:

$$\text{RMSE}_i = \frac{\left\| x_i^t - \bar{x}_i^a \right\|_2}{\sqrt{m}} \quad (14)$$

where $\bar{x}_i^a$ and x_i^t are the analysis mean and true state, respectively; and $\| \cdot \|$ denotes the Euclidean norm.

The average RMSE at time i_{max} and at repeated 20 ($L = 20$) times is defined as described in Hoteit *et al.* (2015), the $\overline{\text{RMSE}}$ were set

$$\overline{\text{RMSE}} = \frac{1}{L} \sum_{l=1}^{L} \left(\frac{1}{i_{max}} \sum_{i=1}^{i_{max}} \text{RMSE}_i \right)_l \quad (15)$$

Also, to further assess the behavior of the filters, the time evolution of the average ensemble spread (AES) of each filter is monitored, which is then computed at every filtering step as the following (Hoteit *et al.*, 2015):

$$\text{AES}_i = \sqrt{\frac{1}{m} \sum_{k=1}^{m} \sigma_{i,k}^2} \quad (16)$$

where $\sigma_{i,k}^2$ is the ensemble variance of $x_{i,k}$.

4. Experimental results and analysis

In this section, the ETKF, IA, and IT are compared in the condition of the assimilation steps, force parameter, ensemble sizes, and observation information changes, and the filter's robustness and accuracy are investigated. In practice, in order to avoid the cross covariance possibly are becoming unbounded one, we multiply $(1 - \alpha)$ on the left of the SVD matrix $(HX^b)^T R^{-1} HX^b$.

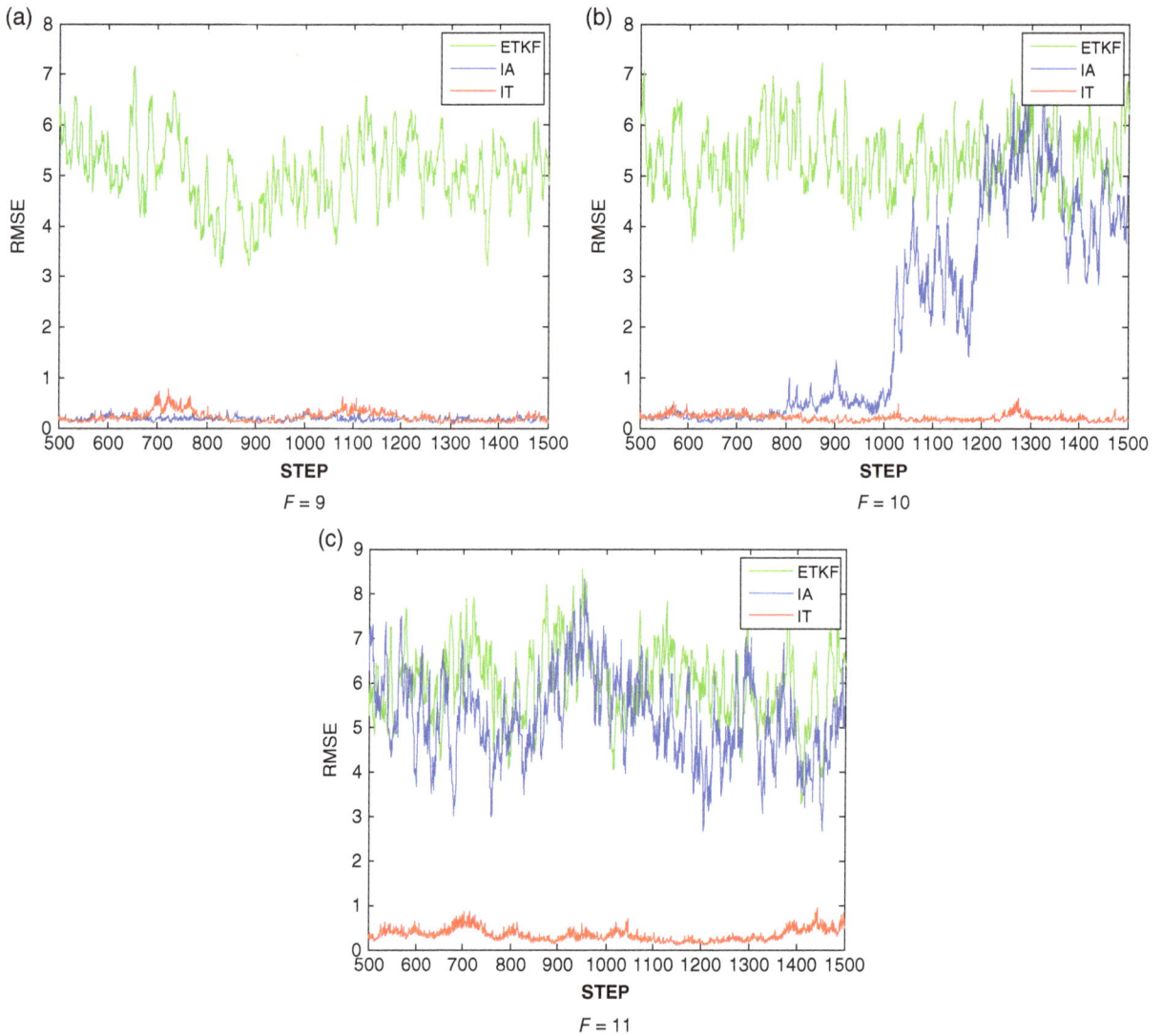

(a)

(b)

(c)

$F = 9$

$F = 10$

$F = 11$

Figure 2. RMSE (using Equation (14)) of the IA and IT in assimilating the L96 model. The values of the parameter F are 9, 10, and 11, respectively.

4.1. Influence of the assimilation steps

Figure 1 illustrates the results when the force parameter $F = 8$; ensemble size $N = 20$; observation error covariance matrix $R = 0.1$ and $\alpha \in [0, 0.1, 0.2, \dots, 0.9]$ at the assimilation time $t = 20$, 25, and 30 (in this experiment, there are 400, 500, and 600 integration steps, respectively), respectively. For each α, we repeat the experiments 20 ($L = 20$) times, each with a randomly drawn initial ensemble.

It can be seen from the figure that: (1) After using the EnTLHF data assimilation method of robust ensemble filter theory, including the inflating of the eigenvalues of analysis error covariance and transform matrix, when $t = 20$, 25, and 30, the time mean RMSE (using Equation (15)) monotonically decreases the functions with respect to α. When $t = 20$, the IT estimation error is less than the IA when $\alpha \in [0.3, \dots, 0.8]$. When $t = 25$, similarly, the IT RMSE is less than the IA in the conditions of $\alpha \in [0.2, \dots, 0.9]$. When $t = 30$, the IT estimate error is lower than the IA for $\alpha \in [0.3, \dots, 0.7]$; (2) For

the same PLC α, the estimation errors of IT and IA are not changed considerably under the three different assimilation time steps. This shows that the change of the short time assimilation steps does not have much influence on the estimation errors of IT and IA; (3) In the three different cases, all of the time mean RMSEs with $\alpha > 0$ are lower than those of the ETKF ($\alpha = 0$). These results indicate that the IA and IT are more robust and have greater filter accuracy than the ETKF. However, the IT is found to be superior to the IA.

4.2. Influence of the force parameter F

Figure 2 illustrates the RMSEs (using Equation (14)) of the ETKF, IA, and IT when the values of the F parameter are 9, 10, and 11. In the experiment, the remaining steps (500–1500) are interrupted during data assimilation, $R = 0.1$, $N = 20$, and PLC $\alpha = 0.6$.

From the results shown in Figure 2, the following can be concluded: (1) When $F = 9$, the RMSE of the ETKF remains large with the increase in assimilation

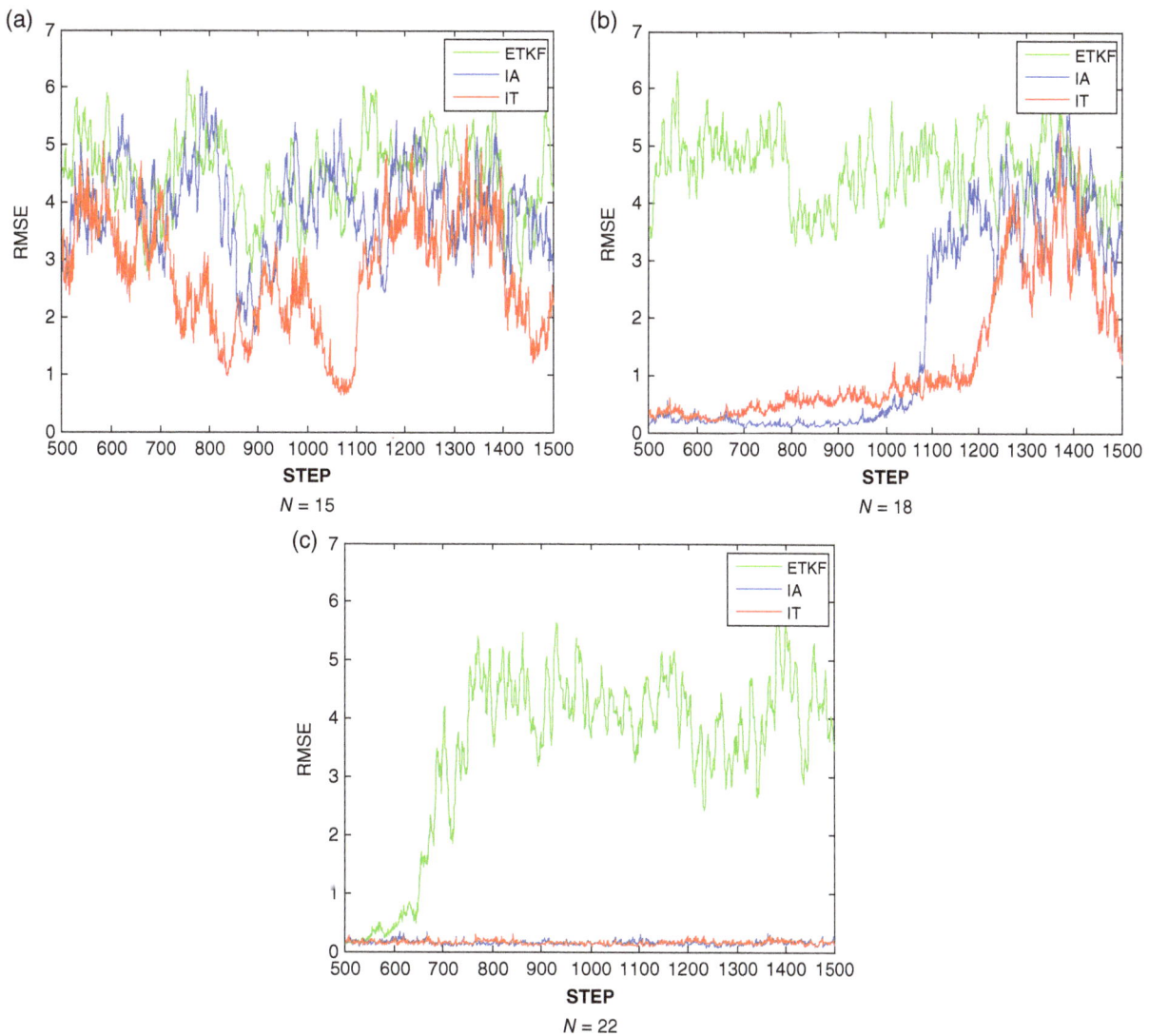

Figure 3. RMSE (using Equation (14)) of the IA, ETKF, and IT in assimilating the L96 model. The values of the parameter N are 15, 18, and 22, respectively.

time. However, the IA and IT have almost indistinguishable values, and the analysis shows that the RMSE remains small. Although the diffidence is not so big in the Lorenz-96 model, IT and IA SVD matrix are $n \times n$ and $m \times n$ (n is the size of ensemble and m is the dimension of state vector), respectively. In this experiment Lorenz-96 is 40 dimension, ensemble numbers are set as 20, so IA has larger SVD matrix than IT (IA is 40×20 and IT is 20×20). As for the calculation costs and matrix dimensions, IT is easier than IA; (2) In the case of $F = 10$, the RMSE of ETKF is also large, and the RMSE of the IA is also small until it reached 1000. After that point, the RMSE rapidly become larger. However, the RMSE of IT remains small with the increase in the assimilation time; (3) The result of $F = 11$ is that the RMSE of the IA and ETKF become large with the increase in assimilation time. However, the RMSE of the IT is found to be small. These results indicate that a filter divergence occurs for the IA, and the fluctuant estimate error of the IT is small. In other words, the IT shows robustness for the

changes of the force parameter. One possible reason for the IT showing a stable phenomenon may be that the IA will directly modify the eigenvectors of the analysis covariance, and each eigenvector has different degrees of inflation. Certain eigenvectors originally may be not so influential to the dynamics of the L96 system, but through inflation, their impacts are over-amplified, especially in the presence of model errors. In general, IT modifies a linear combination of the eigenvectors of the analysis covariance, this may make it slightly less likely to over-amplify the impacts of certain individual eigenvectors that are originally so influential. That is, IT changes the directions of eigenvectors, or in other words, the subspace spanned by the eigenvectors. This is possibly a phenomenon similar to the case of doing the transform $X^a = X^b TU$ in the ETKF, Tödter and Ahrens (2015) reports that using a random centering U is better than using a deterministic U, because a random U tends to rotate into different directions (hence changing the directions of eigenvectors of the analysis covariance).

(a)

(b)

(c)

(d)

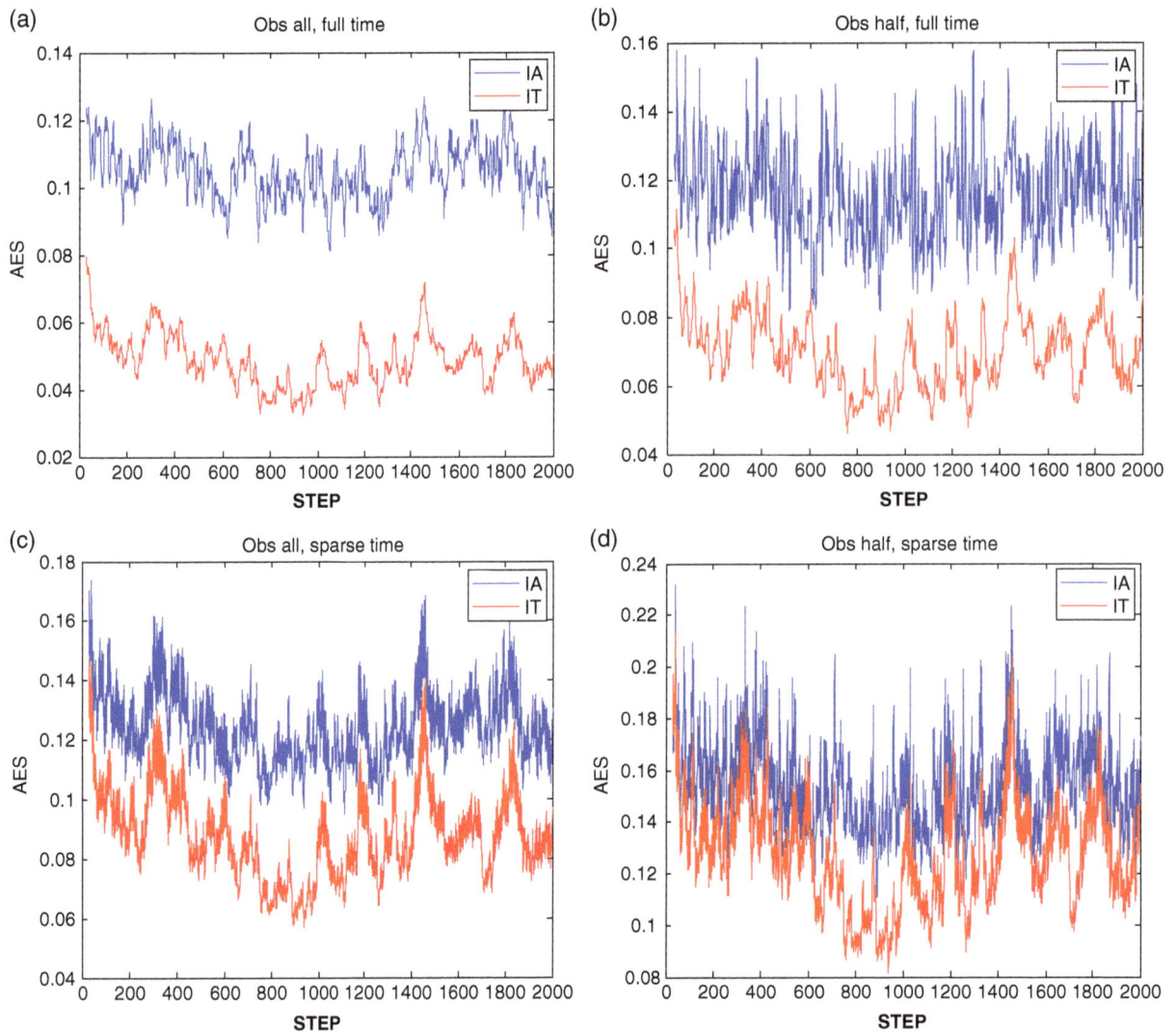

Figure 4. Time evolution of the AES as it results from IA and IT. Full time and sparse time correspond to the assimilation of observations at every model time step and every three model time steps, respectively. Obs all and obs half represent observations which are generated to each numbered element and each odd numbered element of the state vector.

4.3. Influence of the ensemble sizes

Figure 3 shows the RMSEs (using Equation (14)) of the ETKF, IA, and IT with the change of assimilation time for ensemble size $N = 15$, 18, and 22. This experiment's remaining 500–1500 steps are interrupted during data assimilation, $R = 0.1$, $F = 8$, and PLC $\alpha = 0.6$.

Figure 3 shows that in all of the cases, the ETKF have large RMSEs with assimilation steps. In the case of $N = 15$, the RMSEs of the IA and IT assimilation schemes are large when the assimilation step is increased. However, the IT is smaller than that of the IA and ETKF schemes. When $N = 18$, the RMSE of the IA is smaller or equal to that of the IT until the assimilation step reaches 1100. After that point, the RMSE of the IT is smaller than the IA, although both cases are large. However, when $N = 22$, both the IA and IT have almost indistinguishable values of RMSE, which have remained small. The filtering performance of the IT and IA are found to be better than the ETKF method. However, the IT method is superior to

the IA, which is verified by the increase of ensemble members.

4.4. Influence of the observational information

Figure 4 illustrates the time evolution of the AES as it result from the IA and IT implemented with 20 ensemble members, $F = 6$, PLC $\alpha = 0.6$, $R = 0.1$. The full time and sparse time correspond to the assimilation of observations at every model time step and every three model time steps, respectively. The obs all and obs half represent the observations which are generated to each numbered element, and each odd numbered element, of the state vector.

The following can be seen from Figure 4: (1) The representation of the AES of the IT is smaller than the IA during the process of assimilation, when each numbered element of the state vector have observations and assimilation of observations at every model time step; (2) The results of each odd numbered element of the state vector have observations and assimilation

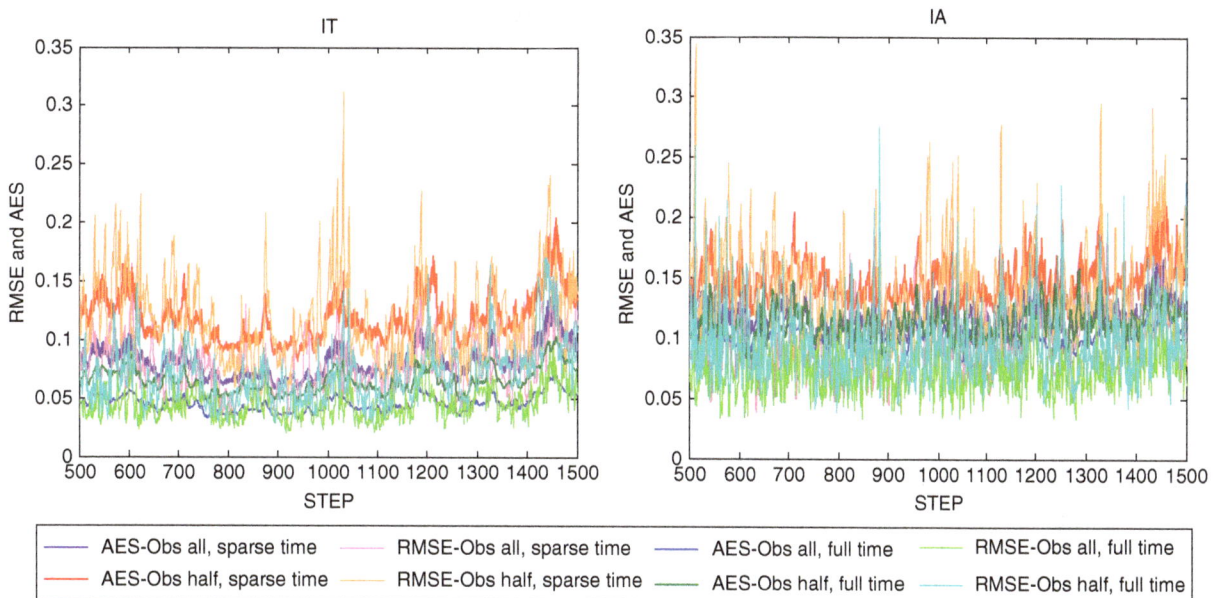

Figure 5. Time evolution of the AES (dark, thick colors) and the corresponding RMSEs (using Equation (14)) (light, thin colors) over the assimilation period with the different observations scenarios. Full time and sparse time correspond to assimilation of observations for every model time step and every three model time steps, respectively. Obs all and obs half represent observations which are generated to each numbered element and each odd numbered element of the state vector.

Table 1. Time average RMSEs (using Equation (15), $L = 1$) and AESs in the IA and the IT with the same observation information as in Figure 5.

	Obs all, full time	Obs half, full time	Obs all, sparse time	Obs half, sparse time
IA RMSE	0.0765	0.1013	0.1011	0.1495
IA AES	0.1056	0.1164	0.1259	0.1563
IT RMSE	0.0492	0.0744	0.0874	0.1553
IT AES	0.0497	0.0712	0.0897	0.1298

of observations at every model time step are that the AES of the IT is less than the IA; (3) Correspondingly, in the cases of each numbered element of the state vector having observations and assimilation of observations at every three model time steps, the AES of the IT is less than the IA in all of the processes of assimilation; (4) Overall, when each odd numbered element of the state vector have observations and assimilation of observations at every three model time steps, then the AES of the IT is less than the IA. This proves that the degree of convergence of the IT is better than that of the IA, whether or not there are observations; that is, the IT has better robustness than the IA.

To analyze the filters' spreads in relationship to their corresponding RMSEs, the time evolution of the spreads and RMSEs (as they resulted from the same single run) are plotted for the best performance of each filter with the different observational scenarios, as shown in Figure 5.

Figure 5 shows that for the IA, all AES curves except for obs half and sparse time, merge together, whereas for IT, it is more distinguishable with the iteration.

On the other hand, the ensemble spread of the IT and the IA continuously exhibits U-turn behavior with the iteration. The ensemble spreads of the IA are higher than those of the IT. The spreads and RMSEs have good comparability for the IT, with the exception of the IA spread, which seems to overestimate in comparison with the corresponding RMSE. One possible reason for this overestimation phenomenon may be that the different inflation methods of the IA and IT have different impacts on the background ensemble at the next assimilation cycle. It may be that the IA has the larger background ensemble discrete degree, leading to the analysis ensemble spread degree also being large, while the RMSEs are determined by the distances between the analysis average value and the truth, and thus are not necessarily as large as the corresponding ensemble spreads.

Table 1 shows the time average RMSEs and AESs in the IT and the IA with the same observation information as in Figure 5.

5. Discussion and conclusions

Using numerical experiments, the relative robustness of the specific forms of the IT are verified in comparison with the IA and ETKF without covariance inflation. The following conclusions are reached: (1) The time mean RMSE of the IT is a monotonically decreasing function with respect to α, for the three types of assimilation time steps. Overall, the IT is found to be better than the IA, in that the verified IT improve the accuracy and robustness of the filter. In the short time of the assimilation process, with the increase of the PLC α, the inflation of the eigenvalues of the transform matrix is indirectly

(IT indirectly inflates the eigenvalues of analysis covariance matrix) equivalent to the inflation analysis error covariance matrix; (2) The estimate error of the ETKF becomes a large value with increases in the model error, and filter divergence occurs in the IA method. However, in the IT, it remains small, which indicates that the IT have strong robustness to the variations of the model error; (3) The IT with the few ensemble members (like 15) have less estimation errors than the other two methods, and the IA and IT schemes have almost indistinguishable values of the estimation error with the increase in ensemble members. Therefore, according to the above theory, the IT is superior to the IA (besides Figure 3(b) shows that the IT results in smaller RMSE up to 1000); (4) The observation information plays an important role in the process of assimilation, and the AES of both the IT and IA increases with the reduction of observational information with fixed assimilation steps. However, the AES of the IT is less than that of the IA. The spreads and RMSEs show good comparability for the IT filter with the different observational scenarios. However, the IA spread seems to overestimate the corresponding RMSEs.

Overall, the IT method shows good robustness with the different conditions, when compared with the ETKF and IA methods. In this study, a robust filtering theory is introduced, which do not need to make assumptions regarding the statistical properties of the model and observations, and through the estimation error growth rate scale, a new data assimilation method of inflation transform matrix eigenvalues is proposed. The variable parameters do not affect the performance of the filter due to the robustness. This method can potentially be widely used in the data assimilation of other nonlinear systems.

Acknowledgements

We are very grateful to the three anonymous reviewers for their constructive suggestions and useful advices. This work is supported by the National Natural Science Foundation of China (NNSFC) project (grant numbers: 41461078, 41061038, and 41561080).

References

Altaf MU, Butler T, Luo XD, Dawson C, Mayo T, Hoteit I. 2013. Improving short-range ensemble Kalman storm surge forecasting using robust adaptive inflation. *Monthly Weather Review* **141**: 2705–2720.

Anderson JL, Anderson SL. 1999. A Monte Carlo implementation of the non-linear filtering problem to produce ensemble assimilations and forecasts. *Monthly Weather Review* **127**: 2741–2758.

Bishop CH, Etherton BJ, Majumdar SJ. 2001. Adaptive sampling with the ensemble transform Kalman filter. Part I: theoretical aspects. *Monthly Weather Review* **129**(3): 420–436.

Evensen G. 2003. The ensemble Kalman filters: theoretical formulation and practical implementation. *Ocean Dynamics* **53**: 343–367.

Hoteit I, Pham DT, Blum J. 2002. A simplified reduced order Kalman filtering and application to altimetric data assimilation in Tropical Pacific. *Journal of Marine Systems* **36**: 101–127.

Hoteit I, Pham DT, Gharamti ME, Luo XD. 2015. Mitigating observation perturbation sampling errors in the stochastic EnKF. *Monthly Weather Review* **143**: 2918–2936.

Julier SJ. 2003. The spherical simplex unscented transformation. *American Control Conference* **3**: 2430–2434.

Livings DM, Dance SL, Nichols NK. 2007. Unbiased ensemble square root filters. *Physica D: Nonlinear Phenomena* **7**(1): 1026505–1026506.

Lorenz EN, Emanuel KA. 1998. Optimal sites for supplementary weather observations: simulation with a small model. *Journal of the Atmospheric Sciences* **55**: 399–414.

Luo XD, Hoteit I. 2011. Robust ensemble filtering and its relation to covariance inflation in the ensemble Kalman filter. *Monthly Weather Review* **139**: 3938–3953.

Luo XD, Hoteit I. 2013. Covariance inflation in the ensemble Kalman filter: a residual nudging perspective and some implications. *Monthly Weather Review* **141**: 3360–3368.

Luo XD, Hoteit I. 2014. Ensemble Kalman filtering with residual nudging: an extension to state estimation problems with nonlinear observation operators. *Monthly Weather Review* **142**: 3696–3712.

Sakov P, Bertino L. 2011. Relation between two common localization methods for the EnKF. *Computational Geosciences* **15**: 225–237.

Simon D. 2006. *Optimal State Estimation: Kalman, H Infinity, and Nonlinear Approaches*. John Wiley & Sons, Inc.: Hoboken, New Jersey and Canada; 333–365.

Tödter J, Ahrens B. 2015. A second-order exact ensemble square root filter for nonlinear data assimilation. *Monthly Weather Review* **143**(6): 1347–1367.

Wang D, Cai X. 2008. Robust data assimilation in hydrological modeling – a comparison of Kalman and H-infinity filters. *Advances in Water Resources* **31**: 455–472.

4

Recent changes on land use/land cover over Indian region and its impact on the weather prediction using Unified model

Unnikrishnan C.K.,[1,*] Biswadip Gharai,[2] Saji Mohandas,[1] Ashu Mamgain,[1] E. N. Rajagopal,[1] Gopal R. Iyengar[1] and P. V. N. Rao[2]

[1]ESSO, MoES, National Centre for Medium Range Weather Forecasting, Noida, India
[2]Atmospheric and Climate Sciences Group, Earth & Climate Science Area, National Remote Sensing Centre, ISRO, Hyderabad, India

*Correspondence to:
U. C.K., ESSO, MoES,
National Centre for Medium
Range Weather Forecasting,
A50, Noida 201309, India.
E-mail:
unnikrishnan@ncmrwf.gov.in

Abstract

This study compares the changes of land use/land cover (Lu/Lc) or the surface type in last decades over India. Recent surface-type fractions show few major regional changes over India. There is a decrease in vegetation fraction, increase in urban and bare soil fractions over India. The Unified Model coupled with Joint UK Land Environment Simulator land surface model was used to investigate the recent Lu/Lc impact on weather prediction. Preliminary results show improvement in weather prediction by the incorporation of the recent Lu/Lc data. This highlights the need to incorporate more realistic Lu/Lc in the dynamical models for better weather prediction.

Keywords: land use land cover; land surface model; weather prediction; India; unified model

1. Introduction

Land surface acts as the lower boundary for the weather prediction models. The land surface forces and modifies the atmosphere above by transferring surface fluxes (latent heat flux, sensible heat flux, momentum and CO_2). The energy, water and carbon balance at surface are characterized by the regional features like topography, land use/land cover (Lu/Lc), soil type, etc. The regional heterogeneity of land surface directly impacts the surface fluxes to atmosphere and its evolution. The surface heterogeneity of land surface is accounted by different types of Lu/Lc data in land surface models. The Lu/Lc plays a key important role in the modulation of regional and local weather. Recent climate studies also suggest that the Lu/Lc changes can have local and remote (teleconnection) impact in dynamical model prediction (Devaraju et al., 2015).

Importance of Lu/Lc change on precipitation was investigated and documented by Pielke et al. (2007). Pielke et al. (2011) had suggested that the intensive Lu/Lc change over regions like India has more direct impact on regional climate. Mahmood et al. (2010) had stressed the importance of global monitoring of Lu/Lc change for both observational and modelling studies. Studies suggest that there are impacts on diurnal changes and mean surface warming as a result of the Lu/Lc change (Kalnay and Cai, 2003). The study of Feddema et al. (2005) suggested that the changes in land cover may influence Hadley and monsoon circulations. Most of these studies had focused on impact on long-term climate.

Studies on impact of land surface processes are limited over Indian region. Unnikrishnan et al. (2013) showed that the weekly satellite-observed vegetation fraction improves land surface parameter prediction over Indian region through better surface flux estimation. Similarly, Kumar et al. (2013) observed that updating vegetation fraction improves regional climate model predictions. Recent study of Xu et al. (2015) noted that Lu/Lc change shows enhanced 2-m air temperature variability in India. It is worth to investigate the impact of recent changes in Lu/Lc over Indian region on weather prediction.

Indian Space Research Organisation (ISRO) has developed meso-scale models compatible Lu/Lc data over Indian region derived from Advanced Wide Field Sensor (AWiFS) (Biswadip, 2014). The National Centre for Medium Range Weather Forecasting (NCMRWF) Unified Model (NCUM) uses by default the International Geosphere and Biosphere Programme (IGBP) Lu/Lc dataset which is based on National Oceanic and Atmospheric Administration's (NOAA) Advanced Very High Resolution Radiometer (AVHRR) data during 1992–1993 period. This paper investigates the recent changes in the nine surface-type fractions and its impact using NCUM coupled with Joint UK Land Environment Simulator (JULES) land surface model by using both IGBP and ISRO Lu/Lc datasets. Two separate prediction experiments (one wet and one dry) are performed to investigate the impact of ISRO Lu/Lc data on weather prediction.

The details of data and model are provided in Section 2. Section 3 compares the surface-type fraction

Table 1. Look-up table for converting 18 class Lu/Lc to 9 class JULES surface type fraction.

	IGBP Class (n)	Fraction of surface types (Fm)								
		Broad leaf tree	Needle leaf tree	C3 grass	C4 grass	Shrubs	Urban	Water	Bare soil	Ice
1	EN forest	0	70	20	0	0	0	0	10	0
2	EB forest	85	0	0	10	0	0	0	5	0
3	DN forest	0	65	25	0	0	0	0	10	0
4	DB forest	60	0	5	10	5	0	0	20	0
5	Mixed forest	35	35	20	0	0	0	0	10	0
6	Closed shrubs	0	0	25	0	60	0	0	15	0
7	Open shrubs	0	0	5	10	35	0	0	50	0
8	Woody savannah	50	0	15	0	25	0	0	10	0
9	Savannah	20	0	0	75	0	0	0	5	0
10	Grassland	0	0	70	15	5	0	0	10	0
11	Permanent wetland	0	0	80	0	0	0	20	0	0
12	Cropland	0	0	75	5	0	0	0	20	0
13	Urban	0	0	0	0	0	100	0	0	0
14	Crop/natural mosaic	5	5	55	15	10	0	0	10	0
15	Snow and ice	0	0	0	0	0	0	0	0	100
16	Barren	0	0	0	0	0	0	0	100	0
17	Water body	0	0	0	0	0	0	100	0	0

from both IGBP and ISRO. The impact of Lu/Lc on weather prediction is presented in Section 4. Section 5 summaries and conclude the results.

2. Data and model

The NCUM adapted from Met office, UK is used at NCMRWF (Rajagopal et al., 2012) for daily weather prediction. This is a grid point model with approximately 25 km horizontal resolution at mid latitude regions and it has 70 vertical levels. It also uses 4D variational data assimilation for creating model initial conditions. The surface parameters like soil moisture, snow depth, sea ice and SST are assimilated using a surface analysis scheme (SURF). JULES land surface model (Best et al., 2011; Clark et al., 2011) is coupled to the Unified Model. JULES has four vertical levels for soil moisture and temperature prediction. It is a tiled land surface model with sub-grid heterogeneity and computes surface temperatures and fluxes separately for each surface type in a grid-box. It can represent a grid box with nine major Lu/Lc types (surface type fractions) namely broad leaf trees, needle leaf trees, temperate grass, tropical grass, shrubs, urban, inland water, bare soil and land ice. JULES exchanges surface fluxes (latent heat flux, sensible heat flux, and CO_2) and momentum to the atmospheric model at each time step. At the same time atmospheric component of Unified Model forces the evolution of JULES land surface model by precipitation, surface short-wave and long-wave radiation, surface wind speed, pressure and moisture.

The NCUM uses by default the AVHRR-based Lu/Lc data from IGBP with 18 class Lu/Lc data (Loveland and Belward, 1997) to derive nine surface types for JULES land surface scheme. The dataset was derived from AVHRR data covering the period between April

1992 and March 1993 and the data have a resolution of 30 arc-second (~1 km) globally.

Recently ISRO IRS P6 satellite-derived Lu/Lc data over Indian region have become available (Biswadip, 2014). AWiFS sensor on board IRS P6 satellite during 2012–2013 period was used to derive these IGBP 18 surface types with a resolution of 30 s. Over Indian region, data in global IGBP data are replaced by ISRO Lu/Lc data. This global data was further processed using Central Ancillary Program (CAP) utility. CAP is a collection of UM utilities to make necessary ancillary input files for Unified Model-like topography, surface-type fraction, etc. Documentation of CAP is available at https://puma.nerc.ac.uk/trac/UM_TOOLS/wiki/ANCIL/CAPbuild#Introduction. CAP utility is used for converting 18 classes of Lu/Lc to 9 classes of JULES surface-type fractions. The aggregation method is used for the conversion to nine surface-type fraction of the target model grid boxes in CAP utility. The grid box surface types fraction (F_m) is calculated as below:

$$F_m = \Sigma \left(fm * \alpha_{mn} \right)$$

where F_m is the fraction of nine surface types ($m = 1–9$). fm is the fraction of each 18 IGBP class m and α_{mn} is the fraction of each nine surface-type m in each IGBP class n. The look up table used for α_{mn} in CAP is shown in Table 1.

The comparison of surface-type fractions in both datasets are discussed in the next section.

3. Surface-type fraction comparison

The recent period surface-type fraction from ISRO Lu/Lc data shows changes in type fractions compared to the IGBP data. The spatial pattern and area average of all surface types (broad leaf trees, needle leaf

Table 2. Area average fraction of surface types over Central India (70–85°E and 17–28°N) and throughout India.

	Central India		All India	
	IGBP	ISRO	IGBP	ISRO
Broad leaf tree	0.0784	0.1008	0.0981	0.1160
Needle leaf tree	0.0077	0.0218	0.0178	0.0253
C3 grass	0.5458	0.5038	0.4274	0.4021
C4 grass	0.0631	0.0598	0.0745	0.0738
Shrubs	0.0761	0.0343	0.1021	0.0793
Urban	0.00073	0.00479	0.0008	0.0034
Water	0.00914	0.0184	0.0149	0.0200
Bare soil	0.2187	0.2560	0.2640	0.2796
Ice	0	0	0	0

trees, temperate grass, tropical grass, shrubs, urban, inland water, bare soil and land ice) are compared in this section. Table 2 shows the changes Lu/Lc type in both datasets. Figure 1 shows the spatial variations of all nine surface-type fractions. The Figure 1(a) shows the surface-type fractions from IGBP and Figure 1(b) shows same from ISRO data. There is no land ice over the region. The average fractions of surface types are calculated over Central India region (17–28°N and 70–85°E) and all India is shown in Table 2. We can see from Figures 1 and Table 2 that there is a decrease in total vegetation type fractions (5.05%) over central India. Another major change is seen in area average bare soil fraction, which has increased in recent period (+3.72%), and area average urban fraction (+0.93%) over Central India. Similar change is also observed in all India. The increase in total average urban fraction, bare soil fraction and reduction in vegetation fraction are results of the anthropogenic activities during last two decades.

4. Impact on weather prediction

The two weather events are selected based on India Meteorological Department (IMD) weather daily/weekly report. The heavy rainfall event over Jammu and Kashmir on 3 September 2014 led to floods over the region. This event was selected for the wet case study. IMD-NCMRWF satellite-merged rainfall (Mitra et al., 2009) was used for the comparison of model rainfall forecasts. This daily rainfall data are available at 0.5° resolution from IMD (www.imdpune.gov.in). The 26 March 2014 was selected as the dry case, on that day there were above normal temperatures (3–4 K anomalies as per IMD weather report) over Western Ghats in Kerala. We have evaluated only the first 24-h model forecasts in this study.

The above normal maximum 2-m temperature was reported over Western Ghats, Kerala, India on 26 March 2014. The NCUM forecasts could reproduce the spatial pattern of above normal maximum 2-m temperatures over Western Ghats. Figure 2 shows the comparison of 2-m temperature model prediction using both Lu/Lc datasets. Even though the model is not able to predict

the actual maximum 2-m temperature reported by IMD (40.3 °C), it is seen that there is an increase of air temperature up to 1–2 K with the use of ISRO Lu/Lc data. The bias of model is also reduced by 1–2 K over the case study region. Lu/Lc types have different albedo, roughness length and surface conductance in the land surface model. The Lu/Lc type can directly impact surface fluxes in the model. The albedo and surface fluxes directly impact the surface energy budget and temperature forecast in the model. The change in Lu/Lc contributes to the change in temperature bias in the experiments.

A heavy rainfall was observed on 3 September 2014 over the Jammu and Kashmir region. Figure 3 shows comparison of model simulated rainfall with observations. It is seen from the figure that the use of the new Lu/Lc dataset has resulted in improved prediction of regional rainfall pattern. The rainfall biases (observation-model) from experiments are shown in Figure 4. The rainfall bias is reduced with ISRO Lu/Lc experiment. The observed average rainfall over the region (74.5–78°E and 33–36.5°N) was 19.2 mm, the model predictions with IGBP Lu/Lc gave an average of 8.9 mm while the ISRO Lu/Lc gave 11.1 mm. There is an improvement of rainfall by 2 mm day^{-1} (~20% of model rainfall) over the region. Lu/Lc types can impact the surface fluxes and lower boundary layer stability, any change in the surface evaporation and lower stability is reflected in the rainfall prediction in the model. This contributes to the change in the rainfall prediction in these experiments.

5. Summary and conclusion

The comparison of surface-type fraction from old IGBP data and recent ISRO data shows major regional changes. The major changes observed in the recent period are reduced total vegetation fractions and an increase in urban and bare soil fractions. These changes are the result of both anthropogenic activity and natural interannual variability of monsoon. The ISRO Lu/Lc data over Indian region were incorporated into NCUM and tested for two cases.

The preliminary results of both wet and dry weather case study of prediction show improvement in forecast by incorporating the ISRO Lu/Lc. Above normal temperature was improved by 1–2 K. This also raises the question whether the maximum 2-m air temperature over Indian region is increasing due to recent the Lu/Lc changes? This question should be addressed with a set of ensemble multi-model predictions along with strong observational evidences. This result also matches with the observation of Sertel et al. (2010), who reported that the incorporating of recent land cover dataset produced more accurate temperature simulation in a regional model over Turkey.

The prediction of Jammu and Kashmir rainfall event by incorporation of ISRO Lu/Lc data showed increased rainfall of 2 mm day^{-1} (around 20% of model rainfall).

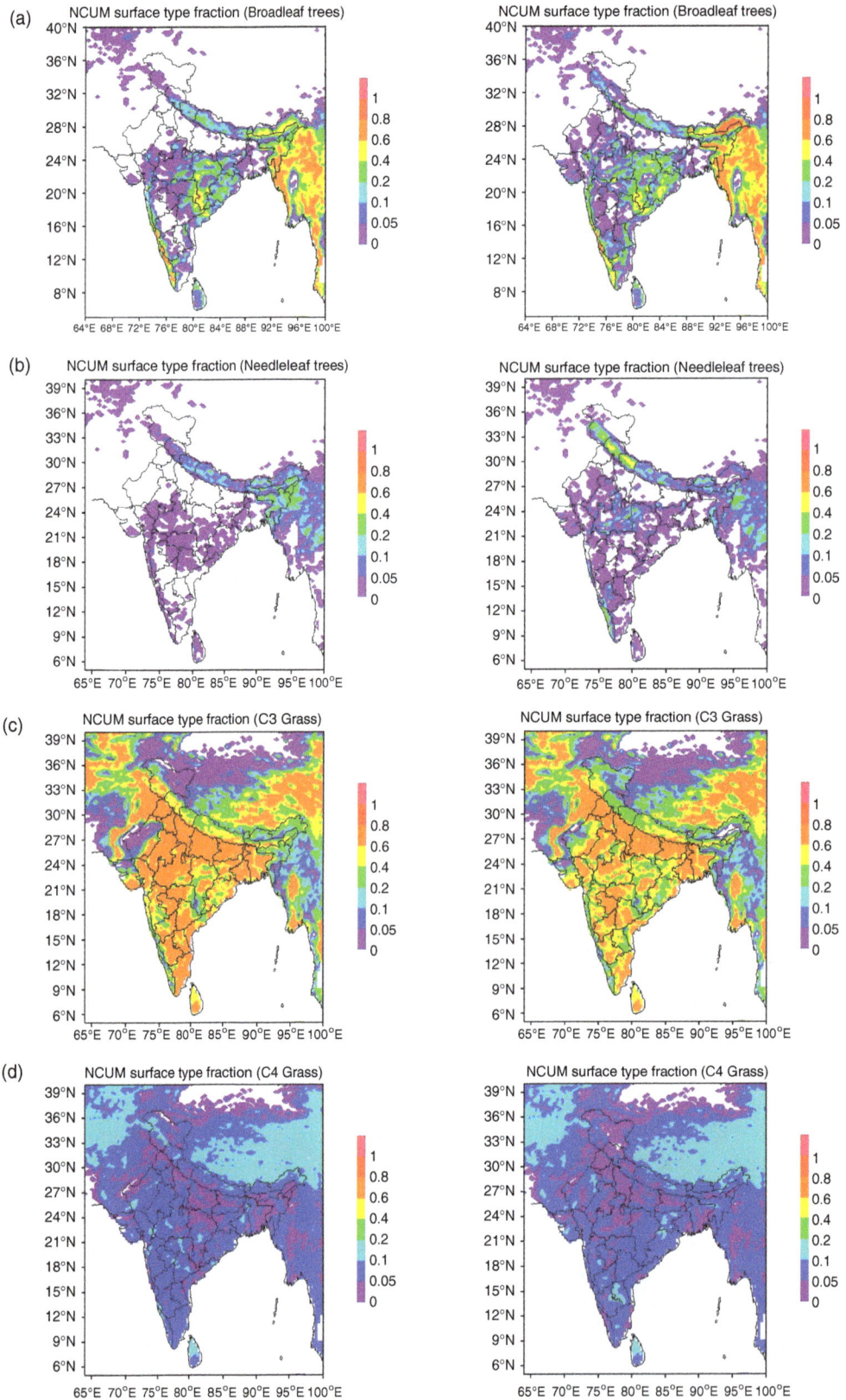

Figure 1. The surface fractions of (a) broad leaf trees, (b) needle leaf tree, (c) c3 grass, (d) c4 grass, (e) shrubs, (f) urban, (g) inland water and (h) bare soil fraction. Left panel shows the IGBP data and right panel shows NRSC ISRO data.

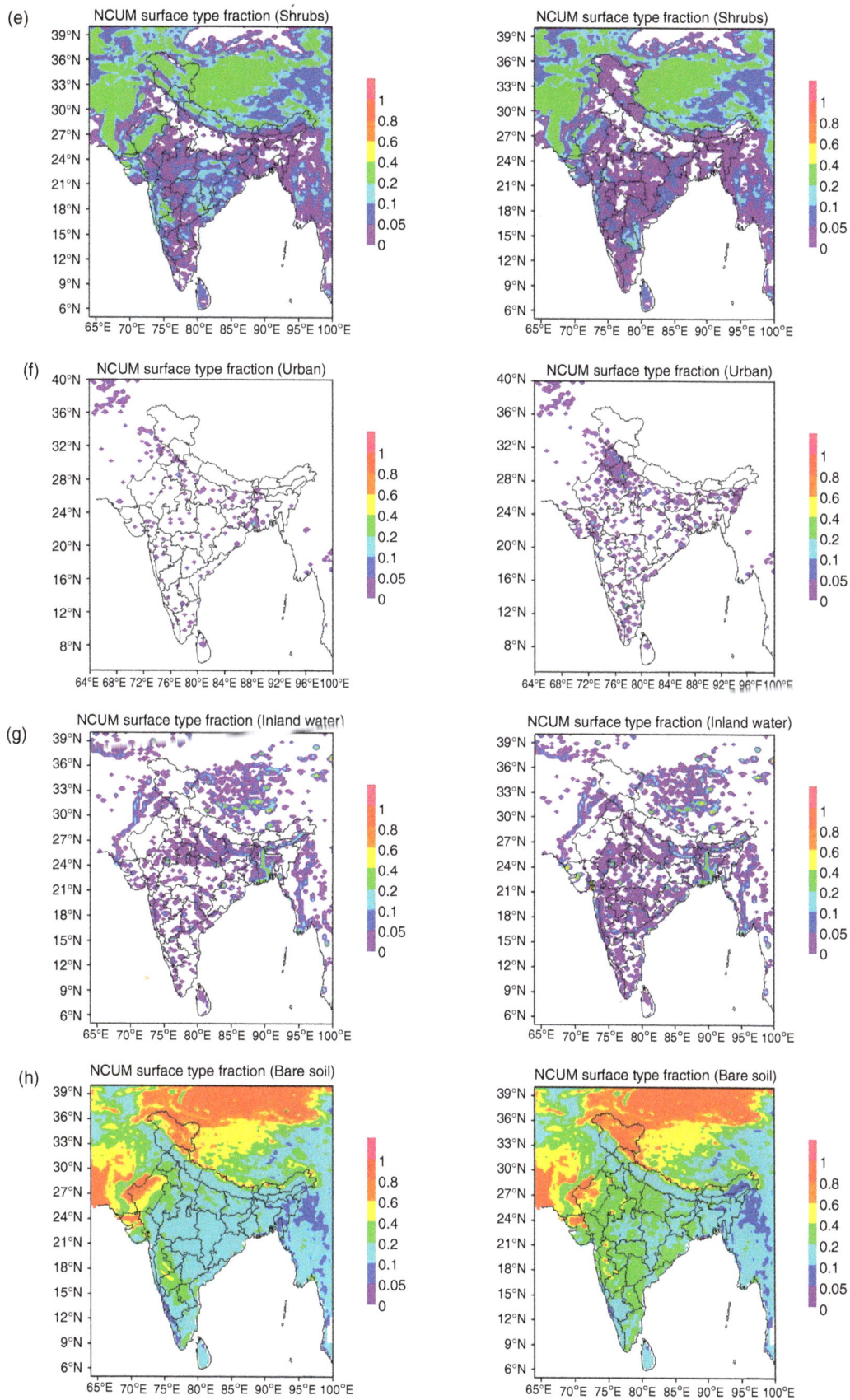

Figure 1. Continued.

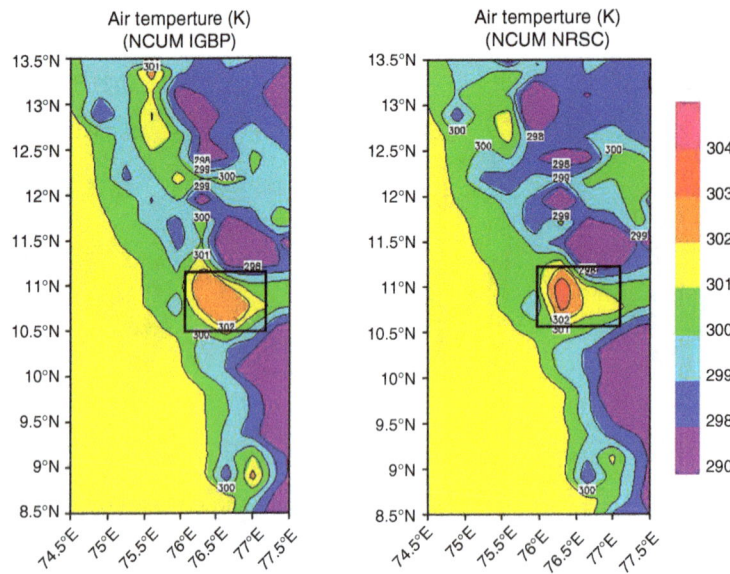

Figure 2. The comparison of 2 m temperature simulations using both Lu/Lc data sets. (a) The simulation with IGBP data and (b) simulation using ISRO data.

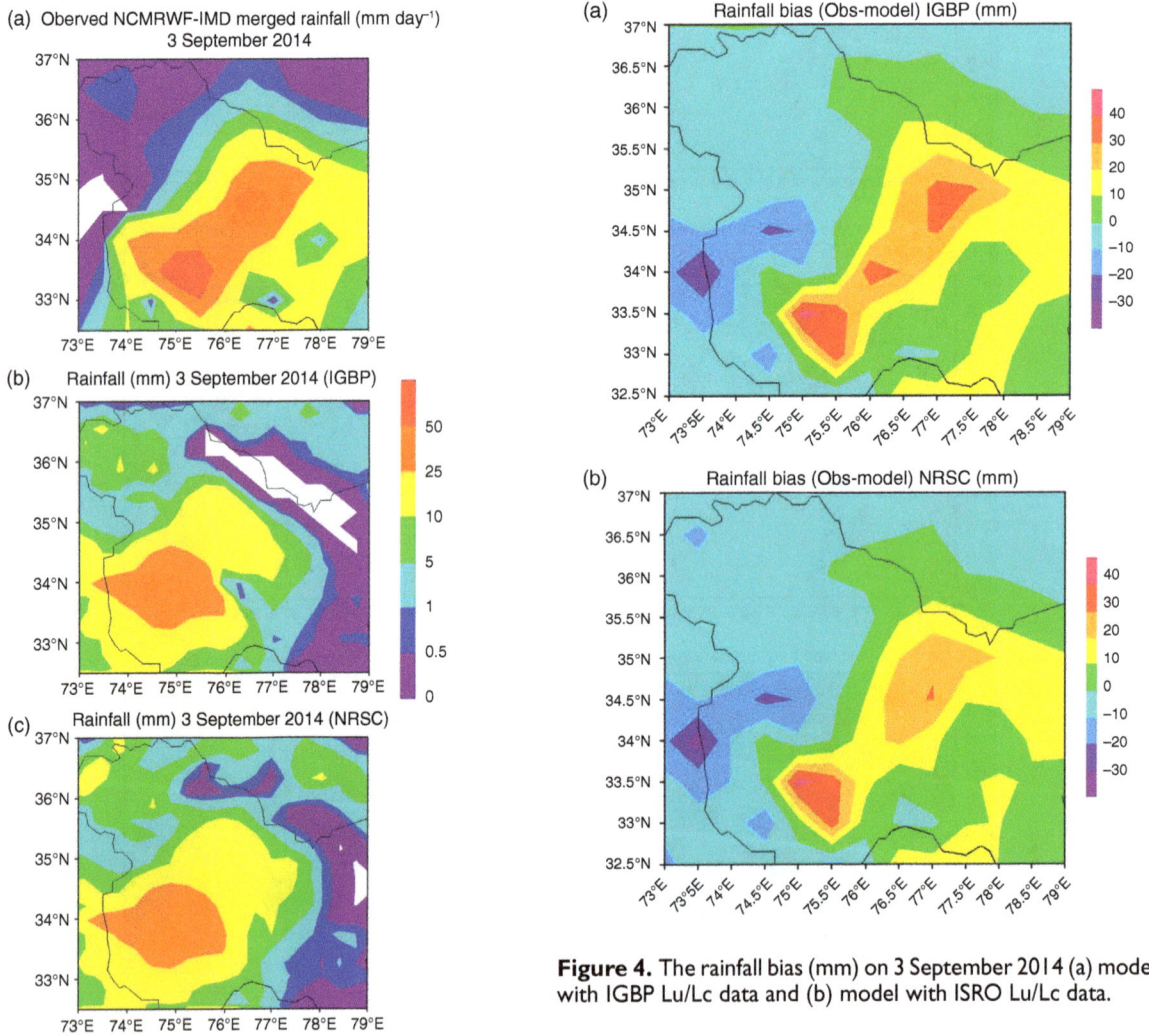

Figure 3. The comparison of model simulated rainfall (mm day^{-1}) with observation on 3 September 2014. (a) Observed rainfall, (b) model using IGBP data and (c) model using ISRO Lu/Lc data.

Figure 4. The rainfall bias (mm) on 3 September 2014 (a) model with IGBP Lu/Lc data and (b) model with ISRO Lu/Lc data.

This result is also consistent with previous experiment of Xie *et al.* (2014), who found that Lu/Lc can impact on precipitation simulation in WRF regional model over Beijing. Our preliminary results suggest that the use of more realistic Lu/Lc data can improve the weather predictions.

Acknowledgements

Unified Model is used at National Centre for Medium Range Weather Forecasting under a MoU between Ministry of Earth Sciences, Government of India and Met Office, UK to collaborate for developing a seamless numerical modelling system for prediction over different time ranges and spatial scales. Authors would like to thank all data sources used in this paper. We also thank IMD for providing the weather report. We would like to thank Ministry of Earth Sciences, Government of India for their support. Authors thank the anonymous reviewers for their constructive comments, which helped to improve the manuscript.

References

Best MJ, Pryor M, Clark DB, Rooney GG, Essery RLH, Ménard CB, Edwards JM, Hendry MA, Porson A, Gedney N, Mercado LM, Sitch S, Blyth E, Boucher O, Cox PM, Grimmond CSB, Harding RJ. 2011. The Joint UK Land Environment Simulator (JULES), model description – Part 1: Energy and water fluxes. *Geoscientific Model Development* **4**: 677–699.

Biswadip G. 2014. IRS-P6 AWiFS derived gridded land use/land cover data compatible to Mesoscale Models (MM5 and WRF) over Indian Region. NRSC Technical Document No. NRSC-ECSA-ACSG-OCT-2014-TR-651, 1–11.

Clark DB, Mercado LM, Sitch S, Jones CD, Gedney N, Best MJ, Pryor M, Rooney GG, Essery RLH, Blyth E, Boucher O, Harding RJ, Huntingford C, Cox PM. 2011. The Joint UK Land Environment Simulator (JULES), model description – Part 2: carbon fluxes and vegetation dynamics. *Geoscientific Model Development* **4**: 701–722.

Devaraju N, Bala G, Modak A. 2015. Effects of large-scale deforestation on precipitation in the monsoon regions: remote versus local effects. *Proceedings of National Academy of Sciences USA* **112**: 3257–3262, doi: 10.1073/pnas.1423439112.

Feddema JJ, Oleson KW, Bonan GB, Mearns LO, Buja LE, Meehl GA, Washington WM. 2005. The importance of land-cover change in simulating future climates. *Science* **310**: 1674–1678, doi: 10.1126/science.1118160.

Kalnay E, Cai M. 2003. Impact of urbanization and land-use change on climate. *Nature* **423**: 528–531.

Kumar P, Bhattacharya BK, Pal PK. 2013. Impact of vegetation fraction from Indian geostationary satellite on short-range weather forecast. *Agricultural and Forest Meteorology* **168**: 82–92.

Loveland TR, Belward AS. 1997. The IGBP-DIS global 1 km land cover data set. DISCover: first results. *International Journal of Remote Sensing* **18**(15): 3289–3295, doi: 10.1080/01431 1697217099.

Mahmood R, Quintanar AI, Conner G, Leeper R, Dobler S, Pielke RA Sr, Beltran-Przekurat A, Hubbard KG, Niyogi D, Bonan G, Lawrence P, Chase T, McNider R, Wu Y, McAlpine C, Deo R, Etter A, Gameda S, Qian B, Carleton A, Adegoke JO, Vezhapparambu S, Asefi S, Nair US, Sertel E, Legates DR, Hale R, Frauenfeld OW, Watts A, Shepherd M, Mitra C, Anantharaj VG, Fall S, Chang H-I, Lund R, Treviño A, Blanken P, Du J, Syktus J. 2010. Impacts of Land Use/Land Cover Change on Climate and Future Research Priorities. *Bulletin of the American Meteorological Society* **91**: 37–46.

Mitra AK, Bohra AK, Rajeevan MN, Krishnamurti TN. 2009. Daily Indian precipitation analyses formed from a merge of rain-gauge with TRMM TMPA satellite derived rainfall estimates. *Journal of Meteorological Society of Japan* **87A**: 265–279, doi: 10.2151/jmsj.87A.265.

Pielke RA, Adegoke J, Beltran-Przekurat A, Hiemstra CA, Lin J, Nair US, Niyogi D, Nobis TE. 2007. An overview of regional land-use and land-cover impacts on rainfall. *Tellus B* **59**: 587–601, doi: 10.1111/j.1600-0889.2007.00251.x.

Pielke RA, Pitman A, Niyogi D, Mahmood R, McAlpine C, Hossain F, Goldewijk KK, Nair U, Betts R, Fall S, Reichstein M, Kabat P, de Noblet N. 2011. Land use/land cover changes and climate: modeling analysis and observational evidence. *WIREs Climate Change* **2**: 828–850.

Rajagopal EN, Iyengar GR, George JP, Das Gupta M, Mohandas S, Siddharth R, Gupta A, Chourasia M, Prasad VS, Aditi, Sharma K, Ashish A. 2012. Implementation of unified model based analysis-forecast system at NCMRWF, NCMRWF Technical Report No. NMRF/TR/2/2012, 1–46.

Sertel E, Robock A, Ormeci C. 2010. Impacts of land cover data quality on regional climate prediction. *International Journal of Climatology* **30**: 1942–1953, doi: 10.1002/joc.2036.

Unnikrishnan CK, Rajeevan M, Vijaya Bhaskara Rao S, Kumar M. 2013. Development of a high resolution land surface dataset for the South Asian monsoon region. *Current Science* **105**(9): 1235–1246.

Xie Y, Shi J, Lei Y, Xing J, Yang A. 2014. Impacts of land cover change on simulating precipitation in Beijing area of China. IEEE International Geoscience and Remote Sensing Symposium (IGARSS), Quebec City, Canada, 13–18 July 2014, 4145–4148, doi: 10.1109/IGARSS.2014.694740.

Xu Z, Mahmood R, Yang ZL, Fu C, Su H. 2015. Investigating diurnal and seasonal climatic response to land use and land cover change over monsoon Asia with, the Community Earth System Model. *Journal of Geophysical Research – Atmospheres* **120**: 1137–1152.

5

Analysis of Rocky Mountain mesoscale convective system initiation location clusters in the Arkansas-Red River Basin

Elisabeth Callen[1,*] and Donna F. Tucker[2]

[1]Department of Geological and Atmospheric Sciences, Iowa State University, Ames, IA, USA
[2]Department of Geography and Atmospheric Science, University of Kansas, Lawrence, KS, USA

*Correspondence to:
E. Callen, Department of
Geological and Atmospheric
Sciences, Iowa State University,
3014 Agronomy, Ames, IA
50011, USA.
E-mail: ecallen@iastate.edu

Abstract

Mesoscale convective systems (MCSs) are the largest precipitation producers in terms of total accumulated precipitation and are, therefore, the focus of this study. For this analysis, the MCSs forming within the Rocky Mountain portion of the Arkansas-Red River Basin were studied in April through September in 1996–2006. Once variables favourable for the initiation were determined, statistical analyses were performed on the variables. Presented are the general results of an analysis of the multiple linear regressions and principal component analyses showing that in each area where MCSs initiate, the most favourable conditions for initiation are unique.

Keywords: mesoscale convective system; statistical analysis; mountain meteorology

1. Introduction

Mesoscale convective systems (MCSs) and mesoscale convective complexes (MCCs; Maddox, 1980; Zipser, 1982) are prolific rain producers during the warm season (Fritsch *et al.*, 1986), potentially causing flash flooding (Doswell *et al.*, 1996 – hereafter D96; Schumacher and Johnson, 2005). The orogenic MCSs (Tripoli and Cotton, 1989) compound flooding issues typically associated with MCSs due to the limited terrain (e.g. valleys) the floodwaters can occupy. While, Tucker and Crook (1998) studied initiation characteristics for an orogenic MCS, which occurred just east of Denver, Colorado (CO), Maddox (1980) and Fritsch *et al.* (1986) determined a portion of MCCs developed on the Rocky Mountains' eastern slope. Also, Carbone *et al.* (2003) indicated that diurnal forcing in the Rocky Mountains plays an important role in the initiation and propagation of MCSs.

Many studies (Cotton *et al.*, 1983; Banta and Schaaf, 1987 – hereafter BS87; Banta, 1990 – hereafter B90; Tucker and Crook, 2001 – hereafter TC01; Tucker and Crook, 2005 – hereafter TC05) have examined thunderstorm mountain initiation and arrived at conclusions concerning the initiation locations and wind speeds and directions and are the starting point for MCS mountain initiation consideration. The discussions from Cotton *et al.* (1983), BS87, B90, TC01, and TC05 indicate that there is a potential for wind speed and direction at ridgetop height to predict the initiation location of a thunderstorm.

While not pertaining to orogenic MCSs in particular, several studies provide valuable information for variable selection. Jirak and Cotton (2007 – hereafter

JC07) and McAnelly and Cotton (1986 – hereafter MC86), among others, went into extensive detail on the MCS precursor environment. Rotunno *et al.* (1988 – hereafter R88) was pivotal in determining shear needed for squall lines (subset of MCSs). Johnson and Mapes (2001 – hereafter JM01) discussed that local barriers do affect time and place of initiation.

As a precursor to this analysis, Tucker and Li (2009 – hereafter TL09) studied the single cell systems, multicell systems, and MCSs within the entire Arkansas-Red River Basin (ARB), specifically looking at precipitation, size, and the number of each system type for the years 1996 through 2006. The study described here uses a small subset of the TL09 database when analysing the initiation (appearance of first precipitation) characteristics for the Rocky Mountain MCSs, which was not carried out in TL09. The study uses precipitation that initiated within the Rocky Mountains that will eventually become the MCSs. These Rocky Mountain MCSs were chosen since three fourths of the ARB precipitation fell from MCSs (TL09) and mountainous terrain can increase the chances of flooding from large precipitation events. Section 2 details the steps taken to determine the MCS initiation variables. Section 3 contains the resulting statistical runs of the multiple linear regressions (MLRs) and principal component analysis (PCAs). Trends for the most significant variables are included to show the overall change from 6 h prior to initiation through the initiation hour with no comparison against null cases. Conclusions follow in section 4. We hypothesize that a single set of variables cannot be applied to the Rocky Mountain portion, parts of CO and New Mexico (NM), of the ARB to determine MCS initiation location. The hypothesis is consistent with Moninger *et al.* (1991), who indicated that one set of criteria in a large domain for initiation does not provide meaningful results (no generalized conditions).

2. Data and methods

This study utilizes the multi-sensor precipitation data (stage 3; Young *et al.*, 2000) used in TL09's analysis. Delineation of the multi-sensor precipitation data into systems using MATLAB occurred as part of TL09. The MCS criteria, met within the ARB, were any system having continuous precipitation for at least 6 h and had a footprint (number of cells occupied) of at least 21 precipitation cells (TL09) at its maximum, where a precipitation cell is approximately 4 km by 4 km (ARM: Climate Research Facility, 2011).

For determining MCSs, only the ones initiating (first precipitation on the hourly digital precipitation array) west of 104°W, in the warm season (April to September), and in the years 1996–2006 in the ARB were used for mountain initiation. These MCSs were located in the westernmost area of the ARB. For these MCSs, only the appearance of the first precipitation (mountain initiation) had to occur by 104°W. The MCSs did not have to meet the MCS criteria by 104°W; the criteria had to be met within the TL09 domain. MATLAB was used to determine the MCSs, change the

coordinate system into Polar Stereographic coordinates, and transform the local time in TL09's database into UTC. Once the transformations occurred, a cluster analysis, using the latitude and longitude of the first appearance of precipitation on the hourly digital precipitation array for each MCS (MCS initiation location), was run using between-groups linkage and Euclidean distance. Between-groups linkage is computed by calculating the smallest average distance between each set of points and then combining the closest points into a group until the desired clusters are obtained. The assumption made is that the MCSs close together in latitude and longitude have similar initiation characteristics leading to individual cluster domains centred on individual peaks/one specific area. Only clusters with 30 or more members were to be singled out for analysis, but the overall results (all the clusters combined) included the values from all clusters with 20 or more members. Figure 1 shows the clusters that were used in this analysis. The line at 104°W indicates the eastern boundary of the Rocky Mountains. No null cases were included due to the available information in the TL09 database.

Once the cluster analyses were completed, two North American Regional Reanalysis (NARR) model runs before and one model run after MCS initiation were downloaded for analysis (ex: initiation hour: 17Z; models downloaded/analysed: 12Z, 15Z, 18Z) from the NOAA National Operational Model Archive and Distribution System (NOMADS). NARR was used

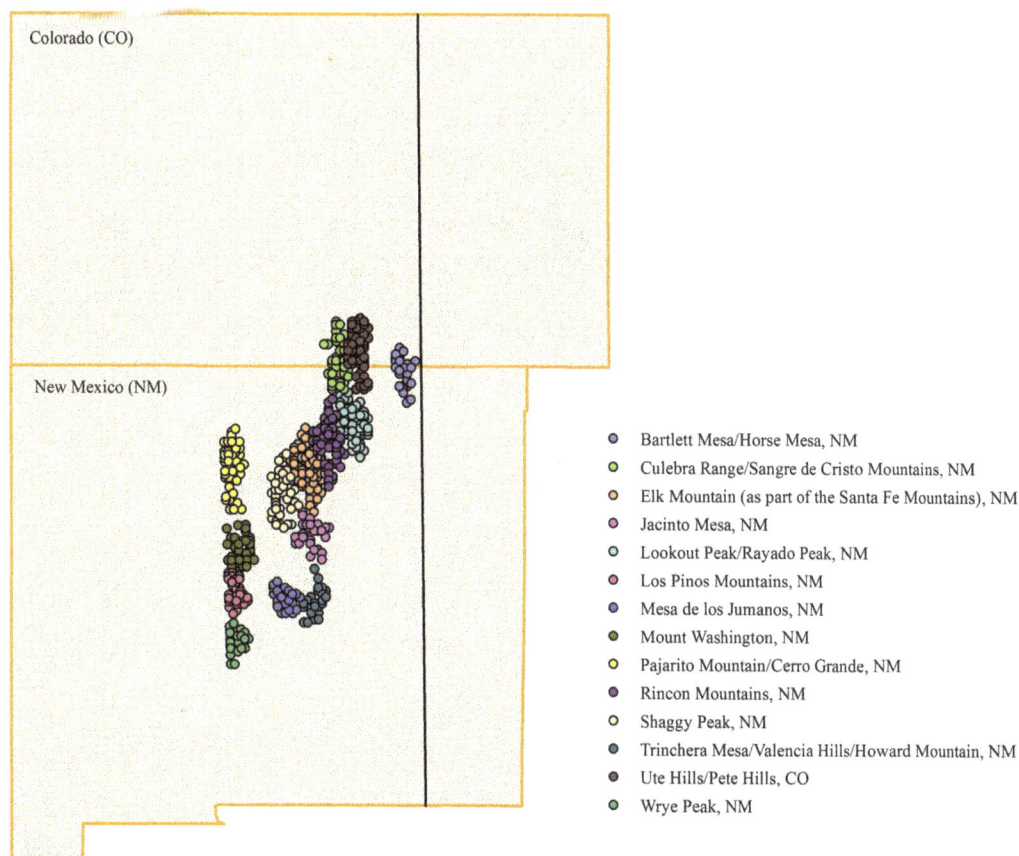

Colorado (CO)

New Mexico (NM)

- Bartlett Mesa/Horse Mesa, NM
- Culebra Range/Sangre de Cristo Mountains, NM
- Elk Mountain (as part of the Santa Fe Mountains), NM
- Jacinto Mesa, NM
- Lookout Peak/Rayado Peak, NM
- Los Pinos Mountains, NM
- Mesa de los Jumanos, NM
- Mount Washington, NM
- Pajarito Mountain/Cerro Grande, NM
- Rincon Mountains, NM
- Shaggy Peak, NM
- Trinchera Mesa/Valencia Hills/Howard Mountain, NM
- Ute Hills/Pete Hills, CO
- Wrye Peak, NM

Figure 1. Cluster locations. Black line indicates 104°W.

over other models as it contained no missing data and was at a constant resolution over the entire period. NARR was also run every 3 h since January 1979 and was run on 45 vertical layers with a horizontal grid spacing of 32 km (Mesinger *et al.*, 2006). Even though the analysis is over a 7-h period and NARR is run every 3 h, there was no temporal interpolation performed to 'fill in' the hourly data between the NARR model runs and the variable values were not 'carried over' into the next hour to fill in the gaps. The NARR data were collected at the initiation location for each MCS.

Integrated Data Viewer (IDV, Murray, 2003) was used to analyse the NARR data to collect the NARR variables listed in Table 1. The polar stereographic coordinates of the systems were re-projected into the correct coordinate system. The derived variables, listed in Table 1, were calculated, using MATLAB, from the variable information (i.e. U and V component wind speed values) collected in IDV (using NARR). No filtering of the MCS initiation environments (e.g. modification by active convection, lifting by various types of fronts) occurred due to the nature of the analysis (observing clusters as a whole) and the size of the smaller

clusters. If the clusters were further broken down by MCS initiation environments, the sample size (especially with the smaller clusters) would be too small to glean any significant results.

MLRs and PCAs were performed on the non-standardized data sets using Statistical Packages for the Social Sciences (SPSS). The data were not standardized because of the variations present from cluster to cluster since the standardization of one cluster is not the same as the standardization of another cluster. MLRs and PCAs were applied to the 3 hourly data spanning the 6 h prior to initiation through the initiation hour at the initiation location, similar to the JC07 analysis. The systems would typically meet MCS criteria once the systems reached the Plains (east of the initiation locations). Fourteen statistical runs for each cluster were obtained – seven for MLRs and seven for PCAs on the 3 hourly data over the 7-h data period.

MLR was used as an abbreviated way to determine the variables that were needed to attain the MCS footprint within each cluster, even if it was not a linear fit. The assumption was the variables needed to produce the footprint were also the variables needed for initiation.

Table 1. Variable information including name, description, units, and the literature reference.

Name	Description	Units	Reference
	NARR		
Thickness	1000–500 hPa thickness	gpm	JC07
PW	Precipitable water	mm	MC86
CAPE	Surface convective available potential energy	$J\,kg^{-1}$	JM01
CIN	Surface convective inhibition	$J\,kg^{-1}$	JM01
SRH	Storm relative helicity (0–3000 m)	$m^2\,s^{-2}$	JC07
GH600	Geopotential height, 600 hPa	gpm	JC07
GH500	Geopotential height, 500 hPa	gpm	JC07
GH300	Geopotential height, 300 hPa	gpm	JC07
GH200	Geopotential height, 200 hPa	gpm	JC07
SH850	Specific humidity, 850 hPa	$kg\,kg^{-1}$	D96
SH800	Specific humidity, 800 hPa	$kg\,kg^{-1}$	D96
SH600	Specific humidity, 600 hPa	$kg\,kg^{-1}$	D96
SH500	Specific humidity, 500 hPa	$kg\,kg^{-1}$	D96
SH300	Specific humidity, 300 hPa	$kg\,kg^{-1}$	D96
SH200	Specific humidity, 200 hPa	$kg\,kg^{-1}$	D96
UC600	U wind component, 600 hPa	$m\,s^{-1}$	B90, TC05
UC500	U wind component, 500 hPa	$m\,s^{-1}$	B90, TC05
UC300	U wind component, 300 hPa	$m\,s^{-1}$	B90, TC05
UC200	U wind component, 200 hPa	$m\,s^{-1}$	B90, TC05
VC600	V wind component, 600 hPa	$m\,s^{-1}$	B90, TC05
VC500	V wind component, 500 hPa	$m\,s^{-1}$	B90, TC05
VC300	V wind component, 300 hPa	$m\,s^{-1}$	B90, TC05
VC200	V wind component, 200 hPa	$m\,s^{-1}$	B90, TC05
T600	Temperature, 600 hPa	°C	JC07
T500	Temperature, 500 hPa	°C	JC07
T300	Temperature, 300 hPa	°C	JC07
T200	Temperature, 200 hPa	°C	JC07
	Derived		
WD600	Wind direction, 600 hPa	°	BS87
WD500	Wind direction, 500 hPa	°	BS87
UWSS500	U component wind shear, 500 hPa to surface	$m\,s^{-1}$	R88
UWS600500	U component wind shear, 500 to 600 hPa	$m\,s^{-1}$	R88
UWSS600	U component wind shear, 600 hPa to surface	$m\,s^{-1}$	R88
VWSS500	V component wind shear, 500 hPa to surface	$m\,s^{-1}$	R88
VWS600500	V component wind shear, 500 to 600 hPa	$m\,s^{-1}$	R88
VWSS600	V component wind shear, 600 hPa to surface	$m\,s^{-1}$	R88

Footprint was used as the dependent variable because it is a part of the MCS criteria. The stepwise method was chosen with an entry value of 0.15 and an exit value of 0.20 with the exit of a variable occurring when the variable was too highly correlated to another variable. One issue was a lack of independence occurring in some of the MLRs, but this dependency was accepted because the variables could be coupled to one another. Cross validation was not performed due to the smaller clusters' sample sizes and due to how the variables were sampled from NARR.

PCA was used as a more in-depth tool than the MLR and also because of its dimension reduction. The variance considered in this analysis is from the overall variance accounted for by the components. The significance of a variable in the PCA was determined by the amount of variance accounted for throughout all components. Higher values of accounted for variance could indicate a stronger association between the variable and the likelihood for initiation.

3. Results

The MLR and PCA results, at the 95% confidence level, are presented for each cluster with 30 or more members, while Callen (2012) presented a cluster by cluster breakdown of all the variable results for clusters with 20 or more members. For the results of the MLRs and PCAs, Table 2 includes the best and worst fits for the analyses from the 3 hourly data over the 7-h data period. The results from Table 2 indicate that as the sample size decreases, the fits were better, which is expected since there were fewer data points to analyse. While, at times, the fits to the data were very poor, these analyses were still used to determine the best variables for potential MCS initiation in the ARB. A second cluster analysis was performed on the original Elk Mountain, NM cluster because the fits

were poor and it was determined the original cluster could be broken down by the 500 and 600 hPa wind directions.

An analysis was also performed on all the clusters together (listed as 'Overall') and was performed to show the individual cluster analyses were a better fit to the data than the overall results. The most significant aspect of the overall analysis was the very poor fits. The poor fits indicated that finding most likely MCS initiation conditions for the entire Rocky Mountain portion of the ARB would not likely succeed.

The individual cluster results could be used in an ingredients-based approach (Johns and Doswell, 1992, D96), with Table 3 showing the trends of the important variables in each cluster over the 7-h analysis window with the trends providing more information about the state of the atmosphere than variable values. Depending on the cluster, 85–90% of the cases have the indicated trends. The ingredients-based approach works since it generalizes the conditions needed for MCS initiation within the various clusters. The importance of a variable was determined by the number of times and how a variable was included in the MLRs and PCAs. Any positive or negative trend shown in Table 3 was considered significant enough to report. As can be seen in the Table 3 trends, no two clusters are the same overall. There are some similarities in the trends from 6 h prior to initiation through the initiation hour. These similarities include: no significant changes in the upper air temperatures, no significant changes in precipitable water (PW), and negative changes in convective inhibition (CIN) with most clusters. However, the other variables differ providing resistance to a common set of variable trends. It should be noted that WD500 and WD600 were not significant in any of the clusters.

Some useful results can be compiled from the information available in Table 3. For example, for the overall Elk Mountain, NM cluster, the first most important variables of a growing GH200, unchanging

Table 2. High and Low R^2 from the MLRs for each cluster and accounted for variance with eigenvalues of one or greater from the PCAs for each cluster. Also, the sample size is included.

Cluster name	N	Highest R^2	Lowest R^2	Highest variance	Lowest variance
Elk Mountain, NM	154	0.794	0.226	89.751	82.145
Elk Mountain wind direction group 1, NM	114	0.893	0.291	91.377	80.198
Elk Mountain wind direction group 2, NM	40	1.000	0.715	99.835	92.294
Ute Hills/Pete Hills, CO	90	0.906	0.226	94.217	86.300
Rincon Mountains, NM	76	0.995	0.457	92.545	86.017
Lookout Peak/Rayado Peak, NM	70	0.996	0.347	93.685	86.564
Pajarito Mountain/Cerro Grande, NM	61	1.000	0.869	94.912	87.970
Culebra Range/Sangre de Cristo Mountains, CO	56	0.996	0.298	93.926	88.737
Shaggy Peak, NM	51	0.999	0.610	95.357	88.736
Los Pinos Mountains, NM	43	1.000	0.687	98.639	89.600
Mount Washington, NM	38	1.000	0.246	96.947	93.801
Wrye Peak, NM	36	1.000	0.178	100.000	92.265
Mesa de los Jumanos, NM	34	1.000	0.613	97.647	93.769
Jacinto Mesa, NM	32	1.000	0.441	100.000	93.055
Bartlett Mesa/Horse Mesa, NM	31	1.000	0.864	100.000	92.515
Trinchera Mesa/Valencia Hills/Howard Mountain, NM	31	1.000	0.416	98.563	93.562
Overall	1165	0.272	0.093	82.979	76.747

Table 3. Trends for each cluster and variable from the 6 h prior to initiation to the initiation hour. P, significant positive trend; M, significant negative trend; N, no significant trend; I, variables are the most significant in that cluster; 2, variables are the second most significant in that cluster; 3, variables are the third most significant in that cluster. The variable's significance is independent of trend and a cluster can have multiple variables with the same significance.

Cluster	Thickness	PW	CAPE	CIN	SRH	GH600	GH500	GH300	GH200	SH850
Elk Mountain, NM	2 P						3 N	2 N	1 P	
Elk Mountain Wind Direction Group I, NM							3 N	2 N	3 P	
Elk Mountain Wind Direction Group 2, NM	3 P									
Ute Hills/Pete Hills, CO		3 P								
Rincon Mountains, NM	3 N	3 N					2 N			
Lookout Peak/Rayado Peak, NM	3 N	3 N					2 N	3 M	3 N	3 M
Pajarito Mountain/Cerro Grande, NM	2 N					1 N	3 N			
Culebra Range/Sangre de Cristo Mountains, CO	2 M					3 N				3 N
Shaggy Peak, NM	1 P	2 N							2 P	
Los Pinos Mountains, NM										
Mount Washington, NM	3 M	3 N		3 M	3 M	3 P	2 P	2 M	2 M	2 M
Wrye Peak, NM			1 P	2 N		1 N	2 N	2 M		
Mesa de los Jumanos, NM							3 P			
Jacinto Mesa, NM	1 M						2 N			
Bartlett Mesa/Horse Mesa, NM	1 P	3 N		3 M		1 P	1 P			3 M
Trinchera Mesa/Valencia Hills/Howard Mountain, NM		2 N		3 P				3 N	1 M	3 M

Cluster	SH800	SH600	SH500	SH300	SH200	UC600	UC500	UC300	UC200	VC600
Elk Mountain, NM	3 M						2 P			
Elk Mountain Wind Direction Group I, NM										
Elk Mountain Wind Direction Group 2, NM							2 M	3 P		3 P
Ute Hills/Pete Hills, CO										
Rincon Mountains, NM										
Lookout Peak/Rayado Peak, NM	3 M		3 N				1 N			
Pajarito Mountain/Cerro Grande, NM										
Culebra Range/Sangre de Cristo Mountains, CO							2 P		3 P	2 M
Shaggy Peak, NM										
Los Pinos Mountains, NM										
Mount Washington, NM	1 N					3 P	3 P	3 N		
Wrye Peak, NM	3 N						3 M	3 P		
Mesa de los Jumanos, NM				2 P			1 N	2 P		
Jacinto Mesa, NM										
Bartlett Mesa/Horse Mesa, NM	3 N	3 N				1 N				
Trinchera Mesa/Valencia Hills/Howard Mountain, NM	3 M	2 N		3 P	2 M	3 P				

Cluster	VC500	VC300	VC200	T600	T500	T300	T200	WD600	WD500	UWSS500
Elk Mountain, NM	1 M						3 N			1 N
Elk Mountain Wind Direction Group I, NM		3 N			3 N	1 N	3 N			1 N
Elk Mountain Wind Direction Group 2, NM		2 N	3 N	3 N	3 N					
Ute Hills/Pete Hills, CO										2 N

Table 3. Continued.

Location	VC500	VC300	VC200	T600	T500	T300	T200	WD600	WD500	UWSS500
Rincon Mountains, NM	3 M			3 N, 3 P		3 N				
Lookout Peak/Rayado Peak, NM					3 M					2 N
Pajarito Mountain/Cerro Grande, NM										
Culebra Range/Sangre de Cristo Mountains, CO										1 P
Shaggy Peak, NM										1 P
Los Pinos Mountains, NM		1 M								
Mount Washington, NM	2 M		1 P	2 M	3 N	2 N				3 N
Wrye Peak, NM	3 N		2 M							3 M
Mesa de los Jumanos, NM	2 M	3 M		3 N			3 N			
Jacinto Mesa, NM										
Bartlett Mesa/Horse Mesa, NM		2 M		3 N	3 N					3 P
Trinchera Mesa/Valencia Hills/Howard Mountain, NM	1 M	1 M	2 P	1 N	3 N		3 N			2 M

Location	UWS600500	UWSS600	VWSS500	VWS600500	VWSS600
Elk Mountain, NM			2 M		3 P
Elk Mountain Wind Direction Group 1, NM			2 M		
Elk Mountain Wind Direction Group 2, NM	2 N	2 P	1 N		
Ute Hills/Pete Hills, CO			3 M		1 N
Rincon Mountains, NM		1 N			
Lookout Peak/Rayado Peak, NM		3 N	2 P		
Pajarito Mountain/Cerro Grande, NM			3 N		
Culebra Range/Sangre de Cristo Mountains, CO	3 P		1 M	3 M	
Shaggy Peak, NM			1 M		
Los Pinos Mountains, NM	1 N	1 P	2 N		3 P
Mount Washington, NM	2 M	3 M	2 M		1 N
Wrye Peak, NM	3 P	1 M	2 P		2 N
Mesa de los Jumanos, NM			3 M		
Jacinto Mesa, NM					
Bartlett Mesa/Horse Mesa, NM	2 M				
Trinchera Mesa/Valencia Hills/Howard Mountain, NM	3 M	2 P	2 M		2 N

UWSS500, and decreasing VC500, indicate if these variables occur, then there is a greater likelihood for MCS initiation within the cluster domain. While more weight is given to the first most significant variables, the second and third most important variables need to be considered as well. This can be done for each cluster. The trends and significant variables show that no two sets of conditions needed for MCS initiation at different peaks are the same. These differences would cause issues for any forecaster trying to predict MCS initiation within the ARB portion of the Rocky Mountains.

For the individual clusters and generalized results, the most significant variables are the wind shear/wind speed, geopotential height, and specific humidity variables, which would be a starting point, although the specific variables differ from cluster to cluster. Observing the wind shear/wind speed, geopotential height, and specific humidity variables alone significantly reduces the number of variables needed for observation. Of the top six most used variables, four were wind shear variables with the wind shear between the surface and an upper level being the most significant. While the geopotential height variables are highly correlated in each cluster, the most significant one is not consistent through the clusters. The specific humidity could be a marker for the depth of the moist layer or for a pocket of moist air present at any level. PW should be considered since MCSs require relatively high amounts, but PW was not often considered one of the most important variables. Another ingredient to consider is T500, included relatively often, because changes could be a proxy for instability as it directly relates to Lifted Index (LI). The needed lift would be provided by the mountains. This variable combination into an ingredients-based approach further reinforces the fact that no two clusters are exactly the same and the ingredients needed for MCS initiation within each cluster domain are different.

The initiation variable differences observed in the clusters could be caused by a variety of factors including topography. The topography affects the flow of all variables causing some areas of have higher variable values than other areas. This effect can be seen especially in the wind variables were ridgetop height winds affect the potential for initiation. The topography-affected flow would cause certain places (i.e. specific peaks) to be more prominent areas of initiation over other areas.

4. Conclusion

While there were similarities present from cluster to cluster in that wind shear, specific humidity, and geopotential height variables occurred often, one variable, overall, is not considered most significant. The hour and cluster dependence of each significant variable is noticeable in Table 3 with the differing trends from 6 h prior to the initiation hour and differing significance of each variable in each cluster. Therefore, one set of criteria for MCS initiation within a large domain will not present meaningful results, consistent with Moninger

et al. (1991). The topography in the area contributes to the difficulty in determining generalized conditions over the Rocky Mountain portion of the ARB since different peaks require different initiation conditions.

In conclusion, there does appear to be certain conditions needed for MCS initiation with the individual clusters like wind shear, specific humidity, and geopotential height which can be used to help predict initiation. It is very difficult to predict MCS initiation within the Rocky Mountain portion of the ARB due to the conditions for MCS initiation in the Rocky Mountains being so diverse. Due to the diverseness of the MCS initiation conditions in the Rocky Mountain portion of the ARB, one forecast model could not be used to predict MCS initiation within this area. The non-generalized conditions would render any generalized forecast model unusable and would call for a much more sophisticated, complex forecast model to account for all the initiation characteristic differences. Many future steps need to be done to verify the analyses to predict the MCSs, including the inclusion of null cases.

References

ARM: Climate Research Facility. 2011. ARM XDC Datastreams. http://www.arm.gov/xdc/xds/abrfc (accessed 1 May 2011).

Banta RM. 1990. The role of mountains flows in making clouds. Atmospheric processes over complex terrain, meteorology monograph. *American Meteorological Society* **45**: 229–282.

Banta RM, Schaaf CB. 1987. Thunderstorm genesis zones in the Colorado Rocky Mountains as determined by traceback of geosynchronous satellite images. *Monthly Weather Review* **115**: 463–476.

Callen E. 2012. A statistical analysis of characteristics of mesoscale convective system mountain initiation location clusters in the Arkansas-Red River Basin. MS thesis, Department of Geography, University of Kansas, 489 pp.

Carbone RE, Tuttle JD, Ahijevych DA, Trier SB. 2003. Inferences of predictability associated with warm season precipitation episodes. *Journal of the Atmospheric Sciences* **59**: 2033–2056.

Cotton WR, George RL, Wetzel PJ, McAnelly RL. 1983. A long-lived mesoscale convective complex. Part 1: the mountain-generated component. *Monthly Weather Review* **111**: 1893–1918.

Doswell CA III, Brooks HE, Maddox RA. 1996. Flash flood forecasting: an ingredients-based methodology. *Weather and Forecasting* **11**: 560–581.

Fritsch JM, Kane RJ, Chelius CR. 1986. The contribution of mesoscale convective weather systems to warm-season precipitation in the United States. *Journal of Applied Meteorology* **25**: 1333–1345.

Jirak IL, Cotton WR. 2007. Observational analysis of the predictability of mesoscale convective systems. *Weather and Forecasting* **22**: 813–838.

Johns RH, Doswell CA III. 1992. Severe local storms forecasting. *Weather and Forecasting* **7**: 588–612.

Johnson RH, Mapes BE. 2001. Mesoscale processes and severe convective weather. Severe convective storms, meteorology monograph. *American Meteorological Society* **50**: 71–122.

Maddox RA. 1980. Mesoscale convective complexes. *Bulletin of the American Meteorological Society* **61**: 1374–1387.

McAnelly RL, Cotton WR. 1986. Meso-β-scale characteristics of an episode of meso-α-scale convective complexes. *Monthly Weather Review* **114**: 1740–1770.

Mesinger F, DiMego G, Kalnay E, Mitchell K, Shafran PC, Ebisuzaki W, Jovic D, Woollen J, Rogers E, Berbery EH, Ek MB, Fan Y, Grumbine R, Higgins W, Li H, Lin Y, Manikin G, Parrish D, Shi W. 2006. North American regional reanalysis. *Bulletin of the American Meteorological Society* **87**: 343–360.

Moninger WR, Bullas J, de Lorenzis B, Ellison E, Flueck J, McLeod JC, Lusk C, Lampru PD, Phillips RS, Roberts WF, Shaw R, Stewart TR, Weaver J, Young KC, Zubrick SM. 1991. Shootout-89, a comparative evaluation of knowledge-based systems that forecast severe weather. *Bulletin of the American Meteorological Society* **72**: 1339–1354.

Murray D. 2003. The Integrated Data Viewer – a web-enabled application for scientific analysis and visualization. In 19th International Conference on Interactive Information Processing Systems, American Meteorological Society, Long Beach, CA, USA.

Rotunno R, Klemp JB, Weisman ML. 1988. A theory for strong, long-lived squall lines. *Journal of the Atmospheric Sciences* **45**: 463–485.

Schumacher RS, Johnson RH. 2005. Organization and environmental properties of extreme-rain-producing mesoscale convective systems. *Monthly Weather Review* **133**: 961–976.

Tripoli GJ, Cotton WR. 1989. Numerical study of an observed orogenic mesoscale convective system. Part I: simulated genesis and comparison with observations. *Monthly Weather Review* **117**: 273–304.

Tucker DF, Crook NA. 1998. The generation of a mesoscale convective system from mountain convection. *Monthly Weather Review* **127**: 1259–1273.

Tucker DF, Crook NA. 2001. Favored regions of convective initiation in the Rocky Mountains. In 9th Conference on Mesoscale Processes, American Meteorological Society, Fort Lauderdale, FL, USA.

Tucker DF, Crook NA. 2005. Flow over heated terrain. Part II: generation of convective precipitation. *Monthly Weather Review* **133**: 2565–2582.

Tucker DF, Li X. 2009. Characteristics of warm season precipitating storms in the Arkansas-Red River Basin. *Journal of Geophysical Research* **114**: D13108, doi: 10.1029/2008JD011093.

Young CB, Bradley AA, Krajewski WF, Kruger A. 2000. Evaluating NEXRAD multisensory precipitation estimates for operational hydrologic forecasting. *Journal of Hydrometeorology* **1**: 241–254.

Zipser EJ. 1982. Use of a conceptual model of the life of mesoscale convective systems to improve very-short-range forecasts. In *Nowcasting*, Browning K (ed). Academic Press; 191–204.

6

Climatology of convective available potential energy (CAPE) in ERA-Interim reanalysis over West Africa

Cyrille Meukaleuni, André Lenouo* and David Monkam

Department of Physics, Faculty of Science, University of Douala, Cameroon

*Correspondence to:
A. Lenouo, Department of
Physics, Faculty of Science,
University of Douala, P.O Box
24157, Douala, Cameroon.
E-mail: lenouo@yahoo.fr

Abstract

Seasonal study of convective available potential energy (CAPE) is done using 6-h ERA-Interim data over West Africa during 35 years (1979–2014). Climatology of CAPE presented in terms of seasonal means, variances and trends shows large values toward 12°–16°N with maxima during summer, according to higher relative humidity due to the arrival of monsoon in West Africa. Spectral analysis in the zone 10°–20°N/20°W–30°E, centered on the latitudes of maxima of CAPE trends at 12°–16°N toward inter-tropical convergence zone (ITCZ) mean position in summer, shows significant power in the 3–5 day within the regions of tropical deep convection in connection with African easterly waves.

Keywords: CAPE; West Africa; ITCZ; ERA-Interim; trends

1. Introduction

Convection plays a crucial role in the terrestrial climate with the formation of the clouds such as the cumulonimbus. The cumulus and other cumulonimbus are conditioned by the general circulation of the atmospheric air (orientation and force of the wind on ground) and the mode of precipitation (Lenouo *et al.*, 2010). The human activities are directly influenced by their effects: cloud cover, downpours, storm, gust of wind, etc. In order to better understand and envisage the episodes of convection's effects, international program as African Monsoon Multidisciplinary Analyses (AMMA) organized a series of measurements on the ground to study the atmospheric stability over West Africa.

The convective available potential energy (CAPE) has become an instability index widely used in the past few decades to evaluate the convective potential of the atmosphere. It is calculated by means of an integral of a vertical profile of cloud buoyancy and has been used for different kinds of studies. For example, CAPE was the appropriate parameter to analyze the conditional instability in the tropical atmosphere (Williams and Renno, 1993) for studies on tropical 'Hot Towers' (Williams *et al.*, 1992) and for other research projects on atmospheric convection (Renno and Ingersoll, 1996). The importance of the CAPE in all these projects has led to a more precise way of calculating their values, and to certain approximations and corrections that may be included in the general formula. Later research into the CAPE also inquired into the possible relationships between this and other parameters that characterize atmospheric conditions (Blanchard, 1998; Monkam, 2002). Brooks *et al.* (2003, 2007) presented global CAPE climatology derived from 7 years using National Centre for Atmospheric Research (NCAR)/National Centres for Environmental Prediction (NCEP) reanalysis. For the United States and Europe, many researches have been done, sometimes through the use of several coupled atmosphere and ocean models (Trapp *et al.*, 2007; Riemann-Campe *et al.*, 2009, 2010) or by using 30 year climatology of CAPE and Convective Inhibition (CIN; Romero *et al.*, 2007) based on ERA-40 from the European Centre for Medium-Range Weather Forecast (ECMWF). Riemann-Campe *et al.* (2010) also provided a global climatology of CAPE and CIN and their relation to convective precipitation, using 1979–2001 ERA-40 reanalyses data and 1979–2009 ECHAM5/MPI-OM model and analyzed both parameters in terms of trends and how they change in a warmer climate. Such statistical analyses which can characterize the climatology of a region, specifically the West Africa region that has a large zone of deep convection deserve to be done.

Therefore, we propose to complete the CAPE analyses from the West Africa climatology based on 35 years (1979–2014) of ERA-Interim data, to provide seasonal ensemble means, a trend analysis and its link to African easterly waves (AEWs). The outline of the article, is as follows: the data and methods of analysis are presented in the Section 2. In Section 3, we present the result of the climatology of CAPE, trends and relationship with AEW. Section 4 presents conclusions and the outlook on future research on higher order statistics.

2. Data and methods

In this document, we have calculated the different variables by using the 0.75° grid ERA-Interim data firstly because of its high quality even if differences sometimes occur due to the varying density of observations (as in the southern hemisphere, and over oceans), and secondly because it is the most recent reanalysis from the ECMWF started in 1979. By taking into account the satellite observations, the quality and quantity of data have improved considerably (Uppala *et al.*, 2005), even if the change in the measurement system leads to an artificial warming trend in the global mean temperature of the lower troposphere (Bengtsson *et al.*, 2004). The differences between observed and ERA-Interim temperature trends were also pointed out by Simmons *et al.* (2004). However, ERA-Interim data yield the general trend signal with an improving performance after 1979. Thus, trends in CAPE can be analyzed for the whole period of time and additionally before and after 1979.

CAPE is calculated between the level of free convection (LFC) and the Level of Neutral Buoyancy (LNB) as measures of the bottom and top of the cloud, respectively. As the effect of moisture on buoyancy is taken into account, the virtual temperature T_V is used and the CAPE is given by R_d gas constant for dry air (287.05 J kg^{-1} K^{-1}):

$$\text{CAPE} = \int_{\text{LFC}}^{\text{LNB}} R_d \left(T_{\text{vp}} - T_{\text{ve}} \right) d\ln(P) \qquad (1)$$

where P, T_{vp}, T_{ve}, LNB and LFC are, respectively, the pressure, the virtual temperature of the air parcel, the virtual temperature of the environment of the air parcel, the LNB and the LFC.

The air parcel rises dry adiabatically from the surface to the lifting condensation level (LCL). Above the LCL, the parcel raises pseudo adiabatically which means that any condensates will immediately fall out of the parcel as rain (Riemann-Campe *et al.*, 2009). Between cloud bottom and cloud top the parcel rises freely as the temperature of the parcel is higher than the temperature of its environment. The pseudo equivalent potential temperature θ_{ep} is used to calculate the temperature of the rising parcel (Emanuel, 1994).

$$\theta_{\text{ep}} = T \left(\frac{P_{sfc}}{P} \right)^{0.2854(1-0.28r)}$$
$$\times \exp \left[r(1 + 0.81r) \left(\frac{3376}{T_{\text{LCL}}} - 2.54 \right) \right] \qquad (2)$$

with the mixing ratio r of dry to moist air and the temperature T_{LCL} of the parcel at the LCL, where the ascent of the air parcel changes from dry adiabatic to pseudo adiabatic and P_{sfc} is the pressure at the sea surface (1000 hPa). The expression of the saturation temperature T_{LCL} can be approximated by (Bolton, 1980):

$$T_{\text{LCL}} = \frac{2840}{3.5 \log T - \log E - 4.805} + 55 \qquad (3)$$

with the vapor pressure E.

Local seasonal trends in CAPE are analyzed by the Mann–Kendall trend test which is a robust trend estimator applicable for any theoretical distribution. As noted by Wilks (2011), investigating the possible trend through time of the central tendency of a data series is of interest in the context of a changing underlying climate,

among other settings. The usual parametric approach to this kind of question is through regression analysis with a time index as the predictor, and the associated test for the null hypothesis that a regression slope is zero. This particular test is used here, as CAPE is not normally distributed in general. The trend in CAPE is calculated on a seasonal mean basis. Only positive CAPE values are considered in the calculation. Grid points with less than two positive CAPE values per season are neglected. The Mann–Kendall score indicates a given trend being positive or negative, which is supplemented by a two sided p-value to provide the probability of a detected trend (Riemann-Campe *et al.*, 2009). Trends are only included in the analysis if their probability exceeds the 95% significance level. The magnitude of a given trend is estimated by linear regression, although the error is rarely normally distributed in CAPE.

The wavelet analysis method is applied here to identify the dominant synoptic oscillation modes and to isolate the synoptic oscillation components. In this work, it is used on daily CAPE time series over West Africa for the period 1979–2009 in order to evaluate the seasonality of the variance of the synoptic time scale and of the related AEW signal. Previous studies have shown that dominant modes of monsoon synoptic variability are characterized by strong and reproducible 3–5 day oscillations (Sultan and Janicot, 2003). Hence, CAPE spectra are calculated on each individual June–September period using wavelet analysis method. A red noise background spectrum is computed from the formula of Gilman *et al.* (1963). The 95% confidence limits about this red noise spectrum are determined using F-statistic (Wilks, 2011).

3. Results and discussions

3.1. Means and variability

The global distribution of seasonally averaged CAPE (Figure 1) follows basically the distribution of the wind at 850 hPa over West Africa. CAPE generally increases from Sahara which is the arid zone with values of about 500 J kg^{-1} to the equator where they can reach 2000 J kg^{-1}. CAPE minima are observed in regions of cold water upwelling and where currents are colder than the ambient ocean temperatures, and in arid regions. CAPE also moves with the inter-tropical convergence zone (ITCZ) as show in Figure 1. However, the northward evolution of monsoon around 10°N during the months of April–May–June (AMJ, Figure 1(b)) to about 15°N in July–August–September (JAS, Figure 1(c)) before decrease southward at October–November–December (OND, Figure 1(d)) suggested that good correlation can exist between CAPE and the seasonal migration of the West Africa monsoon where sufficient moisture are available. The location and intensity of convective systems occurring during JAS are more frequently north of the ITCZ. This suggests the possible influence of AEW *versus* CAPE in their development during

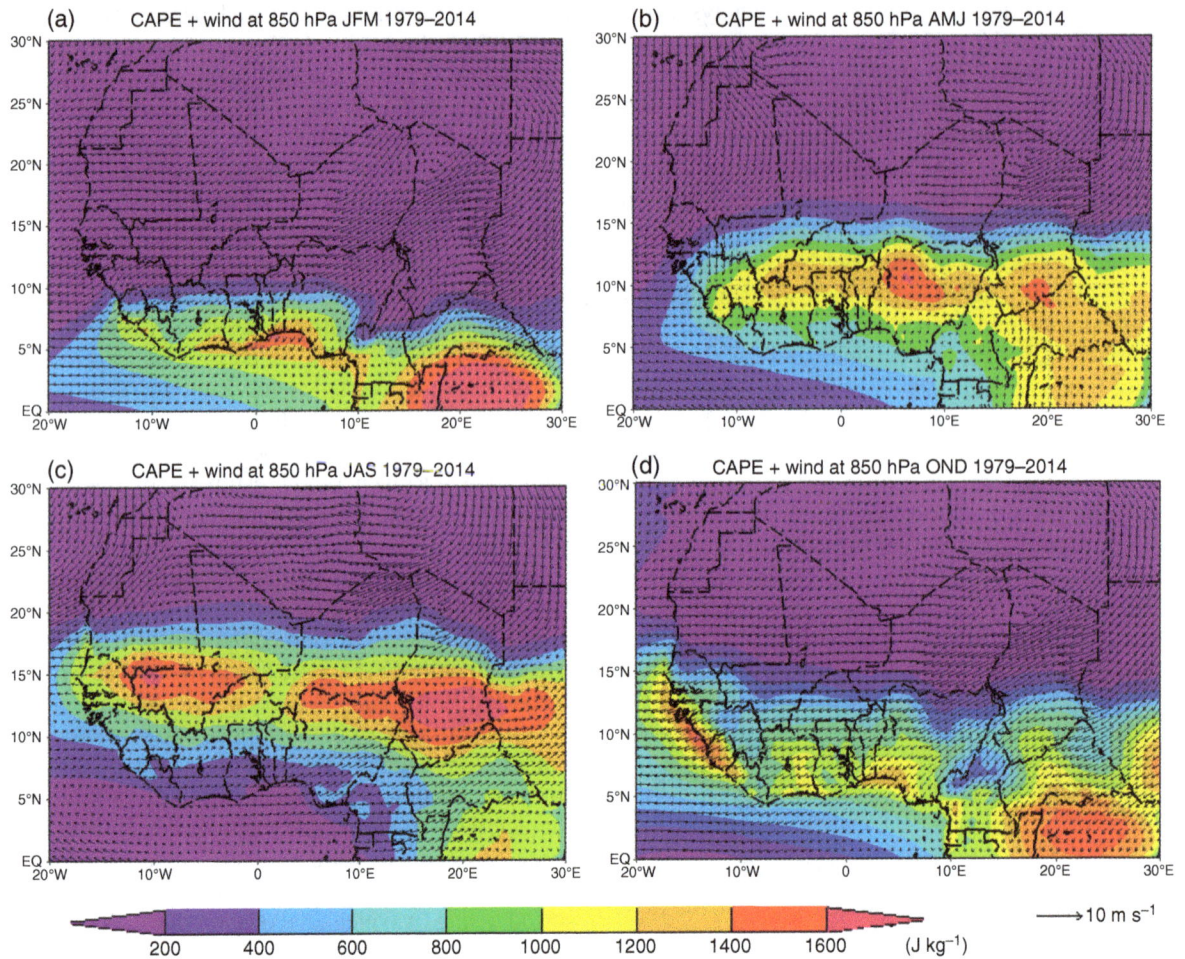

Figure 1. Climatological (1979–2014) seasonal mean of CAPE (in J kg^{-1}) and wind at 850 hPa (in m s^{-1}) averaged during JFM (a), AMJ (b), JAS (c) and OND (d); the country contours are represented by dashed lines. The interval of CAPE in the color scale is 200 J kg^{-1}.

AMJ and JAS. In other way, the maximum CAPE over Congo basin or in Guinea coast can be viewed as the development of Kelvin wave versus CAPE during JFM and OND.

Interannual variance of the seasonal CAPE means varied from 500 to about 6.5×10^7 (J kg^{-1})^2, with mean values around 1.3×10^5 (J kg^{-1})^2 (Figure 2). The zone of maximum variance corresponds to the growth in relative humidity. The geographical distribution of the seasonal differences is similar to those of the seasonally averaged mean. In general, larger values are observed in summer (AMJ, Figure 2(b) and JAS, Figure 2(c)) according to higher relative humidity due to the arrival of the monsoon over West Africa. In the dry season (JFM), high variance is present around the Congo basin and the Benin coast where relative humidity is 80 and 65%, respectively, and in OND (Figure 2(c)) around the Guinean coast where the relative humidity remains around 65%.

3.2. Trends

CAPE trend magnitudes are displayed (Figure 3) at the level of significance exceeding 95%. Trends are

shown for the full 35 years time series (1979–2014). Calculated trends before 1979 might correspond to changes in measurements instead of changes in climate (Bengtsson *et al.*, 2004). Significant trends in CAPE occur in most parts of the West Africa except in dry region. Regions with a positive trend outnumber the regions of negative trends considerably with magnitudes varying in the time periods considered. Trend magnitudes range from about −800 J kg^{-1} to about 1000 J kg^{-1} per decade during 1979–2014. The largest increase in CAPE of about 1600 J kg^{-1} per decade occurs in the Congo basin during JFM (Figure 1(a)) and around the AEW axis (15°N) in the south of Chad and at the borders of Mali and Senegal during JAS (Figure 1(c)). The change of sign yields in a net decrease by 200–600 J kg^{-1} per decade from the dry period (OND and JFM) to the wet season (AMJ and JAS) in the band of latitudes 10°–20°N can be due to the migration of the ITCZ which is also linked to the southward decrease of CAPE during the two periods.

3.3. Link with AEW

Figure 4 shows the relationship between CAPE and AEW through spectral analysis of CAPE during the

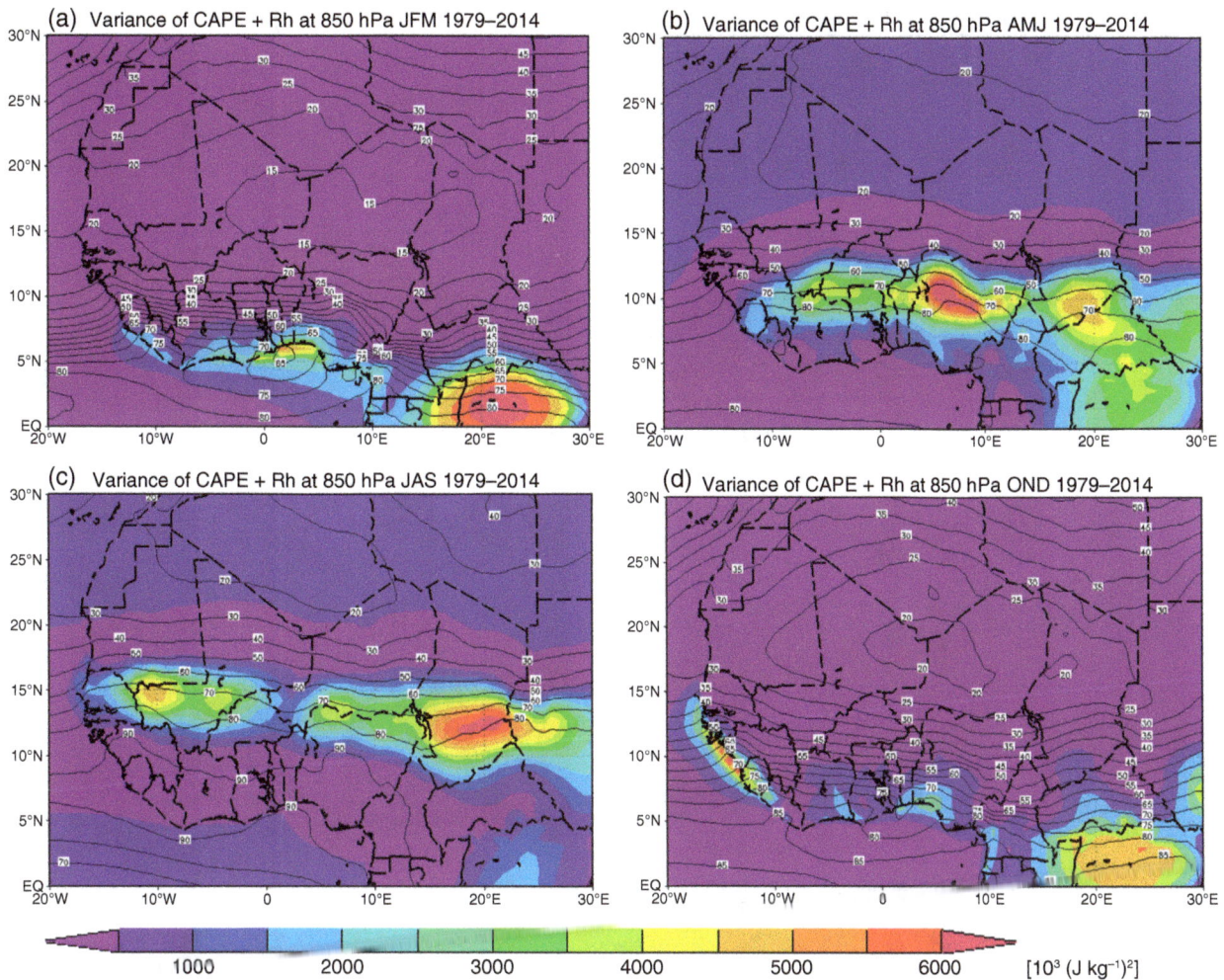

Figure 2. Climatological (1979–2014) inter annual variance of the seasonal CAPE in $(J\,kg^{-1})^2$ and relative humidity at 850 hPa (in %) averaged during JFM (a), AMJ (b), JAS (c) and OND (d); the country contours are represented by dashed lines.

period JAS of 1979–2014. Wavelet analysis is a common tool for decomposing a time series into a time-frequency space and detecting time-frequency variations. The wavelet transform allows the treatment of a signal as a wavelet function called a mother wavelet (here the Morlet wavelet is used; Sultan and Janicot, 2003; Mohr and Thorncroft, 2006). We have tested different mother wavelets and the results look similar. Because the wavelet transform is a band pass filter with a known response function (the wavelet function), it is also a powerful filtering technique. The wavelet analysis method is used here to identify the dominant synoptic oscillation modes and to isolate the synoptic oscillation components. The mean seasonal cycle and interannual variability were removed before computing the spectrum. At the right (Figure 4), the wavelet significance with the red noise background spectrum and the 95% confidence level are shown.

Spectral analysis using high resolution data shows significant power in the 3–5-day band period over tropical West and Central Africa in the box $10°–20°N$ and $20°W–30°E$, where CAPE is high as shown in Figure 1. Within the regions of tropical deep convection, the 3–5-day time scale variance accounts for about

25–35% of the total variance. The 3–5-day convective variance has similar amplitudes from east to west in the band $10°–20°N$, while dynamic measures of AEW activity show stronger amplitudes in the west. Weak AEW activity in the east is consistent with initial wave development there. AEWs are initiated by the convection triggered on the western sides of the Darfur Mountains (western Sudan) and Ethiopia. The subsequent development and growth of AEWs are associated with stronger coherence with convection there (Mohr and Thorncroft, 2006).

4. Summary and conclusions

By using the ERA-Interim data, with a 0.75° grid over West Africa, it was found that CAPE minima are observed in arid regions over West Africa and move with the ITCZ. The evolution of the monsoon around $10°N$ during the months of AMJ to about $15°N$ in JAS prior to a southward decrease in OND suggests good correlation between CAPE and the seasonal migration of West Africa monsoon where sufficient moisture are available. Larger values of CAPE variance are observed

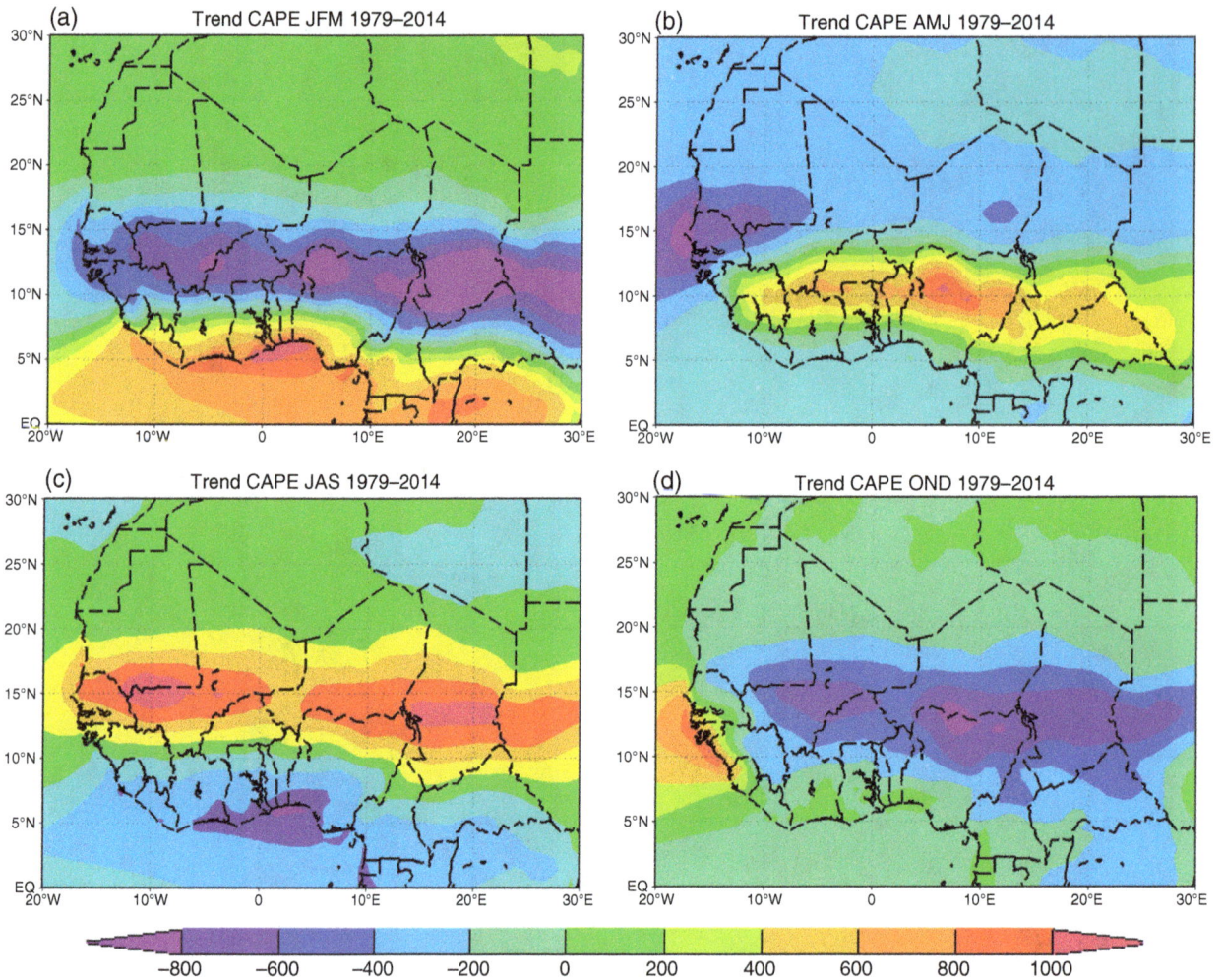

Figure 3. Seasonal trends of CAPE in J kg^{-1} per decade, significant at the 95% level over West Africa during JFM (a), AMJ (b), JAS (c) and OND (d) from 1979 to 2014. The country boundaries are represented by a dash.

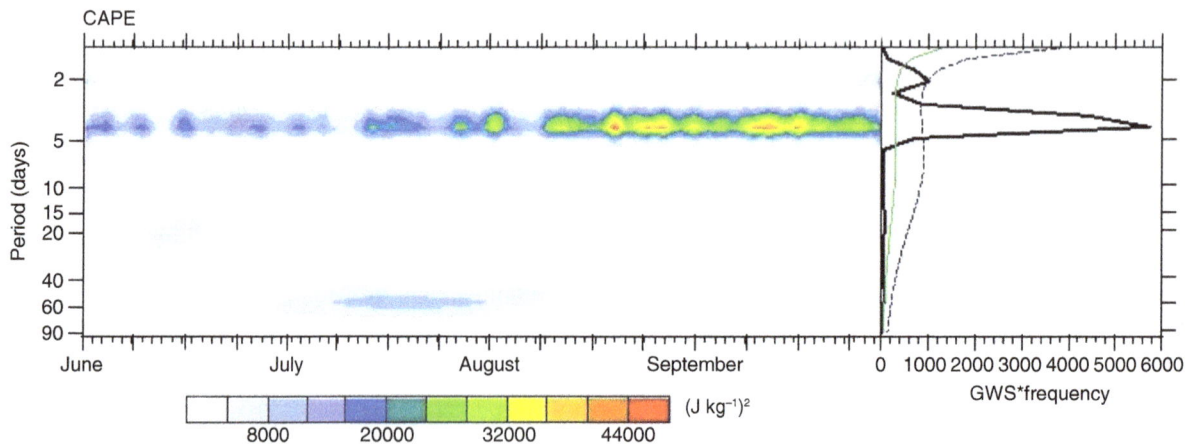

Figure 4. Diagrams of mean wavelet variance of daily unfiltered CAPE over West Africa for 10°–20°N/20°W–30°E during the summer (JAS) in the period 1979–2009. The mean seasonal cycle and interannual variability were removed before computing the spectrum. At the right, the wavelet significance with the red noise backgrounds spectrum and the 95% confidence level.

in the summer according to higher relative humidity due to the arrival of the monsoon in the continent. We found that significant trends in CAPE occur in most parts of the West Africa except in dry region. Trend magnitudes range from about −800 J kg^{-1} to 1000 J kg^{-1} per decade during 1979–2014. The largest increase in CAPE of about 1600 J kg^{-1} per decade occurs in the Congo basin during JFM and around the AEW axis (15°N) in the south of Chad and at the borders of Mali and Senegal during JAS. The change of sign yields decrease from

200 to $600\,\mathrm{J\,kg^{-1}}$ per decade between the dry period (OND and JFM) and the wet season (AMJ and JAS) in the band of latitudes $10°-20°$N. Spectral analysis using high resolution data shows significant power in the $3-5$-day range in the box $10°-20°$N and $20°$W$-30°$E, where CAPE is high. Within the regions of tropical deep convection in the band of latitude $10°-20°$N, the $3-5$-day time scale variance accounts for about $25-35\%$ of the total variance. The $3-5$-day convective variance has similar amplitudes from east to west of this band, while dynamic measures of AEW activity show stronger amplitudes in the west.

Similar intra-seasonal CAPE fluctuations occur also over the Congo basin, especially during the dry season (OND and JFM). This point must be investigated further. Another point would be to quantify the contribution of such intra-seasonal variability to the interannual variability. It will be also interesting to investigate the links with atmospheric modes of this zone since Kamsu-Tamo *et al.* (2014) showed that corresponding regressed deseasonalized atmospheric fields highlight an eastward propagation of patterns consistent with convectively coupled equatorial Kelvin wave dynamics.

Acknowledgements

This research is supported by ICTP, Trieste Italy through the Associate and Federation Schemes Program and 'Programme Pilote Régional Forêts Tropicales Humides d'Afrique Centrale (PPR FTH-AC)' of IRD. We also appreciated the comments of the anonymous reviewers which helped to improve the manuscript.

References

Bengtsson L, Hagemann S, Hodges KI. 2004. Can climate trends be calculated from reanalysis data? *Journal of Geophysical Research* **109**: D11111.

Blanchard DO. 1998. Assessing the vertical distribution of convective available potential energy. *Weather and Forecasting* **13**: 870–877.

Bolton D. 1980. The computation of equivalent potential temperature. *Monthly Weather Review* **108**: 1046–1053.

Brooks HE, Lee JW, Craven JP. 2003. The spatial distribution of severe thunderstorm and tornado environments from global reanalysis data. *Atmospheric Research* **67**: 73–94.

Brooks HE, Anderson AR, Riemann K, Ebbers I, Flachs H. 2007. Climatological aspects of convective parameters from the NCAR/NCEP reanalysis. *Atmospheric Research* **83**: 294–305.

Emanuel KA. 1994. *Atmospheric Convection.* Oxford University Press: New York, NY.

Gilman DL, Fuglister FJ, Mitchel JM Jr. 1963. On the power spectrum of red noise. *Journal of the Atmospheric Sciences* **20**: 182–184.

Lenouo A, Monkam D, Mkankam Kamga F. 2010. Variability of static stability over West Africa during northern summer 1979-2005. *Atmospheric Research* **98**: 353–362.

Mohr KJ, Thorncroft CD. 2006. Intensive convective systems in West Africa and their relationship to the African easterly jet. *The Quarterly Journal of the Royal Meteorological Society* **132**: 163–176.

Monkam D. 2002. Convective available potential energy (CAPE) in Northern Africa and Tropical Atlantic and study of its connections with rainfall in Central and West Africa during summer 1985. *Atmospheric Research* **62**: 125–147.

Kamsu-tamo P-H, Janicot S, Monkam D, Lenouo A. 2014. Convection activity over the Guinean coast and Central Africa during northern spring from synoptic to intra-seasonal timescales. *Climate Dynamics* **43**: 3377–3401, doi: 10.1007/s00382-014-2111-y.

Renno NO, Ingersoll AP. 1996. Natural convection as a heat engine: a theory for CAPE. *Journal of Atmospheric Sciences* **53**: 572–585.

Riemann-Campe K, Fraedrich K, Lunkeit F. 2009. Global climatology of convective available potential energy (CAPE) and convective inhibition (CIN) in ERA-40 reanalysis. *Atmospheric Research* **93**: 534–545.

Riemann-Campe K, Blender R, Fraedrich K. 2010. Global memory analysis in observed and simulated CAPE and CIN. *International Journal of Climatology* **31**(8): 1099–1107.

Romero R, Gaya M, Doswell CA III. 2007. European climatology of severe convective storm environmental parameters: a test for significant tornado events. *Atmospheric Research* **83**: 389–404.

Simmons AJ, Jones PD, da Costa Bechtold V, Beljaars ACM, Kllberg PW, Saarinen S, Uppala SM, Viterbo P, Wedi N. 2004. Comparison of trends and low-frequency variability in CRU, ERA-40, and NCEP/NCAR analyses of surface air temperature. *Journal of Geophysical Research* **109**: D24115.

Sultan B, Janicot S. 2003. The West African monsoon dynamics. Part II: the "preonset" and "onset" of the summer monsoon. *Journal of Climate* **16**: 3407–3427.

Trapp RJ, Diffenbaugh NS, Brooks HE, Baldwin ME, Robinson ED, Pal JS. 2007. Changes in severe thunderstorm environment frequency during the 21st century caused by anthropogenically enhanced global radiative forcing. *Proceedings of the National Academy of Sciences* **104**: 19719–19723.

Uppala SM, Kållberg PW, Simmons AJ, Andrae U, da Costa Bechtold V, Fiorino M, Gibson JK, Haseler J, Hernandez A, Kelly GA, Li X, Onogi K, Saarinen S, Sokka N, Allan RP, Andersson E, Arpe K, Balmaseda MA, Beljaars ACM, van de Berg L, Bidlot J, Bormann N, Caires S, Chevallier F, Dethof A, Dragosavac M, Fisher M, Fuentes M, Hagemann S, Holm E, Hoskins BJ, Isaksen L, Janssen PAEM, Jenne R, McNally AP, Mahfouf J-F, Morcrette J-J, Rayner NA, Saunders RW, Simon P, Sterl A, Trenberth KE, Untch A, Vasiljevic D, Viterbo P, Woollen J. 2005. The ERA-40 reanalysis. *The Quarterly Journal of the Royal Meteorological Society* **131**: 2961–3012.

Wilks DS. 2011. *Statistical Methods in the Atmospheric Sciences.* International Geophysics Series, 3rd ed, Vol. **100**. Academic Press: New York, NY; 669 pp.

Williams E, Renno N. 1993. An analysis of the conditional instability of the tropical atmosphere. *Monthly Weather Review* **121**: 21–36.

Williams ER, Geotis SG, Renno N, Rutledge SA, Rasmussen E, Rickenback T. 1992. A radar and electrical study of tropicalAhot towerB. *Journal of the Atmospheric Sciences* **49**: 1386–1395.

7

Analogous seasonal evolution of the South Atlantic SST dipole indices

Hyacinth C. Nnamchi,[1],*,† ⓘ Fred Kucharski,[2,3] Noel S. Keenlyside[4,5] and Riccardo Farneti[2]

[1]Department of Geography, University of Nigeria, Nsukka, Nigeria
[2]Earth System Physics Section, The Abdus Salam International Centre for Theoretical Physics, Trieste, Italy
[3]Center of Excellence for Climate Change Research/Department of Meteorology, King Abdulaziz University, Jeddah, Saudi Arabia
[4]Geophysical Institute, University of Bergen, Norway
[5]Bjerknes Centre for Climate Research, Bergen, Norway

*Correspondence to:
H. C. Nnamchi, Department of
Geography, University of Nigeria,
Nsukka 410001, Nigeria.
E-mail:
hyacinth.nnamchi@unn.edu.ng

†Current address:
GEOMAR|Helmholtz Centre for
Ocean Research Kiel,
Düsternbrooker Weg 20,
24105 Kiel, Germany.
E-mail: hnnamchi@geomar.de

Abstract

Two variants of sea-surface temperature (SST) dipole indices for the South Atlantic Ocean (SAO) has been previously described representing: (1) the South Atlantic subtropical dipole (SASD) supposedly peaking in austral summer and (2) the SAO dipole (SAOD) in winter. In this study, we present the analysis of observational data sets (1985–2014) showing the SASD and SAOD as largely constituting the same mode of ocean–atmosphere interaction reminiscent of the SAOD structure peaking in winter. Indeed, winter is the only season in which the inverse correlation between the northern and southern poles of both indices is statistically significant. The observed SASD and SAOD indices exhibit robust correlations ($P \leq 0.001$) in all seasons and these are reproduced by 54 of the 63 different models of the Coupled Models Intercomparison Project analysed. Their robust correlations notwithstanding the SASD and SAOD indices appear to better capture different aspects of SAO climate variability and teleconnections.

Keywords: SST dipole; seasonal evolution; CMIP3/5; South Atlantic Ocean

1. Introduction

A dipole structure in sea-surface temperature (SST) anomalies represents the dominant mode of ocean–atmosphere fluctuations in the South Atlantic. The origins of the dipole has been attributed to the interactions of the ocean mixed layer with atmospheric fluctuations via surface heat fluxes (Fauchereau et al., 2003; Sterl and Hazeleger, 2003; Trzaska et al., 2007; Morioka et al., 2011; Nnamchi et al., 2016). The dipole structure is typically oriented in the southwest–northeast direction, and two major variants have been described in the literature – the South Atlantic subtropical dipole (SASD) and the South Atlantic Ocean (SAO) dipole (SAOD).

The monthly standard deviation of the SASD index peaks in austral summer when the SAO mixed layer depth is shallowest (Morioka et al., 2011). The SASD is considered as a subtropical mode, with the northernmost part at ~15°–25°S. However, the empirical orthogonal function (EOF) analysis of seasonally stratified data sets also shows opposite structure of SST anomalies between the equatorial and southwest Atlantic peaking in austral winter (Nnamchi et al., 2011). This dipole pattern is referred to as the SAOD. While the northernmost part of the SAOD extends to the equator and coincides with the Atlantic Niño (ATL3) in both space and time, leading to the notion that the two may represent the same mode of variability (Nnamchi

et al., 2016); by definition, the southern part of the SASD (30°N–40°S, 10°–30°W) falls within that of the SAOD (25°N–40°S, 10°–40°W).

Are SASD and SAOD related or independent modes of ocean–atmosphere variability? To illustrate this question, Figure S1, Supporting Information, shows the first EOF mode of monthly SST anomalies over the SAO; characterised by a dipole associated with a basin-scale cyclonic atmospheric anomalies. There are broad cold anomalies in the southern arms of the SASD and SAOD as well as in the subtropical and equatorial warming peaks leading to SASD and SAOD definitions, respectively. The associated time series displays interannual-to-decadal fluctuations (Venegas et al.,1997; Trzaska et al., 2007; Nnamchi et al., 2016) with robust correlations with the ATL3, SASD, and SAOD indices ($P < 0.001$) as shown in Table S1. The SAOD has similarly significant correlations with the ATL3 and SASD indices ($r \sim 0.70$). The ATL3|SAOD correlation has been described elsewhere (Nnamchi et al., 2016) but that of the SASD and SAOD remains unclear and is therefore the focus of the present study.

2. Data and methods

Using satellite-derived SST data set on $1.0° \times 1.0°$ latitude–longitude horizontal grids (Reynolds et al., 2007) and satellite-era atmospheric reanalysis on

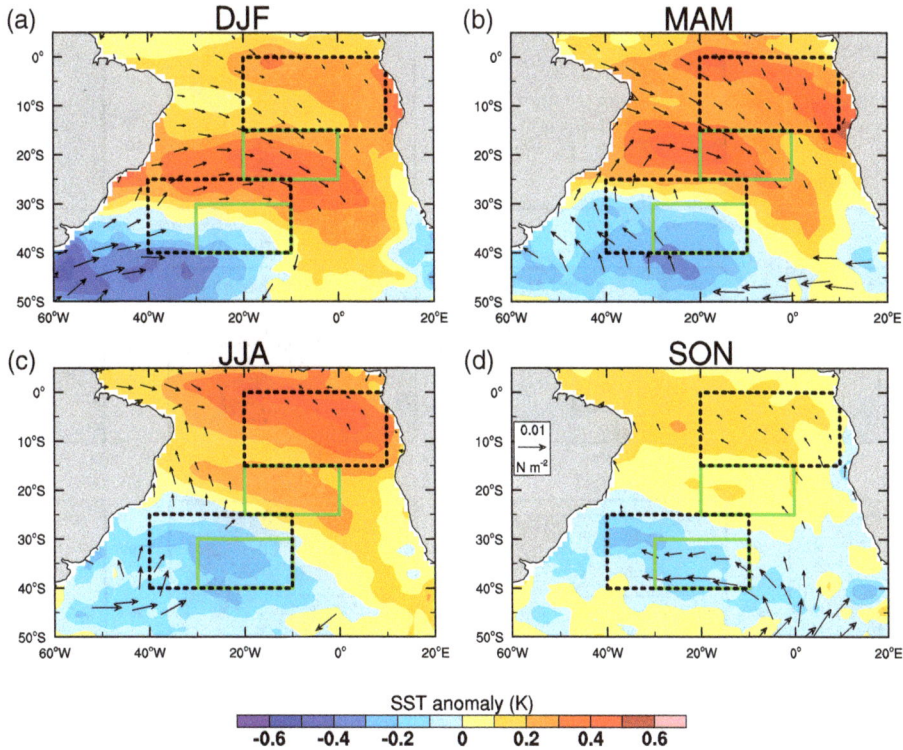

Figure 1. Spatial patterns of the leading S-EOF mode of DJF to SON SST anomalies over the SAO during 1985–2014. Arrows show statistically significant wind stress anomalies at 95% confidence level.

$2.5° \times 2.5°$ grids (Kanamitsu *et al.*, 2002), we analysed the 30-year period from 1985 to 2014 with generally improved observations. We first created SST anomaly by subtracting the mean annual cycle from the original data set and then calculated the deviations from the global-mean warming trend as follows:

$$SSTA'_{x,y,t} = SSTA_{x,y,t} - \beta_t$$

where the subscripts (x, y, t) represent the zonal and meridional directions and time; $SSTA'$ denotes the deviations of the SST anomaly (SSTA) at every grid point (x, y) from the least-squares linear trend coefficient of the global-mean (β). Other variables were linearly detrended at every grid point using the least-squares method. All statistical significance tests are based on two-tailed *t*-test. The warm and cold anomalies of the leading dipole are confined between 5°N and 50°S (Figure S1(a)) and this meridional extent is used for subsequent analyses.

3. Analyses and results

3.1. Seasonal evolution of the interannual SST dipole

Patterns of SST anomalies such as dipole in the SAO primarily represent fluctuations above and below the mean annual cycle in different seasons of the year. Different parts of the equatorial and SAO display marked annual cycle of SST that often exceed the interannual anomalies in amplitude (Burls *et al.*, 2011;

Nnamchi *et al.*, 2016). Thus, the SST anomalies may be strongly modulated by the climatological-mean annual cycles, which denote the background conditions under which the ocean–atmosphere fluctuations evolve. Here, we investigate the seasonal evolution of the dominant dipole mode of SAO SST anomalies using the seasonal-reliant EOF (S-EOF) analysis (Wang and An, 2005). We define a four-season sequence that spans from the austral summer of a year denoted as D(−1) JF(0) to the following spring [September–October–November (SON)] denoted as SON(0). To construct the covariance matrix, the SST anomalies in each seasonal sequence are treated as an integral block [of one time step for the year (0)]. The EOF decomposition is then performed and the expansion coefficient (EC) time series derived for each eigenvector that contains a sequence of four seasonally evolving SST pattern maps from D(−1)JF(0) to SON (0). The major advantage of S-EOF analysis over the conventional method is that the four seasonal maps are directly related to the yearly EC time series and depict the overall seasonal evolution.

The leading S-EOF mode is characterised by the evolution of a meridional dipole structure in SST anomalies over the SAO during the course of the seasons (Figure 1). In DJF, the cold anomalies are mainly confined to the regions around latitude 35°S and further south; the rest of the basin is characterised by warm anomalies. Localised bands of anomalous warm maxima are observed close to the northern pole of the SASD and in the Benguela–equatorial Niño region. The subtropical pattern does not really fit the SASD definition:

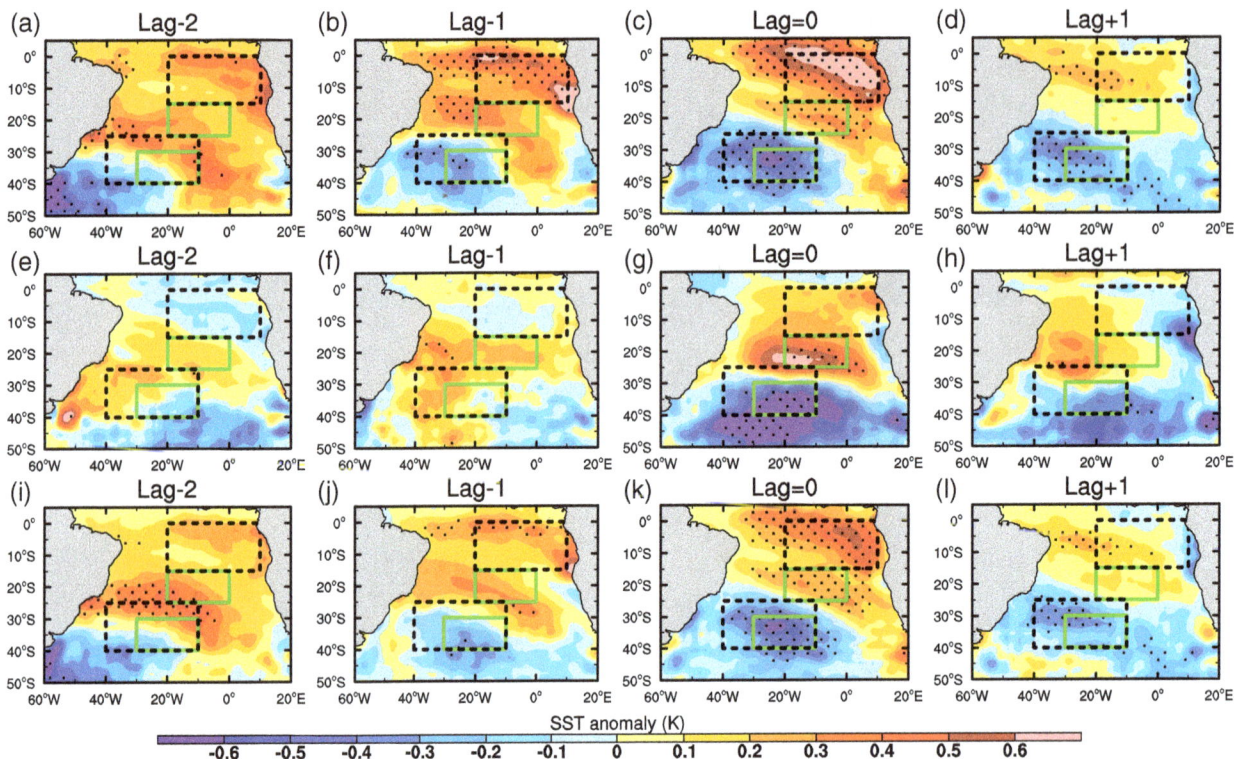

Figure 2. Composite evolution of the SAOD (a–d) and SASD (e–h) SST anomalies. The maps are based on the (±1.0σ) composite difference using the SAOD (SASD) index for JJA (DJF) as lag = 0. The lower panels (i–l) are based on SASD index with JJA as lag = 0. Stipples denote a statistical significance at P ≤ 0.05. The pattern correlations between the corresponding in panels (a–d) and (i–l) are 0.93, 0.88, 0.96, and 0.92 for DJF, MAM, JJA, and SON, respectively.

the cold anomalies are displaced to the west of the SASD box while both the cold and warm anomalies are displaced south by the order of ~5° – 10° of latitude. There are widespread robust wind anomalies suggestive of atmospheric-induced origin of the SST anomalies (Venegas *et al.*, 1997; Sterl and Hazeleger, 2003; Hermes and Reason, 2005) and this pattern tends to lead a similar dipole structure in subtropical Indian Ocean by 1 month (Hermes and Reason, 2005). In the equatorial region, there is a weak Niño-like structure probably reflecting the type II ATL3 described by Okumura and Xie (2006) or an early phase of the Benguela Niño.

In MAM, the SST and wind stress anomalies spread northward such that the cold anomalies reach ~30°S (Figure 1(b)). The cyclonic wind stress anomalies intensify and are associated with enhanced warming anomalies from the northern part of the SASD to the Benguela–equatorial Niño region. This season marks the peak phase of the Benguela Niño and the large-scale cyclonic wind stress anomalies appear as northwesterly perturbations originating from the equatorial region (Florenchie *et al.*, 2003, 2004). Nonetheless, the sequence of the wind stress anomalies from DJF to MAM points to important roles for atmospheric anomalies in the southwest subtropics and extratropics.

By JJA, the dipole structure advances further north (Figure 1(c)); the cold anomalies reach ~20°S with the result that the centre of maximum cold anomalies clearly fit into the southern poles of both the SASD and SAOD. Further north, the Benguela Niño weakens

giving rise to a more zonal structure of warming characteristic of the Atlantic Niño. Compared to MAM, the wind stress anomalies are generally weaker in JJA consistent with previous studies suggesting that peak in wind fluctuations precedes that of SST anomalies (Keenlyside and Latif, 2007; Richter *et al.*, 2013). While the westerly anomalies persist over the western equatorial Atlantic, a reversal (to easterlies) occurs in the eastern basin leading to anomalous convergent motion over the equatorial ATL3 region: 0° – 20°W. Thus, the SST and related wind stress anomalies in JJA may be driven ocean dynamics consistent with the heat budget calculations of Nnamchi *et al.*, (2016). The anomalous easterlies intensify further indicating a strengthening of the trade winds (Figure 1(d)). This may then intensify upwelling and evaporation leading to the decay of the SST anomalies.

3.2. Seasonal variability and correlations of the dipole SST indices with other climate modes

Here, we investigate how well the SST anomalies depicted in Figure 1 could be reproduced using composite difference based on a unit standard deviation (±1.0σ) of the SASD and then SAOD index. For the SASD (SAOD), season (0) is DJF (JJA) on the basis of previous studies suggesting that the patterns peak in these seasons (Morioka *et al.*, 2011; Nnamchi *et al.*, 2011). The composite maps are then lagged in time for the two preceding and one following season. The

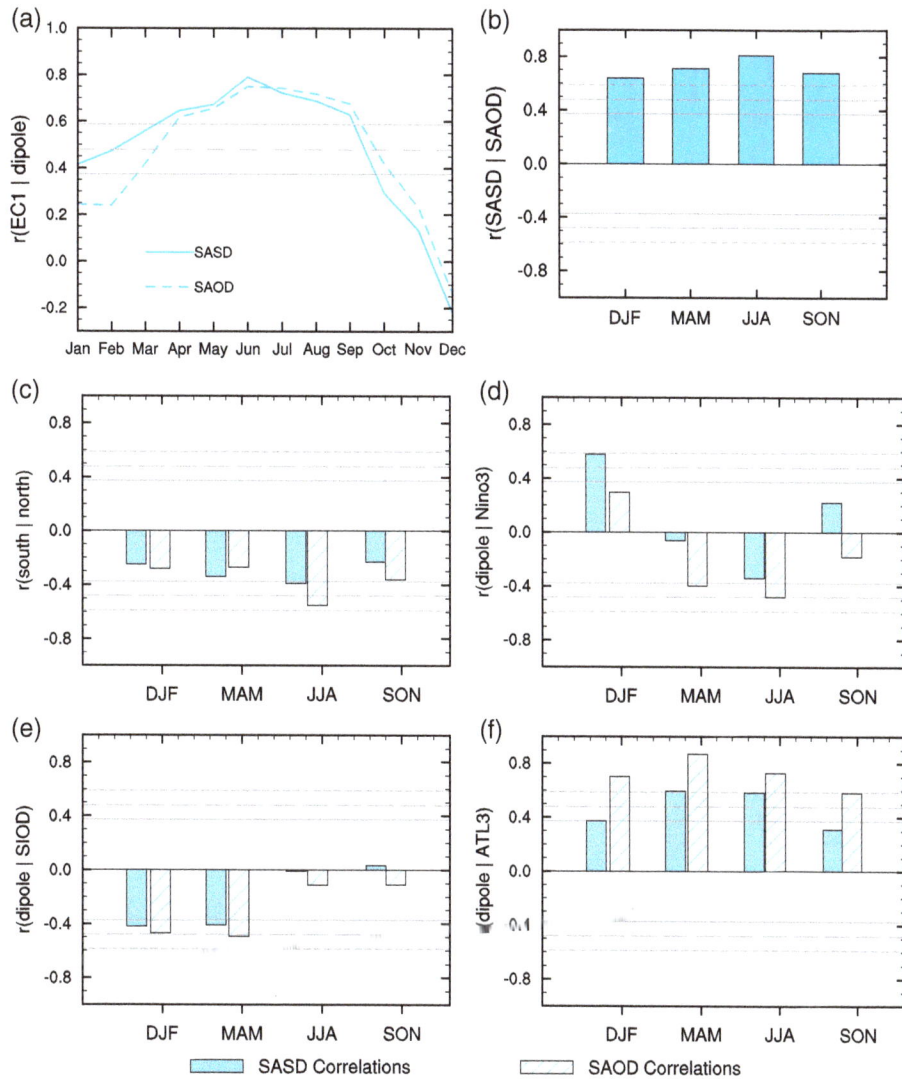

Figure 3. (a) Correlations of the leading S-EOF mode EC time series of SAO SST anomalies with SST-based indices of dipole. (b) Coefficients of correlation between the SASD and SAOD indices and (c) between their northern and southern poles. (d–f) Correlations of the SASD and SAOD indices with Niño-3, SIOD, and ATL3. In all panels, solid ($P = 0.05$); long-dashed ($P = 0.01$); and short-dashed ($P = 0.001$) grey lines denote statistical significance levels. The Niño-3 is defined as SST anomalies averaged in the region 5°N–5°S, 90°–150°W and the SIOD as the difference between averages over (30°–44°S, 74°–44°E) and (19°–35°S, 80°–110°E).

results show robust opposite SST anomalies over the SAO at lag-2, which intensify further at lag-1 leading to the peak phase of the SAOD similar to the S-EOF patterns (Figures 2(a)–(d)). However, such progressive evolution is not found for the SASD (Figures 2(e)–(h)); rather, the SST dipole anomalies are robust only at lag = 0. It is not surprising therefore that the DJF SASD and JJA SAOD time series are not correlated ($r = 0.07$), consistent with the event-based analysis which shows that the $\pm 1.0\sigma$ SST anomalies do not tend to occur in the same year (Table S2). On the other hand, if JJA is used as season (0), the SASD-related anomalies evolve in a fashion analogous to the SAOD (Figures 2(i)–(l)) with spatial correlations in the range of $0.88 \geq r \leq 0.96$.

We further investigate the similarity in temporal evolution of the SASD and SAOD by calculating their month-by-month correlations with EC1 (which

has a constant value for each year; Figure 3(a)). Both correlation curves increase progressively from January, peak in June, and then declines during the remaining months of the year. The SASD curve notably exceeds that of the SAOD curve in the first 3 months, but the two curves converge as from April and by peaking in austral winter seem to describe the SAOD. The SASD and SAOD indices are significantly correlated at $P < 0.001$ in all seasons with a peak in JJA (Figure 3(b)). As a direct measure of the polarity of these dipole indices, the inverse correlation between their southern and northern poles also peak in JJA (Figure 3(c)). Indeed, JJA is the only season for which the correlation is statistically significant for both indices; although the anticorrelation is stronger for the SAOD ($P < 0.01$) than the SASD ($P < 0.05$). The highest absolute correlations in JJA in all panels of Figures 3(a)–(c) appear to confirm the occurrence of the SST dipole peak in this season.

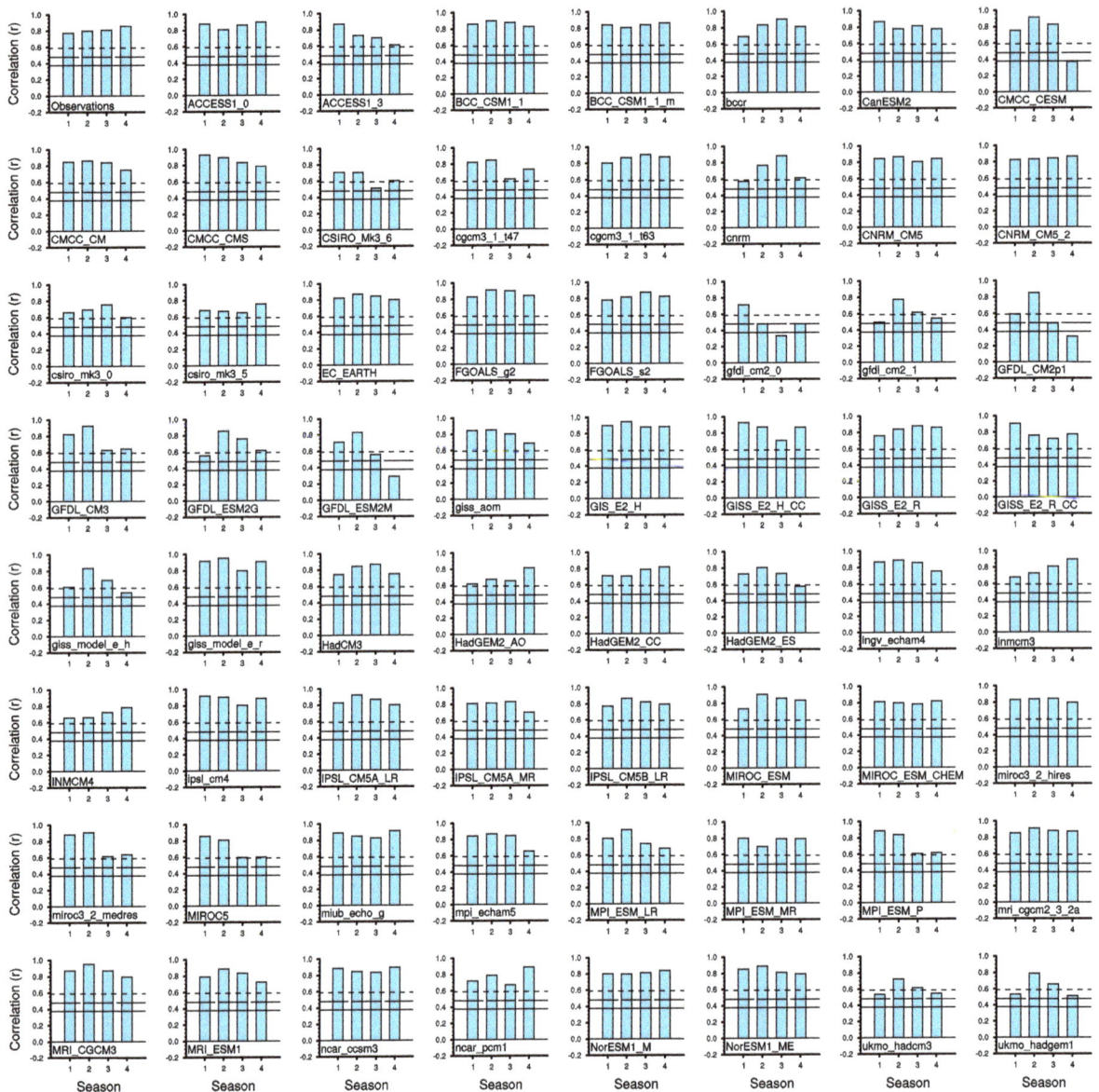

Figure 4. Coefficients of correlation between the SASD and SAOD indices simulated by the CMIP3 and CMIP5 models. The numbers 1, 2, 3, and 4 correspond to DJF, MAM, JJA, and SON, respectively. The panels are arranged in alphabetical order of model names: CMIP3 in lower case and CMIP5 in upper case letters. In all panels, solid ($P = 0.05$); long-dashed ($P = 0.01$); and short-dashed ($P = 0.001$) lines denote statistical significance levels.

Previous studies show that the SAO SST dipole may be related to the Pacific El Niño (Rodrigues *et al.*, 2015), the subtropical Indian Ocean dipole (SIOD) (Hermes and Reason, 2005) and Atlantic Niño (Nnamchi *et al.*, 2016). Here, indices representing these climate modes are correlated with those of the SASD and SAOD for comparison. The SASD (SAOD) index is better correlated with the Niño-3 index in DJF (MAM and JJA) and these seasons depict different phases of El Niño–SAO dipole correlations (Figure 3(d)). The dipole is related to the SIOD in DJF and MAM and to the ATL3 in all seasons; in both cases, the SAOD index exhibits stronger correlations (Figures 3(e) and (f)).

3.3. Representation in coupled climate models

We have so far shown robust correlations between the SASD- and SAOD-related anomalies in space and time.

It is then necessary to understand whether the current generation of state-of-the-art coupled models are able to reproduce these observed correlations or not. Thus, we analysed the SST simulated by 63 different numerical models of the Coupled Models Intercomparison Project phases 3 and 5 (CMIP3/5) (Meehl *et al.*, 2007; Taylor *et al.*, 2012). This represents the 'climate of the 20th-century' experiment of the available 23 CMIP3 (http://www-pcmdi.llnl.gov/ipcc/data_status_tables .htm) and the 40 CMIP5 (http://cmip-pcmdi.llnl.gov/ cmip5/availability.html) models providing first realisation of the 'historical' experiment. The modelled SSTs were bilinearly remapped to a common $1.0° \times 1.0°$ longitude–latitude horizontal grids for comparison with observational data set (Rayner *et al.*, 2003) and both were linearly detrended for the last 30 years available for all models during 1970–1999.

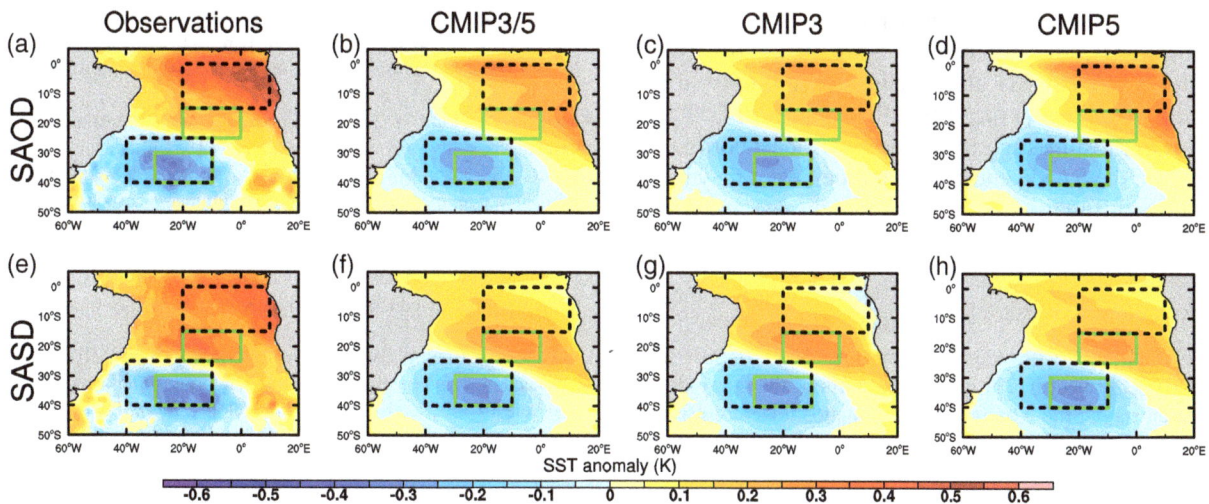

Figure 5. Upper panel: SST anomalies regressed on the SAOD index for JJA based on (a) observational data set, (b) CMIP3/5 multimodel mean, (c) CMIP3 multimodel mean, and (d) CMIP5 multimodel mean. (e–h) Similar to (a–d) but for the SST anomalies regressed SASD index for JJA.

The seasonality of SASD|SAOD correlation may be nonstationary in time as the JJA peak during the 1984–2014 (Figure 3(b)) period is not reproduced during 1970–1999, although the correlations remain robust in all seasons (Figure 4). Consistent with observations, all the CMIP3/5 models simulated positive correlation between the SASD and SAOD indices in all seasons. At $P \leq 0.001$, these correlations are statistically significant in 54 of the 63 models in all four seasons; 58 models in at least three seasons and 59 models in at least two seasons – suggesting that the CMIP3/5 models are able to reproduce the observed robust correlation between the SASD and SAOD indices. We further examined the spatial distribution of the SST anomalies associated with the SASD and SAOD indices in the CMIP3/5 models focusing on JJA during which the dipole has been shown to peak (Figure 5). The maps were constructed by regressing SST anomalies onto the dipole indices for each model and then the multimodel mean calculated for comparison with observations.

Figures 5(a) and (e) show similar dipole SST response to both indices with a pattern correlation of 0.94 in observations. The SASD and SAOD indices correctly reproduce cold anomalies over southwest Atlantic Ocean in the CMIP3/5 models. On the other hand, while the simulated SAOD index is also able to reproduce the broad features of the northern warm anomalies (Figure 5(b)), the SASD shows weak anomalies over the Benguela–equatorial Niño region (Figure 5(f)). The CMIP5 models show overall improvements in the simulation of the SASD compared to CMIP3 – the pattern correlation with observational data set changes from 0.71 to 0.80 but the SAOD pattern remains largely unchanged.

4. Summary and discussions

A comparative analysis of the SASD and SAOD indices using observational data sets (1985–2014) shows a peak during the austral winter consistent with the SAOD definition. Either the SASD or SAOD index may be used to represent the dipole mode: the two indices yield closely matched patterns with spatial correlations of seasonally stratified composite maps in the range of 0.88–0.96. The indices are equally strongly correlated in time in all seasons ($P < 0.001$). However, the SAOD index seems to have stronger polarity and more robust correlations with SIOD and ATL3. Nonetheless, the peak phase of El Niño in DJF is correlated with SASD and it may therefore be more useful for characterising El Niño teleconnections with the SAO in this season. On the other hand, El Niño is correlated with the SAOD in MAM and JJA.

Warm and cold events of the cold tongue tend to occur on a large-scale encompassing the Benguela and equatorial Niño regions (Florenchie *et al.*, 2004; Lübbecke *et al.*, 2010; Lutz *et al.*, 2013). Whereas previous studies linked the origins of this broad warming to equatorial northwesterly atmospheric fluctuations, our results suggest important roles for wind stress anomalies in the southwest SAO subtropics and extratropics starting from the austral summer. The wind anomalies are underlain by opposite phase in SST giving rise to an emerging dipole structure (Colberg and Reason, 2007; Nnamchi *et al.*, 2011, 2016). The dipole evolves progressively and appears to propagate northward during the subsequent months with a peak phase in winter consistent with the seasonality of Atlantic Niño.

The present day generation of coupled models appear to be capable of reproducing the observed robust correlation between the SASD and SAOD indices: 54 of the 63 different CMIP3/5 models analysed have significant coefficients at $P < 0.001$ in all seasons. However, the multimodel mean SST anomalies associated with the SAOD index better represents the observed amplitudes in the Benguela–equatorial Niño region, often considered vital for climate fluctuations over tropical Atlantic and further afield (Losada *et al.*, 2012; Syed and Kucharski, 2016; Kucharski and Joshi, 2017). Nonetheless, the subtropical SAO variability represented by the SASD affects climate variability in this region (Morioka *et al.*, 2011; Rodrigues *et al.*, 2015). Thus, our results suggest that despite robust correlation between the SASD and SAOD indices, each may better capture different aspects of SAO climate variability and teleconnections.

Acknowledgements

HCN and NSK were supported by the EU FP7/2007–2013 PREFACE Project. HCN carried out parts of the study during research stay at the Abdus Salam International Centre for Theoretical Physics through the Associateship Scheme. NSK also acknowledges support from the Research Council of Norway (233680/E10).

Supporting information

The following supporting information is available:

Figure S1. Leading EOF mode of SST anomalies (January 1985–December 2014) over the South Atlantic Ocean: 5°N–50°S, 20°E–60°W. Shown are the (a) spatial pattern and (b) EC time series. In panel (a), the associated variance is shown in the top right corner; solid green boxes (15°–25°S, 0°–20°W and 30°–40°S, 10°–30°W) indicate the SASD domains (Morioka *et al.*, 2011), the dashed black boxes (7°–15°S, 10°E–20°W and 25°–40°S, 10°–40°W) show the SAOD domains (Nnamchi *et al.*, 2011).

Table S1. Correlation of the monthly S-EOF EC1 with equatorial and southern Atlantic SST-based indices, 1985–2014.

Table S2. Years of occurrence of ±1.0σ of the SASD and SAOD indices, 1985–2014.

References

Burls NJ, Reason CJC, Penven P, Philander SG. 2011. Similarities between the tropical Atlantic seasonal cycle and ENSO: an energetics perspective. *Journal of Geophysical Research: Oceans* 116: C11010. https://doi.org/10.1029/2011JC007164.
Colberg F, Reason CJC. 2007. Ocean model diagnosis of low frequency climate variability in the South Atlantic. *Journal of Climate* 20: 1016–1034.
Fauchereau N, Trzaska S, Richard Y, Roucou P, Camberlin P. 2003. Sea-surface temperature co-variability in the southern Atlantic and Indian oceans and its connections with the atmospheric circulation in the southern hemisphere. *International Journal of Climatology* 23: 663–677. https://doi.org/10.1002/joc.905.
Florenchie P, Lutjeharms JRE, Reason CJC, Masson S, Rouault M. 2003. The source of Benguela Niños in the South Atlantic Ocean. *Geophysical Research Letters* 30: 1505, 10. https://doi.org/10.1029/2003GL017172.
Florenchie P, Reason CJC, Lutjeharms JRE, Rouault M, Roy C, Masson S. 2004. Evolution of interannual warm and cold events in the southeast Atlantic Ocean. *Journal of Climate* 17: 2318–2334.
Hermes JC, Reason CJ. 2005. Ocean model diagnosis of interannual coevolving SST variability in the south Indian and South Atlantic oceans. *Journal of Climate* 18: 2864–2882.

Kanamitsu M, Ebisuzaki W, Woollen J, Yang S, Hnilo J, Fiorino M, Potter G. 2002. NCEP–DOE AMIP-II reanalysis (R-2). *Bulletin of the American Meteorological Society* 83: 1631–1643. https://doi.org/10.1175/BAMS-83-11-1631.
Keenlyside NS, Latif M. 2007. Understanding Equatorial Atlantic Interannual Variability. *Journal of Climate* 20: 131–142.
Kucharski F, Joshi MK. 2017. Influence of tropical South Atlantic sea surface temperatures on the Indian summer monsoon in CMIP5 models. *Quarterly Journal of the Royal Meteorological Society* 143: 1351–1363.
Losada T, Rodríguez-Fonseca B, Kucharski F. 2012. Tropical influence on the summer Mediterranean climate. *Atmospheric Science Letters* 13: 36–42.
Lübbecke JF, Böning CW, Keenlyside NS, Xie S-P. 2010. On the connection between Benguela and equatorial Atlantic Niños and the role of the South Atlantic anticyclone. *Journal of Geophysical Research – Atmospheres* 115: C09015. https://doi.org/10.1029/2009JC005964.
Lutz K, Rathmann J, Jacobeit J. 2013. Classification of warm and cold water events in the eastern tropical Atlantic Ocean. *Atmospheric Science Letters* 14: 102–106.
Meehl GA, Covey C, Delworth T, Latif M, McAvaney B, Mitchell JFB, Stouffer RJ, Taylor KE. 2007. The WCRP CMIP3 multi-model dataset: a new era in climate change research. *Bulletin of the American Meteorological Society* 88: 1383–1394.
Morioka Y, Tozuka T, Yamagata T. 2011. On the growth and decay of the subtropical dipole mode in the South Atlantic. *Journal of Climate* 24: 5538–5554.
Nnamchi HC, Li J, Anyadike RNC. 2011. Does a dipole mode really exist in the South Atlantic Ocean? *Journal of Geophysical Research: Atmospheres* 116: D15104. https://doi.org/10.1029/2010JD015579.
Nnamchi HC, Li J, Kucharski F, Kang I-S, Keenlyside NS, Chang P, Farneti R. 2016. An equatorial–extratropical dipole structure of the Atlantic Niño. *Journal of Climate* 29: 7295–7311.
Okumura Y, Xie S-P. 2006. Some overlooked features of tropical Atlantic climate leading to a new Niño-like phenomenon. *Journal of Climate* 19: 5859–5874.
Rayner NA, Parker DE, Horton EB, Folland CK, Alexander LV, Rowell DP, Kent EC, Kaplan A. 2003. Global analyses of sea surface temperature, sea ice, and night marine air temperature since the late nineteenth century. *Journal of Geophysical Research-Atmospheres* 108: 4407. https://doi.org/10.1029/2002JD002670.
Reynolds RW, Smith TM, Liu C, Chelton DB, Casey KS, Schlax MG. 2007. Daily high-resolution-blended analyses for sea surface temperature. *Journal of Climate* 20: 5473–5496.
Richter I, Behera SK, Masumoto Y, Taguchi B, Sasaki H, Yamagata T. 2013. Multiple causes of interannual sea surface temperature variability in the equatorial Atlantic Ocean. *Nature Geoscience* 6: 43–47.
Sterl A, Hazeleger W. 2003. Coupled variability and air-sea interaction in the South Atlantic Ocean. *Climate Dynamics* 21: 559–571.
Rodrigues RR, Campos EJ, Haarsma R. 2015. The impact of ENSO on the South Atlantic subtropical dipole mode. *Journal of Climate* 28: 2691–2705.
Syed FS, Kucharski F. 2016. Statistically related coupled modes of south Asian summer monsoon interannual variability in the tropics. *Atmospheric Science Letters* 17: 183–189.
Taylor K, Stouffer R, Meehl G. 2012. An overview of CMIP5 and the experiment design. *Bulletin of the American Meteorological Society* 93: 485–498.
Trzaska S, Robertson AW, Farrara JD, Mechoso CR. 2007. South Atlantic variability arising from air–sea coupling: local mechanisms and tropical–subtropical interactions. *Journal of Climate* 20: 3345–3365.
Venegas S, Mysak L, Straub D. 1997. Atmosphere–ocean coupled variability in the South Atlantic. *Journal of Climate* 10: 2904–2920.
Wang B, An S-I. 2005. A method for detecting season-dependent modes of climate variability: S-EOF analysis. *Geophysical Research Letters* 32: L15710. https://doi.org/10.1029/2005GL022709.

8

Recent ENSO–PDO precipitation relationships in the Mediterranean California border region

Edgar G. Pavia,[1,*] Federico Graef[1,2] and Ramón Fuentes-Franco[2]

[1]Centro de Investigación Científica y de Educación Superior de Ensenada (CICESE), Ensenada, Mexico
[2]The Abdus Salam International Centre for Theoretical Physics (ICTP), Trieste, Italy

*Correspondence to:
E. G. Pavia, Centro de
Investigación Científica y de
Educación Superior de Ensenada
(CICESE), PO Box 434844, San
Diego, CA 92143-4844, USA.
E-mail: epavia@cicese.mx

Abstract

The Mediterranean California Border Region (MCBR) rainfall's relationship with El Niño-Southern Oscillation (ENSO) and the Pacific Decadal Oscillation (PDO) is reexamined for the period 1951–2014. When stratifying data by ENSO events we found that strong events of either sign yield the highest ENSO–rainfall correlation; but when the stratification was done by rainfall the wet seasons yield the highest ENSO–rainfall correlation. Most strong ENSO events have the same sign as PDO; but the ENSO–rainfall correlation for all ENSO–PDO same-sign events is almost undistinguishable from the full-record's correlation. Timewise stratification shows that 30-year climatological values (MCBR precipitation, PDO and ENSO) and ENSO–rainfall correlations have decreased in recent years.

Keywords: Baja California rainfall; California rainfall; ENSO PDO rainfall relationships

1. Introduction

The influence of the El Niño-Southern Oscillation (ENSO) phenomenon on the rainfall of southern California is well-known (e.g. Schonher and Nicholson, 1989; Fierro, 2014). The situation is similar in northwestern Baja California as rainfall regimes are comparable on both sides of the United States–Mexico border (Pavia and Graef, 2002). The southwestern California-northwestern Baja California region is part of the Mediterranean Californias (Pavia and Badan, 1998), because it is characterized by dry summers and wet winters, with usually wetter winters during warm ENSO (El Niño) events than during cool ENSO (La Niña) events; therefore, this region hereinafter shall be referred to as the Mediterranean California Border Region (MCBR). Furthermore decadal variability (e.g. Pacific Decadal Oscillation, PDO) is known to modulate the ENSO–rainfall relationship in this region in a constructive interference way (Gershunov and Barnett, 1998; Gershunov and Cayan, 2003; Pavia et al., 2006). However in the last decade the ENSO–rainfall relationship in the MCBR has weakened, for example after the last 1997–1998 strong El Niño wet year, the only two recent wet seasons were the 2004–2005 (a neutral event) and the 2010–2011 La Niña. Climatologically this resulted in a reduction of both: 30-year ENSO–rainfall correlations and 30-year regional precipitation climatological values; the latter coinciding with a similar decline in ENSO and PDO climatologies. Therefore the study of these circumstances is of utmost importance for the MCBR, where several major cities are located and where precipitation averages only around 240 mm year^{-1} (see Figure 1). Within the MCBR the situation is even more critical for Baja California than California, because the former lacks the level of hydraulic infrastructure that the latter possesses (MacDonald, 2007). Elsewhere variations in ENSO–rainfall and ENSO–PDO–rainfall relationships have been noted as well (e.g. Rocha, 1999; Krishna Kumar et al., 1999; McCabe and Dettinger, 1999; Chang et al., 2001; Cai et al., 2001; Sarkar et al., 2004; Gao et al., 2006; Zubair and Ropelewski, 2006; Kucharski et al., 2007); thus in this paper we study the changes in ENSO, PDO and rainfall in the MCBR (Figure 1), focusing on the recent 1951–2014 period. Our goal is not to scrutinize single or particular events, but to examine different subsets of our record (stratified in different ways) searching for rather general relationships of the MCBR rainfall with ENSO and PDO.

2. Data and methods

We construct ENSO, PDO and MCBR rainfall indices to obtain baseline correlations among them (for the entire period) and subset correlations for different stratifications (by index values and timewise). We also compute 30-year climatological values for these three variables plus 30-year ENSO–rainfall and PDO′–PDO correlations; where PDO′ is an 'ENSO-forced' model (see Newman et al., 2003; Pavia, 2009). Finally we construct global maps of SST–ENSO and SST–regional precipitation correlation patterns for different periods to substantiate our results.

2.1. Original data

For ENSO we select the monthly El Niño 3.4 index from NOAA's Earth System Research Laboratory/Physical

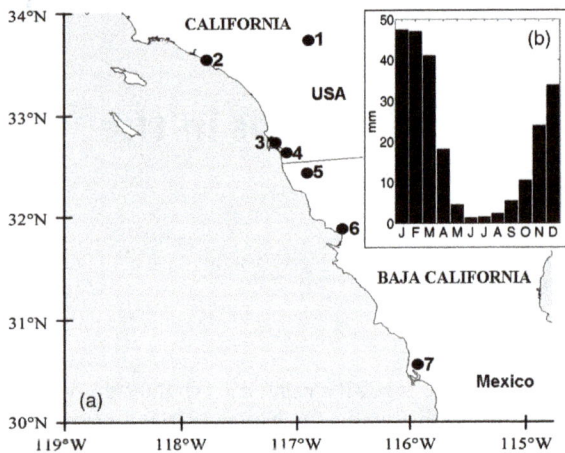

Figure 1. (a) The region of study. Numbers besides a solid circle indicate the stations considered: 1. Hemet, 2. Laguna Beach, 3. San Diego, 4. Chula Vista, 5. Tijuana, 6. Ensenada, 7. San Quintin. (b) Bars indicate the 1951–2014 monthly climatological values of precipitation averaged over the seven stations (annual mean 237 mm year^{-1}).

Sciences Division (ESRL/PSD) because the region of the El Niño 3.4 index is closely related to extra-tropical teleconnections; however, using other ENSO indices yields similar results. For decadal variability we select the PDO index (obtained from the University of Washington's Joint Institute for the Study of the Atmosphere and Ocean). For regional rainfall we select the monthly precipitation data from seven stations: Hemet (PHE), Laguna Beach (PLB), San Diego (PSD) and Chula Vista (PCV), California, obtained from the National Weather Service (NWS) among many others, and from Tijuana (PTJ), Ensenada (PEN) and San Quintin (PSQ), Baja California, obtained from the Mexican Water Authority (CNA) (see Figure 1). We also use global SST data from the Hadley Center (HadISST, Rayner *et al.*, 2003), and gridded precipitation data from Livneh *et al.* (2013) within the region between 115° and 120°W and 30° and 34.5°N to perform further correlation analyses to verify our results. A preliminary analysis of the seven stations and gridded precipitation data gave comparable results; thus the stations dataset is considered typical of MCBR precipitation.

2.2. Data preparation and correlation analyses

We begin by calculating the average of annual total (July to June) precipitation of the seven stations: PHE, PLB, PSD, PCV, PTJ, PEN and PSQ; we consider the standardized average to be representative of our region's rainfall and call it CPI (CPI and a similar index calculated from the gridded data of Livneh *et al.* (2013) are correlated above 0.9). Similarly for the ENSO and PDO indices we use annual averages of the El Niño 3.4 index and the PDO index (also July to June), and call them ENI and PDI (Figure 2). The correlations of ENI and PDI with CPI are performed for the entire 63-year series, as a first assessment of their relationship, and for 30-year running periods. The ENSO–PDO relationship

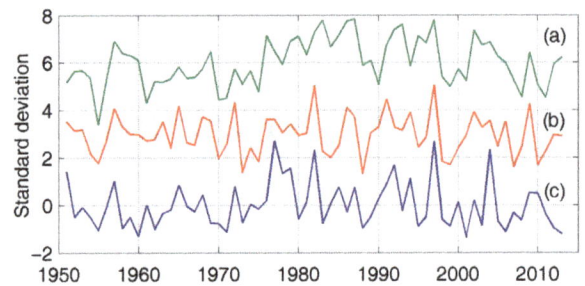

Figure 2. The constructed annual indices: (a) PDO (PDI + 6), (b) ENSO (ENI + 3), (c) MCBR precipitation (CPI).

is assessed thru the use of a model: PDI′ = f(PDI, ENI). We then stratify data by index value to obtain subset correlations; similarly we stratify data timewise by periods and by seasons and calculate climatological values in addition to subset correlations. Finally four series of global SST–ENSO–rainfall spatial correlation maps (1951–1981, 1965–1995, 1983–2013 and 1951–1965 plus 1999–2014) are calculated to examine the details of our results. We selected the mentioned periods in order to show the differences in SST–regional precipitation and SST–ENSO correlation patterns depending on the dominant ENSO phase in each period. The period 1983–2013 includes the two El Niño events with the overall highest amplitude (1983 and 1998), while the 1951–1965 plus 1999–2014 combined-period excludes the four strongest ENSO events during winter (1965–1966, 1972–1973, 1982–1983 and 1997–1998) (see Figure 3).

3. Results

3.1. ENI–CPI correlation

The baseline correlation for our entire record is $R_{\text{all}} = 0.58$; thus only about one third of recent MCBR rainfall variability may be explained by ENSO.

3.1.1. Stratification by ENSO events

To further investigate the correlation between ENI and CPI, we stratified the ENI data into strong ENSO events (|ENI| > 0.8) and neutral events ($-0.8 \leq$ ENI ≤ 0.8). This yields 25 strong ENSO (12 positive) and 38 neutral ENSO events. The correlation $R_{\text{strong}} = 0.83$, is statistically significant as the probability of getting a correlation as large as the observed value by random chance is 10^{-7}, lower and upper bounds for a 95% confidence interval for R_{strong} are [0.64, 0.92]. All but 1 year out of 12 with strong positive ENSO have CPI > 0; that is, rain above the mean. Also all but 2 years out of 13 with strong negative ENSO (ENI < −0.8) have CPI < 0; that is, rain below the mean (see Figure 4). This is the most robust result we could find in these data. With the definition more commonly used in the literature: |ENI| > 1.0, we obtained 20 strong and 43 neutral

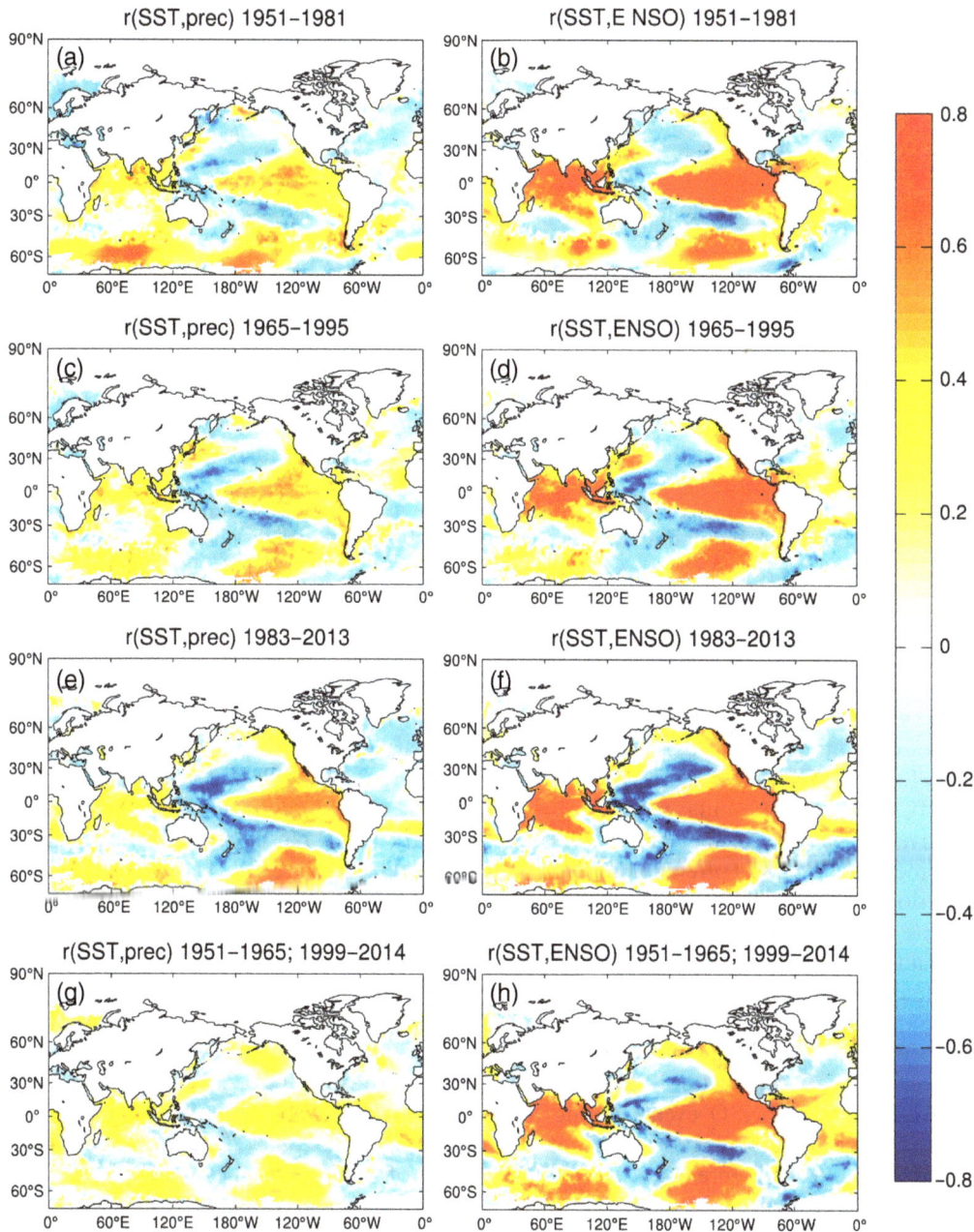

Figure 3. (a) Correlations (value indicated by the color bar) between MCBR annual rainfall and global SST for the period 1951–1981. (b) Correlations between ENSO3.4 index and global SST for the period 1951–1981. (c) and (d) Same as (a) and (b), respectively, but for the 1965–1995 period. (e) and (f) Same as (a) and (b) respectively but for the 1983–2013 period. (g) and (h) Same as (a) and (b) but for the 1951–1965 and 1999–2014 period.

events, and it yields $R_{strong} = 0.82$, $R_{neutral} = 0.41$, correlations which are statistically indistinguishable from those obtained using $|ENI| > 0.8$.

3.1.2. Stratification by regional rainfall events

Similarly when we stratified by wet ($CPI > \varepsilon$), dry ($CPI < -\varepsilon$) and neither ($-\varepsilon \leq CPI \leq \varepsilon$) events (using different values for ε, from 0.5 to 1.0), the correlations were always higher for R_{wet} (~0.5, and significant) than for R_{dry} (~0.0, and non-significant). For $R_{neither}$ we also found values comparable with R_{wet}, but only marginally significant.

3.1.3. Seasonal analysis

We also averaged monthly El Niño 3.4 Index data (ENSO) by season; that is, winter: January, February and March; spring: April, May and June; summer: July, August and September; autumn: October, November and December. For precipitation data (RAIN) we calculated totals for winter and autumn. After an extensive statistical analysis we found that seasonal correlations are in all cases lower than $R_{all} = 0.58$. The highest correlation found was $R(ENSO_{winter}, RAIN_{winter}) = 0.55$; when using ENSO lagged data we found that ENSO's previous year gives $R(ENSO_{autumn\ (n-1)}, RAIN_{winter}) = 0.50$. We did not find

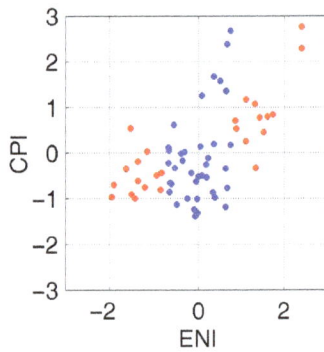

Figure 4. Stratification by ENSO strength: strong ENSO (|ENI| > 0.8) and neutral ENSO (−0.8 ≤ ENI ≤ 0.8), red and blue, respectively.

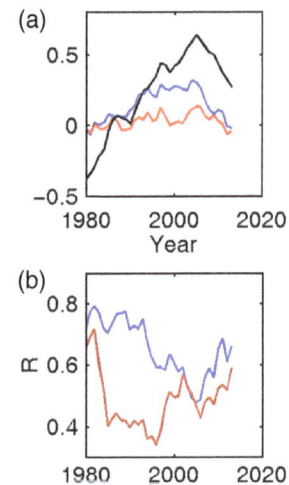

Figure 5. (a) Running 30-year climatologies from 1951–1980 to 1984–2013. (Recall that 2013 means July 2013 to June 2014), PDI (in black), CPI (in blue), ENI (in red). (b) Running 30-year correlations from 1951–1980 to 1984–2013, R(PDI′, PDI) (in blue), R(ENSO, PDO) (in red).

correlations $RAIN_{winter}$ (or $RAIN_{autumn}$) with any other ENSO season higher than 0.50. In general correlations increase as the ENSO season gets closer to $RAIN_{winter}$.

3.1.4. SST–ENSO–rainfall correlation maps

We correlated all regional seasonal winter precipitation with the gridded global SST at all grid points (Figure 3). Similarly we correlated the ENSO index with the gridded global SST. All ENSO–SST correlation maps (Figure 3(b), (d), (f), (h)) show similar patterns among them for positive correlations in the Pacific, in contrast with the precipitation–SST correlation maps. The 1983–2013 period shows the highest correlation patterns (Figure 3(e)), it includes the strongest El Niño events (1983 and 1998), and furthermore this period exhibits the greatest similarity between ENSO–SST correlation (Figure 3(f)) and precipitation–SST correlation (Figure 3(e)). The period with the lowest correlation patterns is 1951–1965 plus 1999–2014, which excludes the four strongest ENSO events during the wintertime within our record: 1965–1966, 1972–1973, 1982–1983 and 1997–1998 (Figure 3(g)), and which also shows the greatest difference with the ENSO–SST correlation patterns (Figure 3(h)).

3.2. The role of the PDO

3.2.1. ENI–CPI correlation

Considering the stratification done in Section 3.1.1 we found that PDI and ENI have the same sign in 76% of strong ENSO events; but the ENI–CPI correlation for all ENSO–PDO same-sign events is $R = 0.62$.

To further investigate the constructive interference of Gershunov et al. (1999) we varied the criterion parameter (φ) that we used to stratify strong and neutral ENSO years from 0.5 to 1.0 and also stratified the years as follows:

(i) PDI ≥ 0 and ENI ≥ φ, (positive phase of PDO and strong El Niño).
(ii) PDI < 0 y ENI < −φ (negative phase of PDO and strong La Niña).
(iii) (i) and (ii).

For example, for $\varphi = 0.5$ (0.8) we get 13 (8), 16 (11) and 29 (19) years for cases (i), (ii) and (iii), respectively. Comparing the corresponding correlation matrices, the correlations (ENI, PDI) and (PDI, CPI) are higher in general (some statistically significantly) in case (iii) than in case (i) but more so than in case (ii). None of the correlations in case ii) is significant and the highest correlation is $R(ENI, PDI) = -0.54$ for $\varphi = 1.0$, with 95% CI being [−0.89, 0.19]. In contrast, all correlations in case iii) are significant. With respect to $R(ENI, CPI)$, it increases from 0.32 for $\varphi = 0.5$ to 0.90 for $\varphi = 1.0$ in case (i), and from 0.72 to 0.82 (not a significant increase) in case (ii); there is also an increase in case (iii), with respect to (i) and (ii), but only for values of $\varphi = 0.5$, 0.6 and 0.7. These results are in line with those found in Pavia et al. (2006): El Niño and the positive phase of PDO show constructive interference, but La Niña and a negative phase of PDO do not. Correlations for negative PDI and negative ENI (23 years) are low and non-significant, compared with the cases positive PDI and positive ENI (20 years).

3.2.2. 30-year analysis

The most notorious feature here is that climatological values have decreased most recently: precipitation, since the 1975–2004 period, PDO since the 1976–2005 period, and ENSO since the 1977–2006 period; while the ENI–CPI correlations ($R_{1957-1986} = 0.53 < R(ENI, CPI) < R_{1970-1999} = 0.73$) have experienced a similar decrease, that is $R_{1984-2013} = 0.54$ (see Figure 5(a)). This may be related to the fact that within our period there was a low-frequency positive PDO phase during the last quarter of the 20th century. There is also tree-ring evidence that during this period the region was unusually (in a 1000-year record) wet (Swetnam and Betancourt, 1998).

3.2.3. PDO ENSO-dependency

In this case we investigated the variability of the ENSO–PDO relationship within our period thru the use of a 'forced' model: $PDI_n' \sim \alpha \times PDI_{n-1} + \beta \times ENI_n$ (see Pavia, 2009) for the entire record and for 30-year running periods. For the entire record we found $R(PDI', PDI)_{all} = 0.70$, and for the running correlations: $R_{1976-2005} = 0.48 < R(PDI', PDI) < R_{1953-1982} = 0.79$ (see Figure 5(b)).

4. Discussion and conclusions

The ENSO influence and the PDO–ENSO constructive interference on annual MCBR precipitation have been confirmed for the period 1951–2014. However the latter is clearer for the case of strong ENSO events of either sign (in 19 out of 25 events ENI and PDI have the same sign and the ENI–CPI correlation is $R_{strong} = 0.83$) than for any other case: neutral, strong-positive, strong-negative, etc. (see Figure 4). Nevertheless the ENI–CPI correlation for all ENI–PDI same-sign events ($R_{same-sign} = 0.62$) is statistically undistinguishable from the baseline correlation ($R_{all} = 0.58$). That is, apart from the strong ENSO case, only about one third of regional annual rainfall variability may be explained by ENSO. Also in this period dry years are significantly less related to ENSO ($R_{dry} = -0.04$) compared with wet years ($R_{wet} = 0.47$), and 85% (92%) of strong negative (positive) ENSO events correspond to years with precipitation below (above) the mean; this is the opposite of what has been suggested: i.e. that La Niña is more consistently and predictably related to dry winters in the region than El Niño is related to wet winters (e.g. Gershunov and Barnett, 1998, Gershunov and Cayan, 2003). Seasonally winter precipitation is better related to winter ENSO than to other season's ENSO values; however, the heavy 1977–1978 regional precipitation (see Figure 2) was not related to a strong positive ENSO event, but rather to a climate shift in the Pacific Ocean (Miller et al., 1994), neither does the recent MCBR drought seem to be associated with La Niña (Seager et al., 2014): the 2010–2011 La Niña was a wet year. This explains the recent decrease in 30-year ENI-CPI correlations (Section 3.2.2), which is almost concurrent with the recent decrease in 30-year climatological values for PDI, ENI and CPI (see Figure 5(a)). This may be related to a somewhat weaken influence of ENSO on the PDO (Figure 5(b)) during the peak of the 30-year PDI value (Figure 5(a)), or perhaps exceptional years, such as 1977–1978 and 2010–2011, weaken the ENSO–MCBR rainfall relationship in general compared with the only strong ENSO case. Furthermore SST–ENSO-regional precipitation maps show that MCBR annual rainfall is clearly influenced by ENSO during periods with strong warm events, as revealed by conspicuous positive correlation patterns (see Figure 3). That is, in the absence of strong El Niño events regional precipitation is not well correlated to ENSO, and frequent rainy seasons usually concur with repeated warm ENSO events, but several years of low precipitation do not always relate with frequent cool ENSO events. This is in line with recent findings, for example: of the last three dry seasons, only the 2011–2012 season was found to be a response to La Niña, whereas the 2012–2013 and 2013–2014 seasons were related to a warm tropical west Pacific sea surface temperature anomaly (Seager et al., 2014). Consequently for the last 60 years or so wet (dry) periods correspond to higher (lower) ENSO–regional rainfall correlations.

Therefore in general, that is irrespective of particular events, strong ENSO events of either sign are good predictors of rainfall in the Mediterranean Californias: strong positive (negative) events implying wet (dry) years; but otherwise ENSO (in general: that is irrespective of particular years) is not a good predictor for rainfall in our region. Similarly constructive interference is only effective for strong ENSO events; otherwise the general coincidence of ENI and PDI signs impacts ENI–CPI correlations very slightly. For dry years the correlations (ENI, CPI) and (PDI, CPI) were found to be negative or not significantly different from zero.

In the last 15 years, besides the 2010 El Niño and the 2011 La Niña, which did not result in expected precipitation anomalies, the lack of strong ENSO events may explain why ENSO has not been a good predictor for MCBR precipitation recently. It can also be that what we call the past positive phase of the PDO during the last quarter of the 20th century was an exceptional period, as shown by a 1000-year tree-ring record (Swetnam and Betancourt, 1998), and the last 15 years is just a return to the normal conditions, similar to the pre-1977 situation.

Finally considering the 2015 El Niño which is expected to influence weather and climate patterns for the 2015–2016 winter: timewise this event is occurring during a dry and lower ENI–CPI correlation period; nevertheless our analysis of ENSO strength stratification indicates that strong events such as the 2015 El Niño are very likely to correspond to a very rainy season in our region. The October 2015 US Winter Outlook, issued by NOAA's Climate Prediction Center (http://www.noaanews.noaa.gov/stories2015/), indicates wetter than average conditions for southwestern California and other southern US states. Also the Mexican Meteorological Service is forecasting precipitations above normal in northwestern Baja California in particular and in northwestern Mexico in general. These statements are both in line with our ENSO-stratification results, with the caveat of our timewise climatological results indicating the recent weakening of the ENSO-rainfall relationship in our region.

Acknowledgements

This work was funded by CONACYT (Mexico). We thank Sasha Gershunov for many insightful suggestions and several brief, but fruitful discussions.

References

Cai W, Whetton PH, Pittock AB. 2001. Fluctuations of the relationship between ENSO and northeast Australia rainfall. *Climate Dynamics* **17**: 421–432.

Chang CP, Harr P, Ju J. 2001. Possible roles of Atlantic circulations on the weakening Indian monsoon rainfall–ENSO relationship. *Journal of Climate* **14**: 2376–2380.

Fierro AO. 2014. Relationships between California rainfall variability and large-scale climate drivers. *International Journal of Climatology* **34**: 3626–3640.

Gao H, Wang YG, He J. 2006. Weakening significance of ENSO as a predictor of summer precipitation in China. *Geophysical Research Letters* **33**: L09807, doi: 10.1029/2005GL025511.

Gershunov A, Barnett T. 1998. Interdecadal modulation of ENSO teleconnections. *Bulletin of the American Meteorological Society* **79**: 2715–2726.

Gershunov A, Cayan DR. 2003. Heavy daily precipitation frequency over the contiguous United States: sources of climatic variability and seasonal predictability. *Journal of Climate* **16**: 2752–2765.

Gershunov A, Barnett T, Cayan DR. 1999. North Pacific interdecadal oscillation seen as factor in ENSO-related North American climate anomalies. *Eos* **80**: 25–30.

Krishna Kumar K, Rajagopalan B, Cane MA. 1999. On the weakening relationship between the Indian monsoon and ENSO. *Science* **284**: 2156–2159.

Kucharski F, Bracco A, Yoo JH, Molteni F. 2007. Low-frequency variability of the Indian monsoon–ENSO relationship and the Tropical Atlantic: the "weakening" of the 1980s and 1990s. *Journal of Climate* **20**: 4255–4265.

Livneh B, Rosenberg EA, Lin C, Nijssen B, Mishra V, Andreadis KM, Maurer EP, Lettenmaier DP. 2013. A long-term hydrologically based dataset of land surface fluxes and states for the conterminous United States: update and extensions. *Journal of Climate* **26**: 9384–9392.

MacDonald GM. 2007. Severe and sustained drought in southern California and the West: present conditions and insights from the past on causes and impacts. *Quaternary International* **173–174**: 87–100.

McCabe GJ, Dettinger MD. 1999. Decadal variations in the strength of ENSO teleconnections with precipitation in the western United States. *International Journal of Climatology* **19**: 1399–1410.

Miller AJ, Cayan DR, Barnett TP, Graham NE, Oberhuber JM. 1994. The 1976–1977 climate shift of the Pacific Ocean. *Oceanography* **7**: 21–26.

Newman M, Compo GP, Alexander MA. 2003. ENSO-forced variability of the Pacific Decadal Oscillation. *Journal of Climate* **16**: 3853–3857.

Pavia EG. 2009. The relationship between Pacific Decadal and Southern Oscillations: implications for the climate of northwestern Baja California. *Geofísica Internacional* **48**: 385–389.

Pavia EG, Badan A. 1998. ENSO modulates rainfall in the Mediterranean Californias. *Geophysical Research Letters* **25**: 3855–3858.

Pavia EG, Graef F. 2002. The recent rainfall climatology of the Mediterranean Californias. *Journal of Climate* **15**: 2697–2701.

Pavia EG, Graef F, Reyes J. 2006. PDO-ENSO effects in the climate of Mexico. *Journal of Climate* **19**: 6433–6438.

Rayner NA, Parker DE, Horton EB, Folland CK, Alexander LV, Rowell DP, Kent EC, Kaplan A. 2003. Global analyses of sea surface temperature, sea ice, and night marine temperature since the late nineteenth century. *Journal of Geophysical Research* **108**(D14): 4407, doi: 10.1029/2002JD002670.

Rocha A. 1999. Low-frequency variability of seasonal rainfall over the Iberian Peninsula and ENSO. *International Journal of Climatology* **19**: 889–901.

Sarkar S, Singh RP, Kafatos M. 2004. Further evidences for the weakening relationship of Indian rainfall and ENSO over India. *Geophysical Research Letters* **31**: L13209, doi: 10.1029/2004GL020259.

Schonher T, Nicholson SE. 1989. The relationship between California rainfall and ENSO events. *Journal of Climate* **2**: 1258–1269.

Seager R, Hoerling M, Schubert S, Wang H, Lyon B, Kumar A, Nakamura J, Henderson N. 2014. Causes and predictability of the 2011–2014 California drought. NOAA Drought Task Force Assement Report, NOAA, Silver Springs, MD, 42 pp. http://cpo.noaa.gov/MAPP/californiadroughtreport (accessed 8 November 2015).

Swetnam TW, Betancourt JL. 1998. Mesoscale disturbance and ecological response to decadal climatic variability in the American Southwest. *Journal of Climate* **11**: 3128–3147.

Zubair L, Ropelewski CF. 2006. The strengthening relationship between ENSO and northeast monsoon rainfall over Sri Lanka and Southern India. *Journal of Climate* **19**: 1567–1575.

9

Land surface–atmosphere interaction in future South American climate using a multi-model ensemble

R. C. Ruscica,[1] C. G. Menéndez[1,2] and A. A. Sörensson[1]*

[1]Centro de Investigaciones del Mar y la Atmósfera, Consejo Nacional de Investigaciones Científicas y Técnicas, Universidad de Buenos Aires, Argentina
[2]Departamento de Ciencias de la Atmósfera y los Océanos, FCEN, Universidad de Buenos Aires, Argentina

*Correspondence to:
A. A. Sörensson, Centro de Investigaciones del Mar y la Atmósfera, Consejo Nacional de Investigaciones Científicas y Técnicas, Universidad de Buenos Aires, Ciudad Universitaria, Int. Guiraldes 2160, Pabellón 2, Piso 2, Ciudad Autónoma de Buenos Aires C1428EGA, Argentina.
E-mail:
sorensson@cima.fcen.uba.ar

Abstract

The land–atmosphere interaction for reference and future climate is estimated with a regional climate model ensemble. In reference climate, more than 50% of the models show interaction in southeastern South America during austral spring, summer and autumn. In future climate, the region remains a strong hotspot although somewhat weakened due to the wet response that enhance energy limitation on the evapotranspiration. The region of the Brazilian Highlands and Matto Grosso appears as a new extensive hotspot during austral spring. This is related to a dry response which is probably accentuated by land surface feedbacks.

Keywords: land–atmosphere interaction; soil moisture; precipitation; coupling; South America; regional climate modeling

1. Introduction

On a monthly to seasonal time scale, soil moisture (SM) has potential to be a low-frequency modulator of climate because of its influence on the partitioning of heat fluxes and its long memory (e.g. Eltahir, 1998). Therefore, forecasts could be improved by including estimates of SM for the regions and seasons where this variable exerts a control on the atmosphere.

For SM to control precipitation (PP), the evapotranspiration (ET) regime has to be limited by SM rather than by radiation. The radiation (or energy)-limited ET regime dominates in wet regions where the moisture stress is low and variability of SM does not affect ET. In dry climates, SM availability is a first-order constraint on ET (e.g. Koster et al., 2004), but in very dry areas, however, SM does not affect PP because ET variability is too weak to induce PP generation. Therefore, regions where SM exerts control on PP tend to appear over transitions zones between dry and wet climates (Koster et al., 2004). A changing climate could alter these regimes due to increasing temperatures and altered PP amounts and temporal distribution.

South America embraces vast areas of both SM-limited regions, such as the Patagonian semi-arid steppe, and energy-limited regions, such as the Amazon rainforest. Southeastern South America (SESA) is one of the regions that have been identified as SM limited (Jung et al., 2010). SESA has also been identified as a region with strong SM–ET and SM–PP coupling during austral summer (DJF) in climate model studies (Wang et al., 2007; Sörensson and Menéndez, 2011; Dirmeyer et al., 2012; Ruscica et al., 2015) and as a region of SM–ET coupling during austral autumn (MAM) and spring (SON, Ruscica et al., 2015). In a reanalysis study by Zeng et al. (2010), subtropical South America was found to be a hotspot of interaction between SM and PP for DJF. Somewhat contrasting to these results, Dirmeyer et al. (2009) used data from several land surface models and identified SESA and the southern Amazon basin in boreal summer (JJA) as well as northeastern Brazil (NeB) and the Altiplano in MAM as regions that combine both land–atmosphere interactions and memory longer than 2 weeks, i.e. regions where forecasts could be improved by including SM information. On the other hand, a great number of studies on the global scale has focused on JJA (Koster et al., 2004; Notaro, 2008; Orlowsky and Seneviratne, 2010; Wei and Dirmeyer, 2010; Zeng et al., 2010; Zhang et al., 2011), showing different locations of high interaction in the South American tropics and agreeing on the absence of SM–ET and SM–PP interactions in subtropical areas.

The purpose of this work is to identify regions of land–atmosphere interaction over South America for all seasons and to identify changes in location and strength of these hotspots in future climate. The study is original in the sense that we use an ensemble of regional climate models (RCMs) simulations, which enhances the spatial resolution in comparison with global climate model (GCM) studies and spans a wider range of uncertainty than single-model studies.

Table I. Basic information on the ensemble members.

	RCA	RegCM3	PROMES	LMDZ
GCM forcing at lateral boundaries and SST	EC5OM-R1, EC5OM-R2 and EC5OM-R3 (Roeckner et al., 2006)	EC5OM-R1 and HadCM3-Q0 (Gordon et al., 2000)	HadCM3-Q0	EC5OM-R3 and IPSL (Hourdin et al., 2006)
Reference	Samuelsson et al. (2011)	da Rocha et al. (2009)	Domínguez et al. (2010)	Hourdin et al. (2006)
Number of vertical atmospheric levels	40	18	37	19
Number of soil moisture levels	2	2	2	2

2. Models and methodology

The CLARIS-LPB project (Boulanger *et al.*, 2011) has provided the first coordinated ensemble of RCM climate change simulations over the South American continent. The ensemble, which covers the period 1961–2100, is designed to span uncertainty of climate change scenarios using several RCMs forced by lateral boundary conditions and sea surface temperature from different GCMs, for the emission scenario A1B (Sánchez *et al.*, 2015). In this study, we use time series of seasonal values of ET and PP for a period of 30 years of reference climate (1961–1990) and future climate (2071–2100) of eight ensemble members based on the combinations of four RCMs forced by three GCMs (Table 1). The land surface schemes of all four RCMs are second generation schemes where stomatal control on ET is parameterized (Sellers *et al.*, 1997). The domains differ slightly among the models due to different types of grids, but all are of continental scale with borders over the surrounding oceans and a horizontal resolution of 0.5° × 0.5°. All models are hydrostatic.

The regions of SM–PP interaction are estimated with the index Γ, described by Zeng *et al.* (2010). Γ is defined as:

$$\Gamma = r_{pp,ET} \cdot \frac{\sigma_{ET}}{\sigma_{PP}}$$

where $r_{PP,ET}$ is the correlation between the time series of seasonal means of PP and ET, and the standard deviations σ_{ET} and σ_{PP} represent their temporal variability. For regions and seasons where the runoff is small in comparison to infiltration, PP can be interpreted as a proxy for SM. If $r_{PP,ET}$ is positive (negative), ET is SM (energy) limited. A necessary, although not sufficient, condition for SM to influence on PP is that the variability of ET is high enough. This criterion surges since, over dry regions, although ET is controlled by SM, the ET anomaly generated by a SM anomaly could be too small to generate a PP anomaly (Guo *et al.*, 2006). In the Γ index, this criterion is represented by the variability of ET normalized by the variability of PP. To avoid spuriously high values of Γ, regions where interannual rainfall variability for each season is less than 0.5 mm day^{-1} are not included in the analysis.

3. Results

The observed and ensemble seasonal mean PP for the reference period are shown in Figure 1(a)–(d) (CRU TS3.1, Mitchell and Jones, 2005) and Figure 1(e)–(h), respectively. In the northeastern corner of Brazil during DJF, the ensemble simulates rainfall of 8 mm day^{-1} in a semiarid region of 1–2 mm day^{-1}. This bias is the result of an artifact of the driving GCMs which, instead of the intertropical and the South Atlantic convergence zones, simulate only one convergence zone, with a maximum over this very dry region (Lin, 2007). When the RCMs of the ensemble are driven by reanalysis (Solman *et al.*, 2013), this bias is not present. In SESA, there is an underestimation of rainfall for all seasons. This dry bias is common in both GCMs and RCMs (Vera *et al.*, 2006; Menéndez *et al.*, 2010) and is also present when the RCMs of the ensemble are driven by reanalysis (Solman *et al.*, 2013).

The ensemble ET is shown in Figure 1(i)–(l). The spatial pattern resembles that of the PP with maximum located approximately over the same regions, except for in very wet situations such as in DJF over central Brazil or in JJA north of the equator. Over dry regions, the magnitude of the two variables is similar, while in wet regions such as the Amazonia, more than half of the PP goes to runoff.

The ensemble mean response to climate change is shown in Figure 2. The response of PP (Figure 2(a)–(d)) shows a dipole pattern of wet response over SESA and dry response north of it, similar to the study by Dirmeyer *et al.* (2014). Over SESA, future PP increases in DJF, MAM and SON with up to 1 mm day^{-1} (10–30% of reference PP), and around the border between Argentina, Paraguay and Brazil in SON, the response is larger than 1 mm day^{-1}. The dry response is strongest in SON and extends over the entire continent north of 20°S. Figure 2(e)–(h) shows that the mean ET response has a similar pattern, although of less magnitude and extension. Over permanent wet regions such as the western Amazon basin or southern Chile, the mean PP decreases without any remarkable changes in mean ET, leading to potential decreased river discharge, similar to the results by Pokhrel *et al.* (2014) and Dirmeyer *et al.* (2014). Worth to notice is that even though the annual temperature of the Amazon region increases with 3°–6° (Sánchez *et al.*, 2015), which

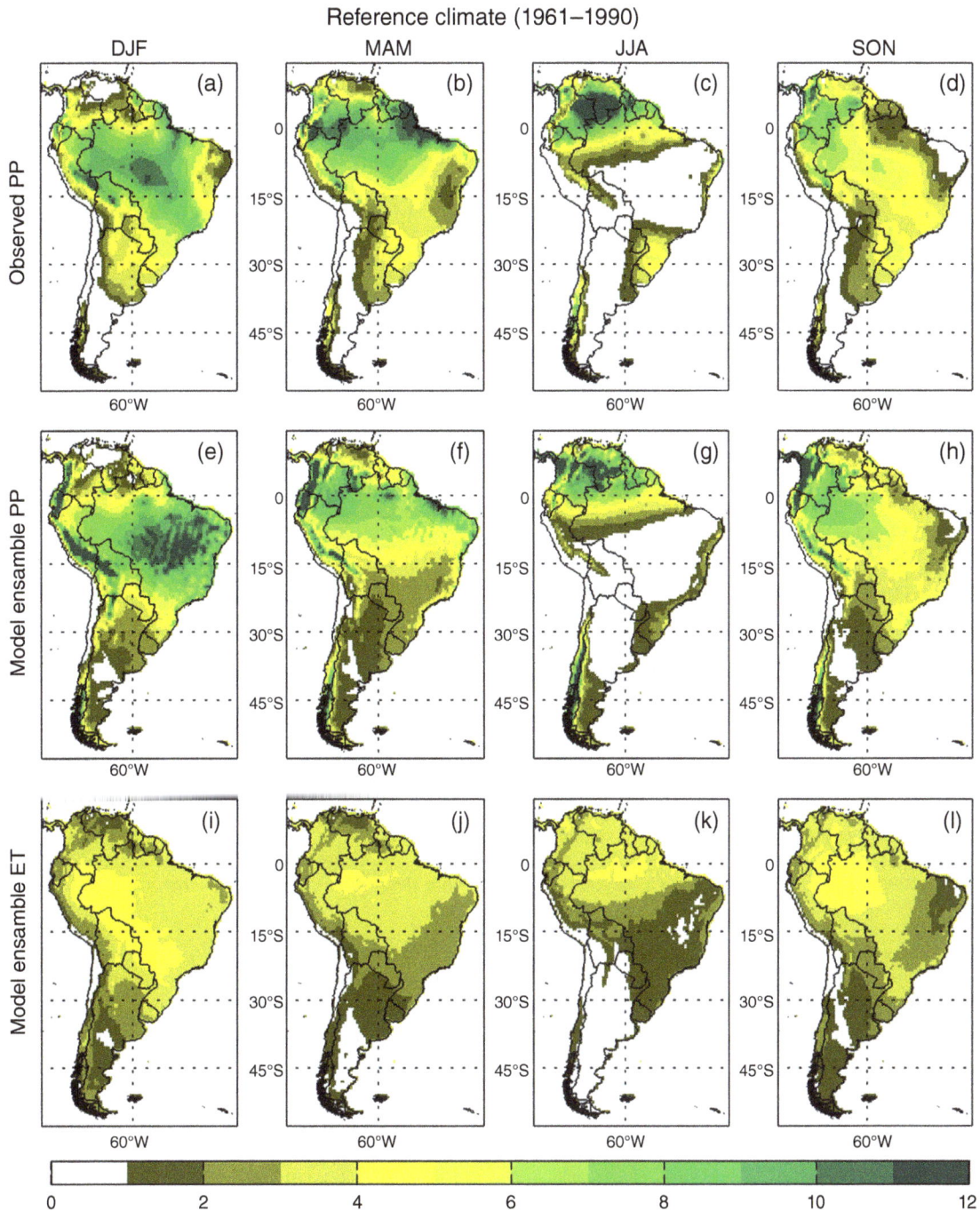

Figure 1. Seasonal means of continental precipitation (PP) and evapotranspiration (ET) during austral summer (DJF), autumn (MAM), winter (JJA) and spring (SON) in the reference climate period (1961–1990). Upper panels (a)–(d): observed PP of the CRU database. Middle panels (e)–(h): ensemble mean PP. Lower panels (i)–(l): ensemble mean ET. Units: mm day^{-1}.

should increase atmospheric demand, the ET does not increase. This could be due to other factors such as decreased wind speed, less insolation due to more clouds or stomata closure due to high temperatures.

Consensus results of land–atmosphere interaction were localized by calculating the number of ensemble members that had a Γ index equal or higher than 0.25. This is shown for the reference climate in Figure 3(a)–(d). There is a clear agreement among members that SESA is a region of land–atmosphere interaction (hotspot) for all seasons, except JJA. The

hotspot is strongest in DJF, where it extends from northern Patagonia in the south to Paraguay in the north and includes Uruguay and southern Brazil. In MAM and SON, the level of inter-member agreement is lower than in DJF, and the hotspot is located more inlands and further north. Overall, this is consistent with Wang *et al.* (2007); Sörensson and Menéndez (2011) and Ruscica *et al.* (2015). In SON, more than 50% of the ensemble members agree on land–atmosphere interaction over NeB. This is reasonable because during the dry JJA (Figure 1(g)), the soil water is depleted over this region,

Response to climate change (2071–2100, 1961–1990)

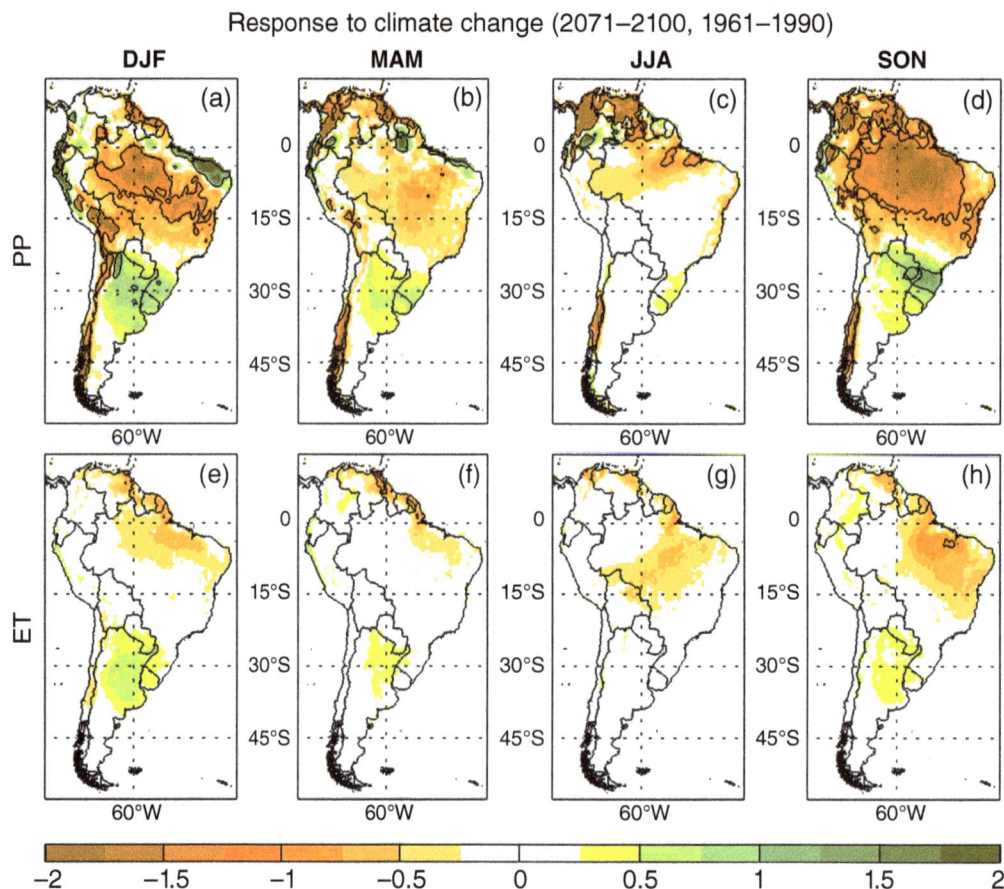

Figure 2. Ensemble response to climate change, defined as the difference of the future and reference climate mean fields (2071–2100 minus 1961–1990). Upper panels (a)–(d): precipitation (PP). Lower panels (e)–(h): evapotranspiration (ET). Units: mm day^{-1}. Contour lines delimit regions with a climate response of more (less) than 1 mm day^{-1} (−1 mm day^{-1}).

and when the wet season starts in SON (Figure 1(h)), the SM responds to the PP and ET responds to SM. It should be recalled that the very eastern corner of NeB has a wet PP bias in SON due to the GCM forcing (Figure 1(d) and (h)). During the same season, SON, half of the members show land–atmosphere interaction over eastern Amazonia.

Figure 3(e)–(h) summarizes the size and location of the hotspots in both periods of analysis. In future climate, SESA is still a region of high interaction (orange zones). This result is similar to the study by Dirmeyer *et al.* (2012), who analyzed the land–atmosphere interaction with a very high-resolution GCM for DJF and JJA for present and future climate. The most notable response to climate change (violet zones) on the continental scale occurs over the Brazilian Highlands and Matto Grosso, where many members agree on strong interaction during SON. This is a consequence of the dryer future climate during this season (Figure 2(h)), transforming the region in a transition zone with better potential conditions for the SM–PP coupling. Over the eastern corner of NeB, the hotspot is lost (green zone) in future climate. This is because Γ is not defined for regions where PP variability is less than 0.5 mm day^{-1}, and this region loses variability as the mean value decreases (Figure 2(d)). Over the Amazon region in JJA,

it can be seen how the double dry–wet transition zone line (green and orange grid points) migrates to a northern single one (violet grid points). This is consistent with the northward shift of the maximum PP gradient during this season (see Figures 1(g) and 2(c)). Central and eastern Amazonia maintain their energy-limited ET regime in spite of decreased PP (Figure 2(a)–(d)), which is in agreement with the Pokhrel *et al.* (2014) study that drive a land hydrology model with future scenario from a GCM.

As SESA in DJF is the most important and well-defined hotspot on the continental scale in both climates, Figure 4 summarizes the response of its size and strength over this region. The red region in the Buenos Aires province is a region where seven to eight members agree on interaction in both reference and future climate (strong hotspot). The violet color over northeastern Argentina and Uruguay and the dark blue over southern Brazil mark regions where the land–atmosphere interaction loses in importance. This is the region of strongest wet response in SESA (Figure 2(a)), and as the PP response is higher than the ET response, the SM should also be higher in the future, being coherent with more energy-limited regime. The green color of the periphery of the hotspot is a moderate hotspot in both climates. This region has

Ensemble consensus of land-atmosphere interaction

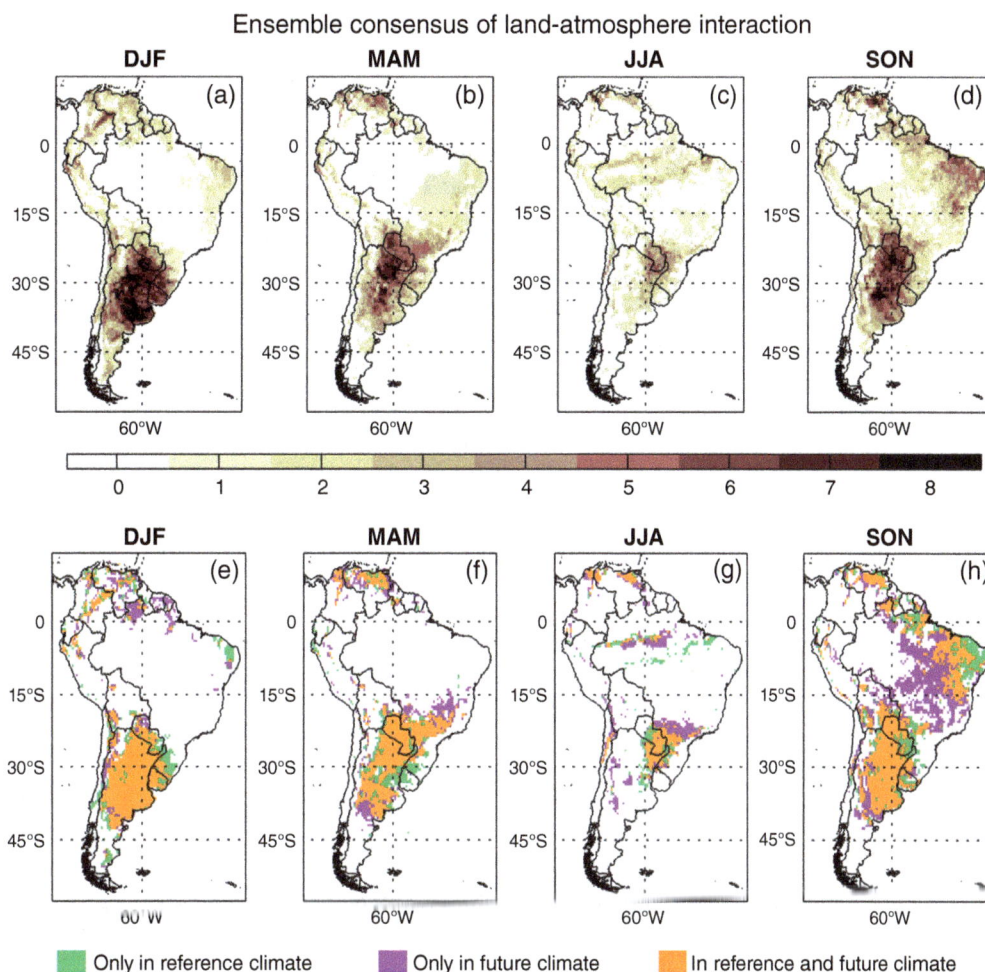

Figure 3. Upper panels (a)–(d): number of ensemble members with land–atmosphere interaction in the reference climate (1961–1990). White land grid points indicate regions without interaction or regions where Γ is not defined (see text for detailed description). Lower panels (e)–(h): extension and location of hotspots of land–atmosphere interaction that exist only in reference climate (1961–1990, green), only in future climate (2071–2100, violet) and that coexist in both climate periods (orange). Hotspot is defined as a region where three or more ensemble members coincide on having land–atmosphere interaction. White continental grid points indicate regions without hotspots in neither period.

a mean response of PP and ET of similar magnitude [less than 0.25 mm day^{-1}, Figure 2(a) and (e)], which is consistent with that the level of SM limitation does not change in the future.

4. Discussion and conclusions

The objectives of this study were to (1) achieve an estimation of land–atmosphere interaction over South America for all seasons and (2) evaluate the response of the interaction to future climate. For this purpose, an index representing land–atmosphere interaction was evaluated for an ensemble of eight RCM–GCM members for both reference and future climate.

The Γ index is advantageous in comparison with others since it is easily computed, does not require ad hoc experiments with climate models and can be used for any consistent ET and PP datasets. However, its main disadvantage is that it does not isolate the causality between two variables (e.g. SM anomalies and

PP anomalies). High Γ values identify regions where land–atmosphere interactions are possible and likely, but, a high correlation between two variables could also be caused by a third one, i.e. a common forcing as SST (Orlowsky and Seneviratne, 2010). However, the results obtained here agree with other modeling studies whose methodologies isolate SM variability as a cause of ET and PP variabilities (Wang *et al.*, 2007; Sörensson and Menéndez, 2011; Ruscica *et al.*, 2014, 2015).

For reference climate (1961–1990), SESA appears as a region of strong land–atmosphere interaction, in particular during austral summer (DJF). This is consistent with other studies (Sörensson and Menéndez, 2011; Dirmeyer *et al.*, 2012; Ruscica *et al.*, 2015), where the most robust hotspot is found during DJF when the variability of ET is high enough to get an effective land–atmosphere interaction. In austral winter (JJA), which is the coldest and driest season in SESA, the region of interaction has a smaller extension and there is less coincidence among ensemble members. However, SESA is simulated too dry in most ensemble

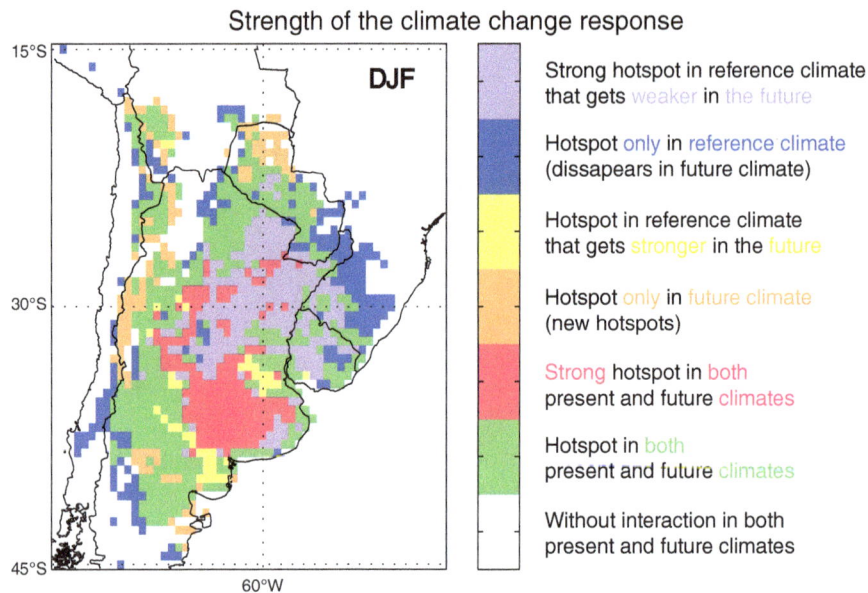

Figure 4. Response of extension and strength of the hotspot over SESA and surroundings in DJF. The definition of hotspot is the same as in Figure 3(e)–(h), and strong hotspot is defined as the agreement of seven to eight ensemble members.

members/seasons, which probably will accentuate the interaction (Ruscica *et al.*, 2015).

More than half of the members agree on strong interaction over NeB and eastern Amazonia during austral spring (SON). In DJF and austral autumn (MAM), the ensemble has a large wet bias over NeB, and therefore it is possible that this region has a strong interaction also in these seasons without being captured by this ensemble.

In future climate, SESA maintains its status as a hotspot, with in general less agreement between members. In regions where PP increases more than ET in DJF, the hotspot loses in strength and extension, while in regions where the response of the two variables is similar, the hotspot maintains its strength. This could be explained by the fact that in the latter case, the region maintains the same level of SM limitation, while in the first case, the ET depends more on atmospheric demand in future climate. In SON, the Brazilian Highlands and Matto Grosso appear as a new, extensive hotspot. This is due to the strong negative response of PP and ET, converting the energy-limited region to a SM-limited region. The dry trend over this region is consistent with the single model results by Sörensson *et al.* (2010). They found that the dry response of SON is associated with a longer dry season in this region and a delay of the monsoon onset. The results here are important because they suggest that this trend can in part be due to, or be amplified by, positive land–atmosphere feedbacks.

Acknowledgements

We acknowledge the European Community's Seventh Framework Program (CLARIS-LPB, Grant Agreement No 212492) and projects LEFE/INSU AO2015-876370 (CNRS, France), PIP No 11220110100932 (CONICET, Argentina) and PICT 2014-0887 (ANPCyT, Argentina).

References

Boulanger JP, Schlindwein S, Gentile E. 2011. CLARIS LPB WP1: metamorphosis of the CLARIS LPB European project: from a mechanistic to a systemic approach. *CLIVAR Exchanges* **16–57**: 7–10.

Dirmeyer PA, Schlosser CA, Brubaker KL. 2009. Precipitation, recycling, and land memory: an integrated analysis. *Journal of Hydrometeorology* **10**: 278–288, doi: 10.1175/2008JHM1016.1.

Dirmeyer PA, Cash BA, Kinter III JL, Stan C, Jung T, Marx L, Towers P, Wedi N, Adams JM, Altshuler EL, Huang B, Jin EK, Manganello J. 2012. Evidence for enhanced land–atmosphere feedback in a warming climate. *Journal of Hydrometeorology* **13**: 981–995, doi: 10.1175/JHM-D-11-0104.1.

Dirmeyer PA, Fang G, Wang Z, Yadav P, Milton A. 2014. Climate change and sectors of the surface water cycle in CMIP5 projections. *Hydrology and Earth System Sciences* **18**: 5317–5329, doi: 10.5194/hess-18-5317-2014.

Domínguez M, Gaertner MA, de Rosnay P, Losada T. 2010. A regional climate model simulation over West Africa: parameterization tests and analysis of land-surface fields. *Climate Dynamics* **35**: 249–265, doi: 10.1007/s00382-010-0769-3.

Eltahir EAB. 1998. A soil moisture-rainfall feedback mechanism. 1.Theory and observations. *Water Resources Research* **34**: 765–776.

Gordon C, Cooper C, Senior CA, Banks H, Gregory JM, Johns TC, Mitchell JFB, Wood RA. 2000. The simulation of SST, sea ice extents and ocean heat transports in a version of the Hadley Centre coupled model without flux adjustments. *Climate Dynamics* **16**: 147–168, doi: 10.1007/s003820050010.

Guo Z, Dirmeyer PA, Koster RD, Sud YC, Bonan G, Oleson KW, Chan E, Verseghy D, Cox P, Gordon CT, McGregor JL, Kanae S, Kowalczyk E, Lawrence D, Liu P, Mocko D, Lu C-H, Mitchell K, Malyshev S, McAvaney B, Oki T, Yamada T, Pitman A, Taylor CM, Vasic R, Xue Y. 2006. GLACE: the global land–atmosphere coupling experiment. Part II: analysis. *Journal of Hydrometeorology* **7**: 611–625, doi: 10.1175/JHM511.1.

Hourdin F, Musat I, Bony S, Braconnot P, Codron F, Dufresne J-L, Fairhead L, Filiberti M-A, Friedlingstein P, Grandpeix J-Y, Krinner K, LeVan P, Li Z-X, Lott F. 2006. The LMDZ4 general circulation model: climate performance and sensitivity to parametrized physics with emphasis on tropical convection. *Climate Dynamics* **27**: 787–813, doi: 10.1007/s00382-006-0158-0.

Jung M, Reichstein M, Ciais P, Seneviratne SI, Sheffield J, Goulden ML, Bonan G, Cescatti A, Chen J, de Jeu R, Dolman JA, Eugster W,

Gerten D, Gianelle D, Gobron N, Heinke J, Kimball J, Law BE, Montagnani L, Mu Q, Mueller B, Oleson K, Papale D, Richardson AD, Roupsard O, Running S, Tomelleri E, Viovy N, Weber U, Williams C, Wood E, Zaehle S, Zhang K. 2010. Recent decline in the global land evapotranspiration trend due to limited moisture supply. *Nature* **467**: 951–954, doi: 10.1038/nature09396.

Koster RD, Dirmeyer PA, Guo Z, Bonan G, Chan E, Cox P, Gordon CT, Kanae S, Kowalczyk E, Lawrence D, Liu P, Lu C-H, Malyshev S, McAvaney B, Mitchell K, Mocko D, Oki T, Oleson K, Pitman A, Sud YC, Taylor CM, Verseghy D, Vasic R, Xue Y, Yamada T. 2004. Regions of coupling between soil moisture and precipitation. *Science* **305**: 1138–1140, doi: 10.1126/science.1100217.

Lin JL. 2007. The double-ITCZ problem in IPCC AR4 coupled GCMs: ocean–atmosphere feedback analysis. *Journal of Climate* **20**: 4497–4525, doi: 10.1175/JCLI4272.1.

Menéndez CG, de Castro M, Boulanger J-P, D'Onofrio A, Sanchez E, Sörensson AA, Blazquez J, Elizalde A, Jacob D, Le Treut H, Li ZX, Núñez MN, Pessacg N, Pfeiffer S, Rojas M, Rolla A, Samuelsson P, Solman SA, Teichmann C. 2010. Downscaling extreme month-long anomalies in southern South America. *Climatic Change* **98**: 379–403, doi: 10.1007/s10584-009-9739-3.

Mitchell TD, Jones PD. 2005. An improved method of constructing a database of monthly climate observations and associated high resolution grids. *International Journal of Climatology* **25**: 693–712, doi: 10.1002/joc.118.

Notaro M. 2008. Statistical identification of global hot spots in soil moisture feedbacks among IPCC AR4 models. *Journal of Geophysical Research* **113**: D09101, doi: 10.1029/2007JD009199.

Orlowsky B, Seneviratne SI. 2010. Statistical analyses of land atmosphere feedbacks and their possible pitfalls. *Journal of Climate* **23**: 3918–3932, doi: 10.1175/2010JCLI3366.

Pokhrel YN, Fan Y, Miguez-Macho G. 2014. Potential hydrologic changes in the Amazon by the end of the 21st century and the groundwater buffer. *Environmental Research Letters* **9**: 084004, doi: 10.1088/1748-9326/9/8/084004.

Roeckner E, Brokopf R, Esch M, Giorgetta M, Hagemann S, Kornblueh L, Manzini E, Schlese U, Schulzweida U. 2006. Sensitivity of simulated climate to horizontal and vertical resolution in the ECHAM5 atmosphere model. *Journal of Climate* **19**: 3771–3791, doi: 10.1175/JCLI3824.1.

Ruscica RC, Sörensson AA, Menéndez CG. 2014. Hydrological links in southeastern South America: soil moisture memory and coupling within a hot spot. *International Journal of Climatology* **34**: 3641–3653, doi: 10.1002/joc.3930.

Ruscica RC, Sörensson AA, Menéndez CG. 2015. Pathways between soil moisture and precipitation in southeastern South America. *Atmospheric Science Letters* **16**: 267–272, doi: 10.1002/asl2.552.

Samuelsson P, Jones CG, Willén U, Ullerstig A, Gollvik S, Hansson U, Jansson C, Kjellström E, Nikulin G, Wyser K. 2011. The Rossby centre regional climate model RCA3: model description and performance. *Tellus A* **63**: 4–23, doi: 10.1111/j.1600-0870.2010.00478.x.

Sánchez E, Solman S, Remedio ARC, Berbery H, Samuelsson P, Da Rocha RP, Mourão C, Li L, Marengo J, de Castro M, Jacob D. 2015. Regional climate modelling in CLARIS-LPB: a concerted approach towards twentyfirst century projections of regional temperature and precipitation over South America. *Climate Dynamics* **45**: 2193–2212, doi: 10.1007/s00382-014-2466-0.

Sellers PJ, Dickinson RE, Randall DA, Betts AK, Hall FG, Berry JA, Collatz GJ, Denning AS, Mooney HA, Nobre CA, Sato N, Field CB, Henderson-Sellers A. 1997. Modeling the exchanges of energy, water, and carbon between continents and the atmosphere. *Science* **275**(5299): 502–509, doi: 10.1126/science.275.5299.502.

Solman SA, Sanchez E, Samuelsson P, da Rocha RP, Li L, Marengo J, Pessacg NL, Remedio ARC, Chou SC, Berbery H, Le Treut H, de Castro M, Jacob D. 2013. Evaluation of an ensemble of regional climate model simulations over South America driven by the ERA-interim reanalysis: model performance and uncertainties. *Climate Dynamics* **41**: 1139–1159, doi: 10.1007/s00382-013-1667-2.

Sörensson AA, Menéndez CG. 2011. Summer soil-precipitation coupling in South America. *Tellus A* **63**: 56–68, doi: 10.1111/j.1600-0870.2010.00468.x.

Sörensson AA, Menéndez CG, Ruscica R, Alexander P, Samuelsson P, Willén U. 2010. Projected precipitation changes in South America: a dynamical downscaling within CLARIS. *Meteorologische Zeitschrift* **19**: 347–355, doi: 10.1127/0941-2948/2010/0467.

Vera C, Silvestri G, Liebmann B, Gonzalez P. 2006. Climate change scenarios for seasonal precipitation in South America from IPCC-AR4 models. *Geophysical Research Letters* **33**: L13707, doi: 10.1029/2006GL025759.

Wang G, Yeonjoo K, Wang D. 2007. Quantifying the strength of soil moisture–precipitation coupling and its sensitivity to changes in surface water budget. *Journal of Hydrometeorology* **8**: 551–570, doi: 10.1175/JHM573.1.

Wei J, Dirmeyer PA. 2010. Toward understanding the large-scale land-atmosphere coupling in the models: roles of different processes. *Geophysical Research Letters* **37**: L19707, doi: 10.1029/2010GL044769.

Zeng X, Barlage M, Castro C, Fling K. 2010. Comparison of land–precipitation coupling strength using observations and models. *Journal of Hydrometeorology* **11**: 979–994, doi: 10.1175/2010JHM1226.1.

Zhang L, Dirmeyer PA, Wei J, Guo Z, Cheng-Hsuan L. 2011. Land–atmosphere coupling strength in the global forecast system. *Journal of Hydrometeorology* **12**: 147–156, doi: 10.1175/2010JHM1319.1.

10

The evolving concept of air pollution: a small-world network or scale-free network?

Linan Sun,[1,2] Zuhan Liu,[3,4,*] Jiayao Wang,[1,2] Lili Wang,[5] Xuecai Bao,[3,4] Zhaoming Wu[3,4] and Bo Yu[6]

[1] School of Resource and Environmental Sciences, Wuhan University, China
[2] Zhengzhou Institute of Surveying and Mapping, China
[3] Jiangxi Province Key Laboratory of Water Information Cooperative Sensing and Intelligent Processing, Nanchang, China
[4] School of Information Engineering, Nanchang Institute of Technology, China
[5] School of Science, Nanchang Institute of Technology, China
[6] College of Architecture and Environment, Sichuan University, Chengdu, China

*Correspondence to:
Z. H. Liu, Jiangxi Province Key Laboratory of Water Information Cooperative Sensing and Intelligent Processing, No. 289 Tianxiang Dadao street, Nanchang 330099, China.
E-mail: lzh512@126.com

Abstract

To analyze the dynamics of air pollution, a homogenous partition of the coarse graining process is employed to transform the daily air pollution index series in Lanzhou into a character series consisting of five characters (R, r, e, d and D). The nodes of the pollution fluctuation network are 125 three-symbol strings (i.e. 125 fluctuation patterns in a duration of 3 days) linked in the network's topology by a time sequence. The network contains integrated information about the interconnections and interactions among the fluctuation patterns of pollution in the network topology. After calculating the dynamical statistics of degree and degree distribution, we find that the distribution follows a three-stage power-law distribution characterized by a scale-free property with hierarchy structure and small-world effect. Therefore, the pollution fluctuation network is not only a scale-free network with hierarchy but also a small-world network. The higher the degree of the node is, the greater the probability that the pollution fluctuation modes will occur. The main nodes of pollution fluctuation networks generally contain the symbols R and r, which demonstrates that the feature of pollution fluctuation is mainly ascending.

Keywords: Lanzhou; complex network; scale-free network; small-world network; degree distribution

1. Introduction

With the rapid development of China's economy, the environment is placed under increasingly heavy pressure. The atmosphere has become pervasive, a condition that is closely related to the health of people everywhere. Lanzhou in China, as an important part of the economy, has been developing rapidly and has a high population density. The region's air pollution and other atmospheric environmental issues have gradually elicited widespread attention (Sanchez-Reyna et al., 2005; Chen et al., 2015) Technical support for air pollution control and air quality amelioration must be provided. The dynamic change in air quality must be systematically analyzed, and the changing laws of atmosphere pollution should be determined.

In recent years, research on atmospheric environmental change mainly focused on SO_2, NO_x (Hamer and Shallcross, 2007; Griffiths and Cox, 2009) and total suspended particulate (Latha and Badrinath, 2004; Sanchez-Reyna et al., 2005) and involved the organic compounds and heavy metals in PM_{10} and $PM_{2.5}$. The research methods employed are mainly analysis methods that combine investigation-sampling and physics–chemistry experiments (West et al., 2005; Hamer and Shallcross, 2007), pollution-load method (Zhang et al., 2014), Spearman's rank correlation coefficient method (Heo and Lee, 2013), environmental Kuznets curve model (Xiao et al., 2011), neural network model (Pasini and Modugno, 2013), fuzzy mathematics method (Li and Liu, 2013), and so on. The development of air pollution is a complex dynamic system, and the air pollution series itself is nonlinear and non-stationary. Hence, using traditional methods to reveal the dynamic characteristics of air pollution is difficult and cannot highlight the acts of nature.

Lanzhou is one of the most heavily air-polluted cities in China. To establish reasonable preventive countermeasures, the pollutant characteristics and the mechanisms of the pollution indexes' temporal evolution need to be understood. In recent years, numerous researchers have devoted a considerable amount of attention on Lanzhou's atmospheric environmental quality and air pollution, including the causes and influencing factors of air pollution (Wang, 1992; Yu et al., 2011), the physical mechanism of atmospheric pollution formation in Lanzhou (Wang, 1992; Hu and Zhang, 1999; Yu et al., 2011). To further explore effective pollution forecasting methods, the meteorological conditions and meteorological data characteristics in Lanzhou and their relationship with pollutant concentrations have been investigated (Wang et al., 2000; Shang et al., 2001). Several researchers established different types of atmospheric pollution diffusion models based on Lanzhou's special geographical location of valley basin (Jiang and Peng, 2002; Xie and Jie, 2010). In these studies, the application of several models to Lanzhou (e.g. the site-optimized model (Dirks *et al.*, 2006)) is limited because of the topography of the valley. The analysis of causes and formation mechanism of atmospheric pollution has not been fully developed. Time variation and dynamism exist in the distribution of the concentration of atmospheric pollutants, and a chaotic time series should be applied to reflect the temporal evolution law of atmospheric pollutants along with the change in time and other factors.

Complex network refers to a network that has one or all properties of self-organization, self-similarity, attractor, small world and scale free (Watts and Strogatz, 1998; Jeong *et al.*, 2000). Since the late 20th century, the number of studies on complex networks has been increasing. Complex networks have been widely utilized in research on social science (Wasserman and Faust, 1994), computer science (Gao and Li, 2009), transportation (Soh *et al.*, 2010) and other fields. However, its application in environmental science is rare.

The relation between motion track and formal language is established by symbol dynamics, and complexity is characterized by grammar complexity theory. The core content is the coarse grain of symbol dynamics and time series. Applying different levels of coarse graining, rounding a small level of details, and making them characteristic quantities help highlight the essential characteristics (Mitran *et al.*, 2012). Hence, studying the change law of atmosphere pollution with a complex network is a new attempt and thus has an important research value.

According to the air pollution index data in Lanzhou obtained on 1 July 2000 to 30 June 2015 from the website of the Chinese Ministry of Environmental Protection (http://www.zhb.gov.cn/), this study examine the daily API data of Lanzhou in China using complex network theory and reveal the dynamic characteristics of air pollution change in Lanzhou from the perspective of a complex network. By using the uniform probability principle in the foundation of a homogenous partition of the coarse graining process, the daily air pollution index series in Lanzhou is transformed into a character series consisting of five characters (R, r, e, d and D). The nodes of the pollution fluctuation network are 125 three-symbol strings linked in the network's topology by a time sequence. The network contains integrated information about the interconnections and interactions between fluctuation patterns of pollution in network topology. To better understand the complex characteristics of the air pollution system, the dynamic statistical characteristics and topological parameters of the network are analyzed, and the inherent law of the pollution fluctuation network is obtained.

2. Data source and data preprocessing

The data source is the daily air quality index of Lanzhou offered on website of the Chinese Ministry of

Environmental Protection (http://www.zhb.gov.cn/); the data were obtained from 1 July 2000 to 30 June 2015. When the air pollution index data of 9 days were unavailable, we made up for it by the arithmetic average value of 2 days' data. If 1 July 2000 is set as number 1, then 30 June 2015 is set as 5477. To eliminate border effects, the data were extended by the cycle method, and the main data processing and wavelet transform were accomplished with Matlab 7.6. The seasons were divided as follows: spring (March to May), summer (June to August), autumn (September to November), and winter (December to February). The time series were filtered with a locally projective nonlinear noise reduction filter (Chelidze, 2013). The method works on the hypothesis that a natural time series is a combination of both a low-dimensional dynamical system and high-dimensional (random) noise. Unlike linear filters, nonlinear ones only remove noisy data points. These points can then be replaced by estimates computed from a nonlinear interpolation process (Fuentes, 2003).

3. Building a pollution network

The creation of a pollution network for the daily air pollution index series reflects the change fluctuation characteristics of air pollution in Lanzhou. We calculated the topology statistics of pollution from the perspective of complex networks. The following is a brief introduction of the main steps implemented to build the pollution network.

The first step is a five-valued coarse graining process. Calculating fluctuation $k(t)$ of the air pollution index series is as follows:

$$k(t) = \frac{P(t + \Delta t) - P(t)}{\Delta t}, \qquad (1)$$

where Δt is the time interval scale of the factor sequence. We mainly focused on $\Delta t = 2$, i.e. pollution index fluctuations in any three consecutive days.

The variation slope k of three consecutive days for air pollution index time series $P(t)$ can be fitted with the least squares method, that is,

$$k(i/3) = \frac{\sum_{t=1}^{i} t \times P(t) - \frac{1}{i}\left(\sum_{t=1}^{i} P(t)\right)\left(\sum_{t=1}^{i} t\right)}{\sum_{t=1}^{i} t^2 - \frac{1}{i}\left(\sum_{t=1}^{i} t\right)^2}$$

$$i = 3, 4, \dots, 5477 \qquad (2)$$

Calculating probability P_k of possible fluctuation values in the air pollution index series is as follows:

$$P_k = \int_{-\infty}^{k} \frac{Num(x)}{N} dx, \qquad (3)$$

where $Num(x)$ is the number of times fluctuation mode x occurred corresponding to an air pollution index. We divided the P_k values into five equal intervals.

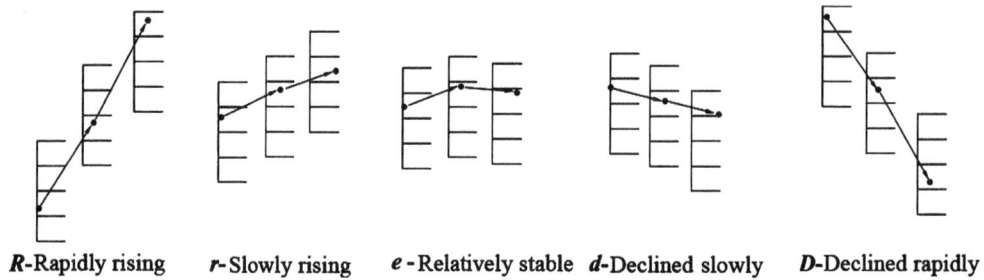

R-Rapidly rising   **r-Slowly rising**   **e-Relatively stable** **d-Declined slowly**   **D-Declined rapidly**

Figure 1. Meaning of the symbols R, r, e, d and D in this study.

Pollution fluctuation $k(t)$, which lies within the five equal intervals, can then be presented as R, r, e, d and D, as shown below.

$$Si = \begin{cases} R, \ 0 < P_k < 0.2, \\ r, \ 0.2 \le P_k < 0.4, \\ e, \ 0.4 \le P_k < 0.6, \\ d, \ 0.6 \le P_k < 0.8, \\ D, \ 0.8 \le P_k < 1.0. \end{cases} \qquad (4)$$

The meaning of R, r, e and d in this study is shown in Figure 1.

Daily air pollution index series $P(t)$ was transformed into a corresponding symbolic series in Lanzhou in the last 15 years.

$$S_P = \left(S_1 S_2 S_3 \ldots \right), S_i \in (R, r, e, d, D) \qquad (5)$$

Transforming time series into symbolic series, i.e. symbolic time series analysis (STSA), is a new analysis method developed from symbolic dynamic theory, chaos theory, and information theory. The basic idea is to transform a data series with a number of possible values into a symbolic series with only a few different values. For the selection of symbols, owing to the difference in symbolic rules (Lehman *et al.*, 1997) the symbols that scholars select are different (Wang *et al.*, 2011). Although having too many symbols can sample the details of information, the amount of operation becomes too large. However, the details of the series may be obscured by small symbols with a very high loss rate of the amount of information; this can change the dynamic characteristics of the system (Chen *et al.*, 2010).

Research on the atmospheric system generally focuses on the change in the process. However, this chapter investigates the fluctuation patterns that represent air pollution index fluctuations in several consecutive days, so meta-patterns are guaranteed to be equal in quantity when the numerical fluctuations are categorized. In this manner, we can ensure the probability that five actors occur equally in the fluctuation patterns.

Time interval scale Δt is an important parameter in the process of transforming an air pollution index numeric series into a symbolic series. If the Δt value is selected differently, the time series resolution will also differ. For example, Zhou *et al.* (2008) set Δt to 1 in a study of temperature change. The number of times, i.e. $N(R)$,

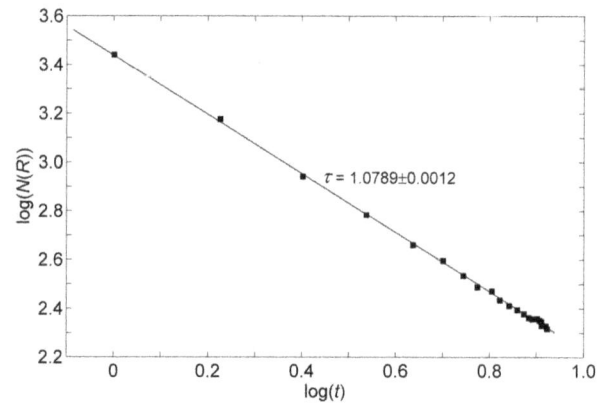

Figure 2. Graphs of symbol statistics log(N) and log(t).

$N(r)$, $N(e)$, $N(d)$ and $N(D)$, which R, r, e, d and D occur may not be the same; hence, the correlation among them is not the same. With regard to the air pollution index series of Lanzhou from 1 July 2000 to 30 June 2005, we analyzed the statistical characteristics of $N(R)$, $N(r)$, $N(e)$, $N(d)$ and $N(D)$ in the symbol series that are built at different time interval scales Δt. We found that the statistics $N(R)$, $N(r)$, $N(e)$, $N(d)$ and $N(D)$ and time interval scale Δt follow the power law $N \sim \Delta t^\tau$. Figure 2 shows the graphs of symbol statistics in double logarithmic coordinate.

Figure 2 reveals that a good linear relationship exists among symbols R, r, e, d and D. This condition indicates that air pollution index fluctuation itself presents a good scale-free characteristic. The scale-free characteristic of fluctuation reflects the optimization in the network evolution process (Watts and Strogatz, 1998; Brede, 2011); it tends to have a small time scale in processing the symbols using five-valued coarse graining (Liu *et al.*, 2007). Hence, it is effective to investigate the transformation from air pollution fluctuation into symbolic series in three consecutive days in this study (Li and Wang, 2003).

The second step is to build the network. We introduced a weighted network to describe the correlation among the fluctuation patterns in the air pollution index series. The nodes of the network are 125 fluctuation patterns with three-symbol strings, and the edge of the network is the previous node points to the next node. That is, one pattern is transformed into the next one, and one pollution process is transformed into another. The weight of the edge between two nodes is the number

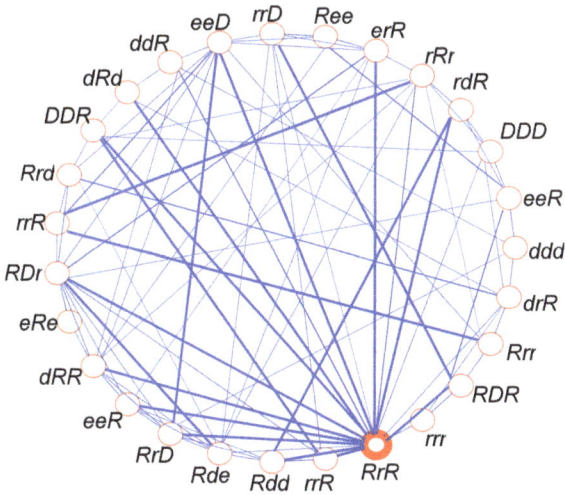

Figure 3. Complex network constructed with partial nodes of the air pollution index series.

Table 1. Sequence of the size of node degree of various fluctuation modes in the pollution fluctuation network.

Node	RrR	rdR	DDR	rrR	rRR	DeR	RDr	rRr	eRe	RRe
Degree	362	247	237	226	209	182	173	151	142	133
Grade	1	2	3	4	5	6	7	8	9	10
Node	eeR	ddd	reR	eDR	Ree	dRd	DDD	–	DDd	–
Degree	130	124	123	111	108	102	97	–	2	–
Grade	11	12	13	14	15	16	17	–	125	–

of multiple disjoint parallel connections between them. For example, in the pollution fluctuation network we built, the symbolic series are *eRdDeRdrdeDDDr eDDDrDedDdDdedrRreeRrreRedrrDdredDrDDedDe reDdDeeRdeeRedrdeDdD* The three-symbol strings are utilized as the nodes of networks, and the directed connection of the nodes is *eRd→DeR→drd→eDD →Dre→DDD→rDe→dDd→Dde→drR→ree→Rrr→e Re→drr→Ddr→edD→rDD→edD→ere→DdD→eeR →dee→Red→rde→ DdD*.

From this, we can build a directed weighted network that shows the interactions among various fluctuation patterns. Figure 3 provides the correlation diagram of the part of nodes.

4. Discussion

With the lucubration of complex networks, many conceptions and measures have been proposed to describe and express the structural characteristics of such networks. Meanwhile, degree and degree distribution are the most important statistical characteristics.

Degree is also called connectivity, and 'degree of nodes' means the edge that is connected to the nodes. Degree has different meanings in different networks. In a social network, degree can represent individual effects and importance. The higher the degree is, the more influence and function exist in the entire system and vice versa. Degree distribution represents the probability distribution function $p(k)$ of node degree, and it means the probability that the node is connected with k edges. The current research involves two common degree distributions. The first one is exponential degree distribution, that is, $p(k)$ attenuates exponentially with increasing k. The second one is power-law distribution, that is, $p(k) \sim k^{\lambda}$, in which λ is the degree exponent; it has various kinetic properties in different networks. In addition, degree distribution has other forms. For instance, it is a two-point distribution in a star network and a one-point distribution in a regular network.

Figure 3 shows the complex network constructed with partial nodes of the daily average air pollution index series. The thickness of the edge lines that connect the nodes reflects the strength of correlation. For instance, the thickest line is between nodes *RrR* and *rdR*, which means the correlation between the two air pollution fluctuation modes is the most intensive; in addition, the two modes have good contact in the long term. Table 1 shows the sequence of the size of the node degree of fluctuation modes in the air pollution fluctuation network. Several node degrees, such as nodes *DDR*, *rrR* and *DeR*, are relatively large. This shows that the network fluctuation modes represented by these nodes play an important direct correlation role in the air pollution fluctuation network.

All the fluctuation modes were transformed into these important modes. The frequency of mode transformation is high, so the extreme pollution event is likely to occur in Lanzhou. In addition, we also derived the character frequency statistics on the node degree of the pollution fluctuation network. Symbol *R* often presents a sharp rise in first 17 nodes of a high degree, whereas *D* seldom declines rapidly, which symbol *R* and *D* appear 18 times respectively. Hence, the pollution fluctuation characterized by a sharp rise appears more often in the pollution change.

Figure 4 shows the degree distribution and cumulative degree distribution of nodes. The degree distribution of nodes in the fluctuation network satisfies the power-law distribution overall and has a heavy tail as a result of a random selection mechanism caused by random connection. However, as long as the amount of certainty is sufficiently large, the random heavy tail of the power-law distribution will be restrained (Patriarca *et al.*, 2006). In addition, the node degree of the network obeys the three-segment power-law distribution. Therefore, the pollution fluctuation network has a scale-free property, but the degree is distributed unevenly, and the difference in the importance of the degree among different pollution fluctuation modes is relatively large. After fitting and statistical calculation, the cut-off points were determined to be 60 and 100. The first segment index $\lambda_1 = -3.418$ ($R^2 = 0.997$), the second segment index $\lambda_2 = -0.235$ (R^2 equals 0.993), and the third segment index $\lambda_3 = -6.157$ ($R^2 = 0.987$).

Scholars Carlson and Doyle (2002) from the University of California, Santa Barbara, explained why the node degree obeyed the power-law distribution by using *HOT* theory, Kitano (2004, 2007) also conducted a similar research. We utilized the pollution system as an

(a)

(b)

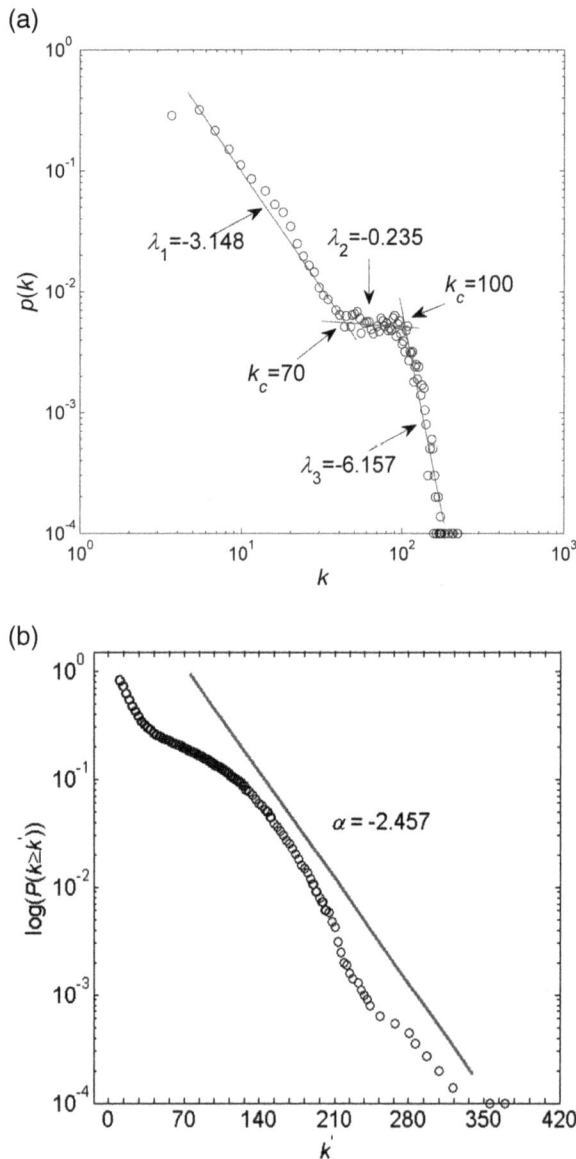

Figure 4. Degree distribution and accumulated degree distribution.

example to explain the cause of the power-law distribution by *HOT* theory. The atmosphere system is composed of many subsystems; large ones include climate and hydrological systems, and small ones include temperature, precipitation, and pollution dynamic systems. If the pollution system can bear significant changes in the robust climatological–hydrological factors, such as temperature, humidity, precipitation, and pollution, but cannot bear the disturbance of small pollution events with some uncertainties, such as exhaust emissions of home kitchens and emissions of a few air pollutants from catering, small factories, and vehicles, then serious pollution events may be created by the SOC behavior of air pollutants (Shi *et al.*, 2008; Shi and Liu, 2009). When the atmosphere pollution system is in a *HOT* state, the system satisfies the power-law distribution.

In the semi-logarithmic coordinate as shown Figure 4, pollution network approximately obey exponential

decay distribution (Han *et al.*, 2009) and can randomly select the network evolution process. The pollution fluctuation mode occurs with some randomness (when and which fluctuation mode will occur are unknown). This scenario further indicates that the air pollution index has chaotic characteristic, but the number of times that all types of pollution fluctuation modes occur abides by certain laws in a long period of time. The range of degree distribution for the fluctuation networks is relatively narrow. Therefore, air pollution index fluctuations obey an index distribution, and this law is similar to the distribution of the number of times that all types of earthquakes occur in the earth system (Abe and Suzuki, 2012). The reason may be that the pollution process follows the maximum entropy principle (Thomas, 1979), which embodies the inherent essential dynamic characteristics of the atmosphere system to a certain extent.

As a result, the air pollution fluctuation network has a scale-free property and a small-world effect and is a small-world network with a scale-free property. A high-speed railway's and an airline's compound network as well as the topological properties of international oil prices have such characteristics. The two characteristics of the air pollution fluctuation network reflect the harmony and unity of certainty and randomness. With chaotic characteristics, natural unity and diversity are further described with the passing of time in the pollution process. Many networks have scale-free characteristics and small-world effect. For instance, the atmospheric network is not only a small-world network but also a scale-free network and indicates a long-range correlation.

5. Conclusions

A pollution fluctuation network was built on the homogenous partition of the coarse graining method by applying the complex network to the daily air pollution index series of Lanzhou. In accordance with the principle of equal probability, the fluctuation of climatological–hydrological factors was transformed into a symbol series. To describe the complexity of the spatial structure in the air pollution system quantitatively based on the structural characteristics, the relevant factor values were inputted into the corresponding network topology.

To obtain the inherent law of the fluctuation network, we specifically analyzed the topological properties of the degree distribution of the fluctuation network and obtained the following conclusions.

(1) The degree distribution of the pollution fluctuation network obeys a three-segment distribution; the network is a small-world network with scale-free characteristics. The frequency of the various fluctuation modes of the pollution fluctuation network follows the maximum entropy principle at a large time scale. This is probably the embodiment of the self-organized criticality of the pollution network.

(2) Several differences exist in the degree value of the nodes in the fluctuation network. The degree of RrR, rdR, DDR, rrR and rRR is markness, these models of temperature fluctuant represent by the five nodes play an important direct correlation role in the process of pollution change, and most of all pollution fluctuation modes have risen or a sharp rise trend.

In conclusion, by adopting the theory of complex network, we analyzed the daily air pollution index in Lanzhou, identified the inherent laws of the pollution fluctuation network, and obtained a series of valuable conclusions that are significant in controlling and reducing the air pollution in Lanzhou. In the future, we plan to enhance our work in following three aspects. First, we will investigate the natural dynamic mechanism of air pollution. Secondly, we will study the causes of the differences among different modes. Third, we will explore why the fluctuation network obeys a three-segment distribution.

Acknowledgements

This work is supported by the Science and Technology Research Project of Jiangxi Provincial Education Department (GJJ151109) and (GJJ151099), the National Natural Science Foundation of China (41271392) and (61401189) and the Natural Science Foundation of Jiangxi, China (20151BAB207010).

References

Abe S, Suzuki N. 2012. Dynamical evolution of the community structure of complex earthquake network. *EPL* **99**: 313–316.

Brede M. 2011. Growth and optimality in network evolution. *Artificial Life* **17**: 281–291.

Carlson JM, Doyle J. 2002. Complexity and robustness. *Proceedings of the National Academy of Sciences USA* **99**: 2538–2545.

Chelidze D. 2013. Smooth projective noise reduction for nonlinear time series. *Topics in Nonlinear Dynamics* **1**: 77–83.

Chen WD, Xu H, Guo Q. 2010. Dynamic analysis on the topological properties of the complex network of international oil prices. *Acta Physica Sinica* **59**: 4514–4523.

Chen B, Lu SW, Li SN, Wang B. 2015. Impact of fine particulate fluctuation and other variables on Beijing's air quality index. *Environmental Science & Pollution Research* **22**: 5139–5151.

Dirks KN, Nanni A, Dirks VI. 2006. Modelling and predicting urban atmospheric pollutants in the Aosta Valley region of Italy using a site-optimised model. *Atmospheric Science Letters* **7**: 15–20.

Fuentes M. 2003. Statistical assessment of geographic areas of compliance with air quality standards. *Journal of Geophysical Research, [Atmospheres]* **108**: 87–107.

Gao S, Li C. 2009. Complex network model for software system and complexity measurement. 2009 World Congress on Computer Science and Information Engineering, *IEEE Computer Society*, Los Angeles, CA, 624–628.

Griffiths PT, Cox RA. 2009. Temperature dependence of heterogeneous uptake of N2O5 by ammonium sulfate aerosol. *Atmospheric Science Letters* **10**: 159–163.

Hamer PD, Shallcross DE. 2007. Modelling the impact of oxygenated VOC and meteorology upon the boundary layer photochemistry at the South Pole. *Atmospheric Science Letters* **8**: 14–20.

Han Q, Brenguier JL, Kuo KS, Naeger A. 2009. A new IR technique for monitoring low cloud properties using geostationary satellite data. *Atmospheric Science Letters* **10**: 115–121.

Heo S, Lee J-T. 2013. Study of environmental health problems in Korea using integrated environmental health indicators. *International Journal of Environmental Research & Public Health* **10**: 3140–3156.

Hu YQ, Zhang Q. 1999. Atmospheric pollution mechanism and prevention countermeasure of the Lanzhou. *China Environment Science* **19**: 119–122.

Jeong H, Tombor B, Albert R, Oltvai Z, Barabasi AL. 2000. The large-scale organization of metabolic networks. *Nature* **407**: 651–654.

Jiang JH, Peng XD. 2002. Numerical study on air pollution concentration over Lanzhou winter under complex terrain. *Plateau Meteorology* **21**: 1–7.

Kitano H. 2004. Cancer as a robust system: implications for anticancer therapy. *Nature Reviews Cancer* **4**: 227–235.

Kitano H. 2007. Towards a theory of biological robustness. *Molecular Systems Biology* **3**: 137.

Latha KM, Badrinath KVS. 2004. Studies on atmospheric turbidity over a tropical urban environment. *Atmospheric Science Letters* **5**: 134–139.

Lehman M, Rechester AB, White RBS. 1997. Symbolic analysis of chaotic signals and turbulent fluctuations. *Physical Review Letters* **78**: 54–57.

Li A, Liu JY. 2013. Efficient environmental air quality evaluation scheme based on the fuzzy mathematics method. In *Informatics and Management Science*, Du W (ed), Vol. **III**. Springer: London; 645–651.

Li P, Wang BH. 2003. DNA series analysis method applied in study of financial data time series. *Journal of the Graduate School of the Chinese Academy of Science* **41**: 200–204.

Liu P, Izvekov S, Voth GA. 2007. Multiscale coarse-graining of monosaccharides. *Journal of Physical Chemistry B* **111**: 11566–11575.

Mitran G, Ilie S, Tabacu I, Nicolae V. 2012. Modeling the impact of road traffic on air pollution in urban environment case study: a new overpass in the city of Craiova. *Environmental Engineering & Management Journal* **11**: 407–412.

Pasini A, Modugno G. 2013. Climatic attribution at the regional scale: a case study on the role of circulation patterns and external forcings. *Atmospheric Science Letters* **14**: 301–305.

Patriarca M, Chakraborti A, Germano G. 2006. Influence of saving propensity on the power-law tail of the wealth distribution. *Physica A* **369**: 723–736.

Sanchez-Reyna G, Wang KY, Gallardo JC, Gallardo JC, Shallcross DE. 2005. Association between PM₁₀ mass concentration and wind direction in London. *Atmospheric Science Letters* **6**: 204–210.

Shang KZ, Da CY, Fu YZ, Wang SG, Yang DB. 2001. The stable energy in Lanzhou city and the relations between air pollution and it. *Plateau Meteorology* **20**: 76–81.

Shi K, Liu CQ. 2009. Self-organized criticality of air pollution. *Atmospheric Environment* **43**: 3301–3304.

Shi K, Liu CQ, Ai NS, Zhang XH. 2008. Using three methods to investigate time-scaling properties in air pollution indexes time series. *Nonlinear Analysis: Real World Applications* **9**: 693–707.

Soh H, Lim S, Zhang T, Zhang TY, Fu XJ, Lee GKK, Hung TGG, Dic P, Prakasam S, Wong L. 2010. Weighted complex network analysis of travel routes on the Singapore public transportation system. *Physica A* **389**: 5852–5863.

Thomas MU. 1979. Technical note – a generalized maximum entropy principle. *Operations Research* **27**: 1188–1196.

Wang JM. 1992. Turbulence characteristics in an urban atmosphere of complex terrain. *Atmospheric Environment* **26**: 2717–2724.

Wang SG, Jiang DB, Yang DB, Shang KZ, Qi B. 2000. A study on characteristics of change of maximum mixing depths in Lanzhou. *Plateau Meteorology* **19**: 363–370.

Wang XM, Zhang B, Qiu DY. 2011. The quantitative characterization of symbolic series of a boost converter. *IEEE Transactions on Power Electronics* **26**: 2101–2105.

Wasserman S, Faust K. 1994. *Social Network Analysis Methods and Applications*. Cambridge University Press: Cambridge.

Watts DJ, Strogatz SH. 1998. Collective dynamics of 'small-world' networks. *Nature* **393**: 440–442.

West LJ, Killworth PD, Cornell VC, Challenor PG. 2005. Remote visualization of Labrador convection in large oceanic datasets. *Atmospheric Science Letters* **6**: 198–203.

Xiao Q, Gao Y, Hu D, Tan H, Wang TX. 2011. Assessment of the interactions between economic growth and industrial waste water discharges using co-integration analysis: a case study for China's Hunan Province. *International Journal of Environmental Research & Public Health* **8**: 2937–2950.

Xie X, Jie LI. 2010. Numerical simulation of diurnal variations of the air pollutants in winter of Lanzhou city. *Climatic & Environmental Research* **15**: 695–703.

Yu B, Huang CM, Liu ZH, Wang HP, Wang LL. 2011. A chaotic analysis on air pollution index change over past 10 years in Lanzhou, northwest China. *Stochastic Environmental Research and Risk Assessment* **25**: 643–653.

Zhang RB, Qian X, Zhu WT, Gao HL, Hu W, Wang JH. 2014. Simulation and evaluation of pollution load reduction scenarios for water environmental management: a case study of inflow river of Taihu Lake, China. *International Journal of Environmental Research & Public Health* **11**: 9306–9324.

Zhou L, Gong ZQ, Zhi R, Feng GL. 2008. An approach to research the topology of Chinese temperature sequence based on complex network. *Acta Physica Sinica* **57**: 7380–7389.

11

Features of vortex split MSSWs that are problematic to forecast

Masakazu Taguchi*

Department of Earth Science, Aichi University of Education, Kariya, Japan

*Correspondence to:
M. Taguchi, Department of
Earth Science, Aichi University of
Education, Hirosawa I,
Igaya-cho, Kariya 448-8542,
Japan.
E-mail:
mtaguchi@auecc.aichi-edu.ac.jp*

Abstract

A companion paper demonstrated that it is more difficult to forecast major stratospheric sudden warmings (MSSWs) of the vortex split type on medium range time scales of about 2 weeks than other MSSWs. As its extension, this study further investigates more specific features of planetary waves for the greater difficulty through a composite analysis using the Japanese 55-year reanalysis data and the Japan Meteorological Agency 1-month hindcast data. Results show that the hindcast data of about two week lead times to the MSSWs largely underestimate the vortex stretching and split at 10 hPa, and the forcing and propagation of planetary wave of zonal wavenumber 2 to the stratosphere. The underestimation in the wave forcing largely reflects a deficiency in simulating a nonlinear, or quadratic, term in wave anomalies (meridional wind and temperature) from the climatology.

Keywords: predictability; major SSWs; planetary waves

1. Introduction

There has been a growing interest in predictability of the extratropical stratospheric circulation, as numerical weather forecast models/systems have been developed to well represent the stratosphere (e.g. Gerber *et al.*, 2012; Tripathi *et al.*, 2015). Examining predictability of extreme stratospheric states, such as major stratospheric sudden warmings (MSSWs), is useful to better understand the nature of the stratosphere–troposphere system. This includes assessment of the possible importance of the stratosphere in extended range weather forecasts as suggested by, e.g. Baldwin and Dunkerton (2001). It will be also useful to improve the numerical forecast models/systems themselves.

Companion papers have extensively investigated predictability of the Northern winter stratosphere using the Japanese 55-year reanalysis (JRA-55) data and the Japan Meteorological Agency (JMA) 1-month hindcast (HC) data. Taguchi (2015) compared stratospheric forecast errors between vortex weakening and strengthening conditions. Some of these conditions correspond to MSSWs (their types were not considered) and vortex intensifications (e.g. Limpasuvan *et al.*, 2005). The study showed that it is more difficult to forecast vortex weakening conditions than strengthening conditions. This reflects greater difficulty in forecasting planetary wave amplifications in the upper troposphere and lower stratosphere leading to weakening conditions.

Taguchi (2016) examined a possible connection of predictability of MSSWs to their types. It was speculated that predictable time limits are shorter for vortex split MSSWs including a large contribution from the zonal wavenumber 2 component (wave 2; Mukougawa and Hirooka, 2004), although the speculation

was not conclusive because it relied on just a few cases. Taguchi (2016) demonstrated that errors of medium range (approximately 2 weeks) forecasts of 21 actual MSSWs are larger when the vortex is highly stretched or split with high aspect ratio and amplified wave 2. It hypothesized that amplified wave 2 events in the troposphere for vortex split MSSWs are more difficult to forecast than wave 1 events, although it left detailed features and mechanisms open.

MSSWs are observed to typically occur in either of the vortex displacement or split type (Charlton and Polvani, 2007; Mitchell *et al.*, 2011; Seviour *et al.*, 2013). The former is characterized by the vortex moving far from the pole, and the latter by the vortex dividing into two separate vortices. Vortex displacement and split events tend to be predominantly associated with amplified wave 1 and wave 2, respectively (e.g. Charlton and Polvani, 2007). Extensive studies showed that these two types have different surface impacts (e.g. Mitchell *et al.*, 2013; O'Callaghan *et al.*, 2014; Seviour *et al.*, 2016), suggesting importance of a better understanding of differences in the predictability of MSSWs of the two types. Maycock and Hitchcock (2015) suggested possible importance of lower stratospheric wind anomalies for diagnosing surface impacts of MSSWs.

As an extension of the companion papers (Taguchi, 2015, 2016), this study further explores more specific features especially of planetary waves for the greater difficulty in forecasting the vortex split MSSWs. To this end, we contrast these characteristic cases to the other cases though a composite analysis using the same reanalysis and HC data.

The rest of the paper is organized as follows. Section 2 explains the data used in this study. Section 3 describes the results. Section 4 provides summary and discussion.

2. Data

This study largely uses the same JRA-55 reanalysis and JMA HC data as in Taguchi (2016).

The real world is represented by daily averages of the JRA-55 reanalysis data (Kobayashi *et al.*, 2015), with $2.5° \times 2.5°$ horizontal grids and 37 levels up to 1 hPa. Twenty-one MSSWs are identified during December–January–February (DJF) from 1978/79 to 2012/13 using the method outlined in Charlton and Polvani (2007). This method identifies a MSSW as when the zonal mean zonal wind [U] at 60°N, 10 hPa reverses from a westerly flow to an easterly flow. Here, U denotes the zonal wind, and squares brackets denote the zonal mean. The day when the zonal wind reverses is referred to as the key or onset day of the MSSW (denoted as lag = 0 day).

As a forecast data set, this study mainly uses the HC data from the March 2014 version of the JMA 1-month HC experiments (The newer version is 'March 2014', which was mistyped in Taguchi (2016) as 'March 2013'.). The experiments employ the JMA global model with a horizontal grid size of about 55 km. A set of five ensemble forecasts is performed for one month from each of the 10th, 20th, and last day of each month from 1981 to 2012.

Whereas Taguchi (2016) use two versions (March 2011 and March 2014) of the HC experiments, for the sake of simplicity this study basically uses only the newer one. Two exceptions are the MSSWs in February 1979 and in February 1980: these are not covered by the newer version, and the older version is used. Taguchi (2016) showed that the differences in stratospheric forecast errors between the two versions are not large, e.g. when compared to case-to-case variability.

This study examines ensemble mean fields in the HC data of lead times of about 2 weeks to the 21 MSSWs. For each of them, we choose only one HC set that is initialized about 2 weeks (ranging from 11 to 20 days) before the onset day.

In order to characterize the vortex geometry around the MSSWs, two vortex moment parameters, centroid latitude (CL) and aspect ratio (AR), are calculated for 10 hPa height Z10 (Seviour *et al.*, 2013). Whereas CL measures the latitude of the vortex center, AR is a measure of vortex stretching. It is noted that the calculation is impossible or will not work well, e.g. when the vortex completely breaks down. Such cases occur for time lags of approximately 5–25 days of some MSSWs, and are excluded from the results. This treatment hardly affects our argument since we are interested in the period before and around the onset day.

3. Results

Figure 1 confirms the main result in Taguchi (2016) that errors of about two week forecasts for the MSSWs are larger for vortex split MSSWs. It is a scatter plot

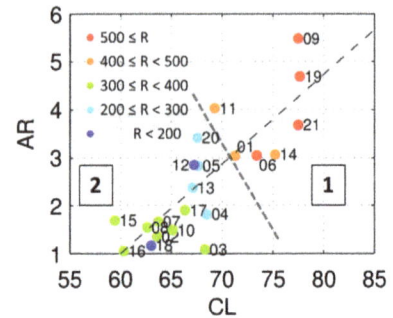

Figure 1. Scatter plot between CL and AR of the JRA-55 Z10 data averaged for lag = ±2 days of the 21 MSSWs. Each data point is colored by RMSE$_{Z10}$ (R) for the 5 days. This figure is essentially identical to Figure 3(a) in Taguchi (2016), but also classifies the 21 MSSWs into two groups for the following analysis.

between CL and AR for the JRA-55 10 hPa height averaged for lag = ±2 days of the 21 MSSWs. The errors are quantified by the root mean square error (RMSE$_{Z10}$) calculated for the 10 hPa height difference field between the JRA-55 and HC data on each forecast day poleward of 20°N.

In order to better understand how/why the larger errors occur for the vortex split MSSWs, we perform a composite analysis by classifying the 21 MSSWs into two groups (Figure 1). The group 1 is the target group for the seven vortex split MSSWs located in the upper right with the larger errors. The group 2 is for the other 14 MSSWs. The classification seeks to maximize the RMSE$_{Z10}$ difference between the two groups. The following results do not depend on a few outliers, since the results hold when a few randomly chosen cases are artificially removed from each group (this is repeated 100 times, not shown).

Figure 2(a)–(c) shows composite time series of CL and AR, as well as [U], for the two groups in the JRA-55 data. For both groups, the polar vortex is characterized by high CL and low AR about 3 weeks before the onset day, with westerly wind around 40 m s^{-1}, before the zonal wind reverses on lag = 0 day.

For the group 1, AR exhibits a peak around the onset day, with CL remaining high. The situation is opposite for the group 2 as both CL and AR are low. These differences in CL and AR between the two groups are statistically significant at the 90% level according to a Student's *t*-test as denoted by magenta squares. Here, the relaxed confidence level of 90% is used since the sample size is not large. These results are consistent with Mitchell *et al.* (2011), who apply a vortex moment analysis to potential vorticity on potential temperature surfaces.

Figure 2(d)–(f) shows forecast errors of these quantities, together with RMSE$_{Z10}$ in Figure 2(g). A notable feature is that the HC data largely underestimate the observed increase in AR for the group 1. This is likely to contribute the larger errors in [U] and Z10. The HC data also underestimate the observed weakening and equatorward displacement of the vortex for the group 2.

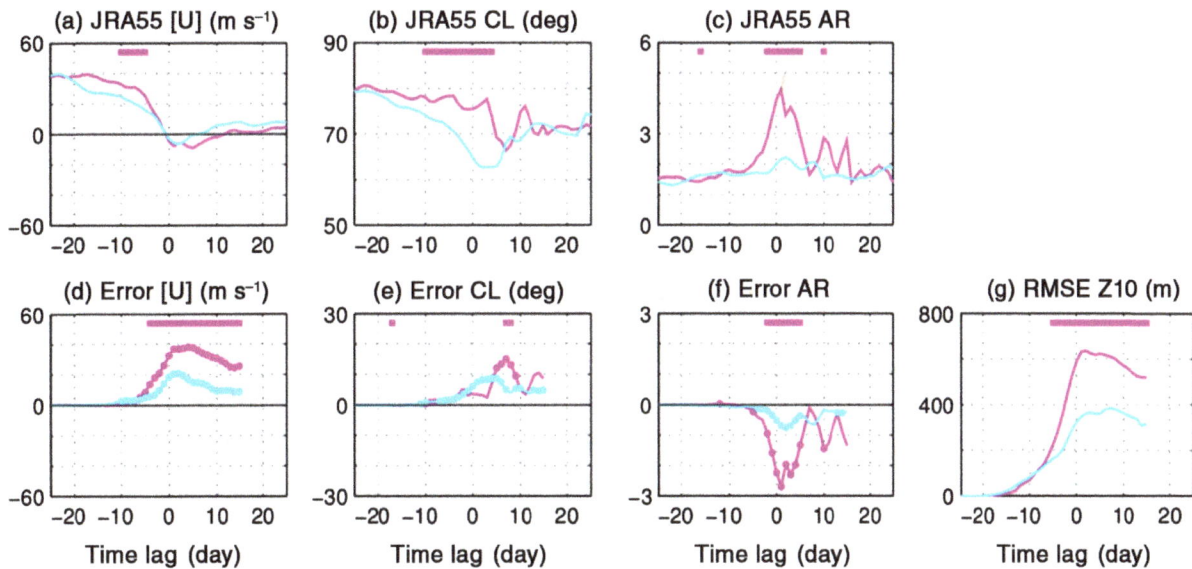

Figure 2. (a)–(c) Composite time series in the JRA-55 data for the two groups: (a) zonal mean zonal wind [U] at 60°N, 10 hPa, (b) CL, and (c) AR. Results for the groups 1 and 2 are shown by magenta and cyan, respectively. Panels (d)–(g) are similar, but for forecast errors. Panel (g) plots RMSE$_{Z10}$. The forecast errors are plotted only up to lag = 15 days due to the limited length of the HC data. Magenta squares near the top in each panel indicate that the composites for the group 1 are significantly different from those for the group 2 at the 90% level. Dots in (d)–(f) indicate that the errors are significantly different from zero at the 90% level.

Since the changes in CL and AR are related to those in planetary waves, Figure 3(a)–(c) shows observed anomalies (deviations from the seasonally varying climatology) of the vertical component (FZ) of the Eliassen-Palm (EP) flux in Northern high latitudes (40–90°N) at 100 hPa. The MSSWs are associated with positive FZ anomalies of waves 1–3 in both groups (Figure 3(a)) as is well known (e.g. Limpasuvan et al., 2004). The FZ anomalies are predominantly contributed by wave 1 and wave 2 for the group 2 and 1, respectively. These features are consistent with the fact that vortex displacement and split MSSWs are contributed by wave 1 and wave 2, respectively (e.g. Charlton and Polvani, 2007).

The HC data tend to underestimate FZ for both groups, but the underestimation is larger in magnitude for the group 1 (Figure 3(d)). This reflects that the underestimation in wave 2 FZ for the group 1 is larger than that in wave 1 FZ for the group 2 (Figure 3(e) and (f)). The large underestimation in wave 2 FZ for the group 1 is consistent with that in AR (Figure 2(f)).

The underestimation in 100 hPa FZ is further examined by calculating budgets of EP flux anomalies in 40–90°N, 100–300 hPa as

$$FZ300 = FZ100 - FY40 + CONV \quad (1)$$

Here, FZ300 and FZ100 denote FZ anomalies at 300 and 100 hPa, respectively, integrated from 40 to 90°N, FY40 denotes FY (meridional component of the EP flux) anomalies at 40°N from 300 to 100 hPa, and CONV denotes the convergence. Equation (1) states that some part of wave forcing from the extratropical troposphere enters the lower stratosphere while the rest propagates to lower latitudes or converges in the region. The budget resolves the lower stratosphere and upper

Table 1. Budgets of composite EP flux anomalies (Eq. (1)) in kg s^{-2} × 10^{10} for the region of 40–90°N, 100–300 hPa averaged from lag = −10 to 0 days. Numbers in parentheses are normalized by relevant FZ300. The sum of FZ100, FY40, and CONV is not equal to FZ300 in some cases due to rounding.

	Group 1, Wave 2			Group 2, Wave 1		
	JRA-55	HC	Error	JRA-55	HC	Error
FZ300	76.7 (100)	36.9 (100)	−39.8	56.3 (100)	39.5 (100)	−16.8
FZ100	32.4 (42.2)	9.0 (24.4)	−23.4	31.8 (56.5)	19.1 (48.4)	−12.7
FY40	−2.2 (−2.9)	−4.8 (−13.0)	−2.6	−0.38 (−0.67)	−2.0 (−5.1)	−1.6
CONV	42.1 (54.9)	23.1 (62.6)	−19.0	24.1 (42.8)	18.3 (46.3)	−5.8

troposphere, since this region modulates the propagation of planetary waves (Chen and Robinson, 1992). Table 1 shows the composite results for the two groups averaged from lag = −10 to 0 days when the FZ anomalies and their forecast errors are large in magnitude to the onset day (Figure 3). The dominant wave component, wave 2 for the group 1 and wave 1 for the group 2, is used.

One sees that the HC data underestimate FZ at 100 and also 300 hPa in both groups and that the underestimation is stronger for the group 1. The difference in the underestimation between the two groups is statistically significant above or near the 90% level (p value is 2.1×10^{-3} at 100 hPa and 0.11 at 300 hPa).

It is also possible to examine the budgets when normalizing them by the relevant FZ300 term. This suggests that the HC data overestimate the equatorward propagation and convergence for both groups, which act to decrease the efficiency of the wave propagation between 300 and 100 hPa. It is also suggested that the underestimation of the propagation is stronger for the group 1. However, such normalization makes it difficult

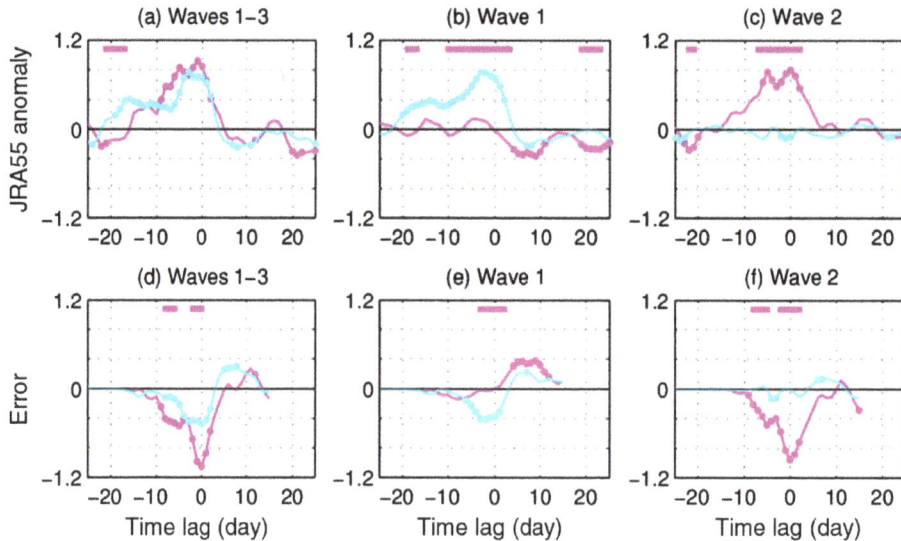

Figure 3. (a)–(c) As in Figure 2(a)–(c), but showing anomalous FZ (in $10^5 \, kg \, m^{-1} \, s^{-2}$) averaged in 40–90°N, 100 hPa for the JRA-55 data: (a) waves 1–3, (b) wave 1, and (c) wave 2. Panels (d)–(f) plot forecast errors in the same unit.

to obtain statistical significance for these errors and their differences between the two groups, as it leads to larger case-to-case variability (not shown).

In order to better understand the underestimation in the tropospheric wave forcing (FZ300), the poleward eddy heat flux is examined when the meridional wind V and temperature T are decomposed into the seasonally varying climatology and anomaly fields (denoted by subscripts c and a, respectively). The heat flux is examined for simplicity using its proportional nature to FZ. Anomalous eddy heat flux on an arbitrary day can be expressed as

$$\left[V^* \, T^*\right]_a = \left[V^* \, T^*\right] - \left[V^* \, T^*\right]_c$$
$$= \left[V_c^* \, T_a^* + V_a^* \, T_c^*\right] + \left[V_a^* \, T_a^*\right]_a \quad (2)$$

(e.g. Nishii *et al.*, 2009). Here, asterisks denote waves. The first and second square bracket terms on the r.h.s. are linear and nonlinear (quadratic), respectively, with respect to anomalies.

Figure 4(a) shows the decomposition results for the two groups. The same time and latitudinal averages are taken for the dominant wave component as in the EP flux budgets (Table 1). The linear term contributes to the positive heat flux anomalies in both groups. Furthermore, the nonlinear term plays a more important role for the group 1 than for the other ($p = 0.11$). These results are consistent with previous studies, as the importance of the linear term in driving SSWs was pointed out by, e.g. Garfinkel *et al.* (2010), Nishii *et al.* (2009) and Smith and Kushner (2012). The important role of the nonlinear term for vortex split SSWs was also claimed by Smith and Kushner (2012).

The calculation based on Eq. (2) is repeated for the HC data to obtain forecast errors (Figure 4(b)). Here, all climatological fields are taken from the JRA-55 data. Both linear and nonlinear terms are underestimated in the two groups. A notable feature is that the nonlinear term plays an important role in the group 1. The

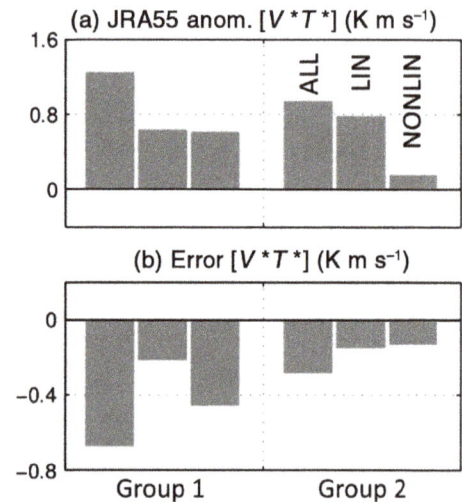

Figure 4. (a) Composite poleward eddy heat flux anomalies over 40–90°N, 100 hPa from lag = −10 to 0 days for the two groups in the JRA-55 data. Contributions from the linear (LIN) and nonlinear (NONLIN) terms, and their sum (ALL) are plotted (Eq. (2)). Panel (b) is similar, but for forecast errors. Results for the dominant component in each group (wave 2 for the group 1, and wave 1 for the group 2) are plotted as in Table 1.

underestimation in the nonlinear term is judged to be statistically different, albeit weakly, between the two groups at $p = 0.15$. The significance for the difference in the linear term underestimation is much weaker or absent.

Synoptic Z300 maps support the underestimation in the linear term for the two groups (Figure 5). The observed anomalies enhance the climatological waves (wave 2 for the group 1 and wave 1 for the group 2) through constructive interference (Figure 5(a)–(c)), consistent with the contribution from the linear term (Figure 4(a)). The significance for the group 1 is not high. This may suggest that the phase of wave

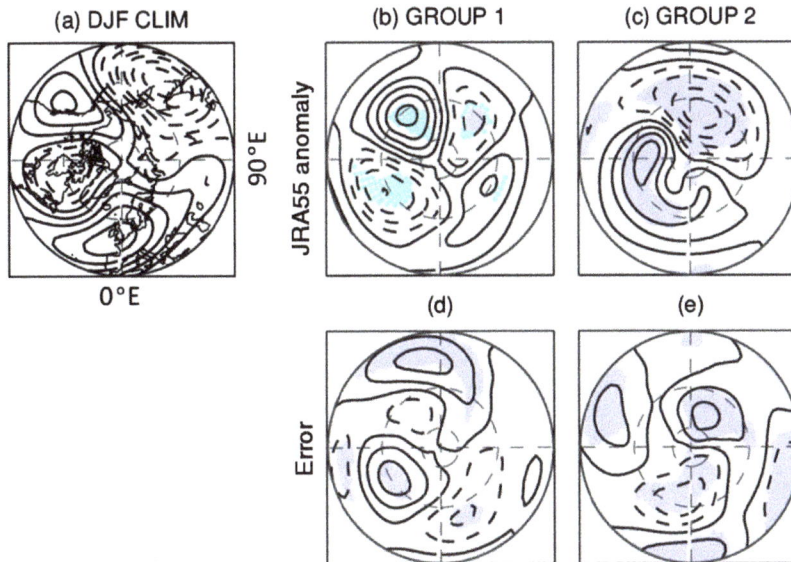

Figure 5. Synoptic Z300 maps of waves 1–3. Panels (b) and (c) plot anomalous fields for the two groups from lag = −10 to 0 days in the JRA-55 data, together with the DJF climatology in (a). Panels (d) and (e) plot forecast errors. Contour interval is 50 m in (a), and 25 m in the others. Blue shades in (b)– (e) show that the anomalies or errors are significantly different from zero at the 90% level. Cyan dots in (b) are similar, but the confidence level is lowered to 80%.

anomalies before these MSSWs are not coherent with the climatology but are variable from one case to another (recall the importance of the nonlinear term in Figure 4(a)). The HC data tend to underestimate the observed Z300 anomalies for both groups (Figure 5(d) and (e)), consistent with the underestimation in the linear term (Figure 4(b)).

Future studies could examine if the observed anomalies and forecast errors of the nonlinear term ($V_a^* T_a^*$, before taking the zonal mean) have distinct zonal distributions.

4. Summary and discussion

This study has explored more specific features especially of planetary waves for the greater difficulty in forecasting vortex split MSSWs (Taguchi, 2016). To this end, we characterize these cases in a composite analysis using the JRA-55 reanalysis and JMA HC data. We have shown that for the vortex split MSSWs, the HC data largely underestimate the vortex stretching and split at 10 hPa, and the wave 2 forcing and propagation to the stratosphere. This result may be expected, because these are just different aspects of the same problem. A large underestimation of the nonlinear term in the 300 hPa heat flux (or FZ) occurs for the group 1. The underestimation of the wave 2 forcing and propagation for the group 1 is suggested to be larger than those of wave 1 for the group 2, although the statistical significance is limited for some quantities.

The present results are relevant to a recent study by Tripathi *et al.* (2016). Tripathi *et al.* examined predictability of the January 2013 MSSW of a vortex split feature in five numerical weather prediction systems. They show that the models have reasonably good skill

in forecasting amplified wave 2 associated with tropospheric blocking features, but have limited skill for amplified wave 2 in the stratosphere. This may be interpreted as difficulty in forecasting the upward propagation of wave 2.

This study may be extended in forcing and propagation of wave 2. Whereas we examined wave 2 forcing for the MSSWs, one can examine such wave events regardless of MSSWs. This will deal with predictability of tropospheric circulation anomalies in more general. We also suggest that wave 2 propagation may be more difficult to forecast than wave 1 counterpart. This may be rephrased as higher sensitivity of wave 2 propagation, e.g. to changes in mean zonal wind and wave forcing. This possibility can be pursued using linear model calculations of planetary wave propagation.

Acknowledgements

The JRA-55 data are developed by the JMA, and obtained from Research Data Archive at the National Center for Atmospheric Research, Computational and Information Systems Laboratory. The HC data are provided by way of the Meteorological Research Consortium, a framework for research cooperation between the JMA and Meteorological Society of Japan. Comments from two reviewers improve the manuscript. This work was supported by JSPS KAKENHI Grant Numbers JP24224011 and JP15K05286.

References

Baldwin MP, Dunkerton TJ. 2001. Stratospheric harbingers of anomalous weather regimes. *Science* **294**: 581–584, doi: 10.1126/science.1063315.

Charlton AJ, Polvani LM. 2007. A new look at stratospheric sudden warming events: part I. Climatology and modeling benchmarks. *Journal of Climate* **20**: 449–469, doi: 10.1175/JCLI3996.1.

Chen P, Robinson WA. 1992. Propagation of planetary waves between the troposphere and stratosphere. *Journal of the Atmospheric Sciences* **49**: 2533–2545, doi: 10.1175/1520-0469(1992)049<2533:POPWBT>2.0.CO;2.

Garfinkel CI, Hartmann DL, Sassi F. 2010. Tropospheric precursors of anomalous Northern Hemisphere stratospheric polar vortices. *Journal of Climate* **23**: 3282–3299, doi: 10.1175/2010JCLI3010.1.

Gerber EP, Butler A, Calvo N, Charlton-Perez A, Giorgetta M, Manzini E, Perlwitz J, Polvani LM, Sassi F, Scaife AA, Shaw TA, Son S-W, Watanabe S. 2012. Assessing and understanding the impact of stratospheric dynamics and variability on the earth system. *Bulletin of the American Meteorological Society* **93**: 845–859, doi: 10.1175/BAMS-D-11-00145.1.

Kobayashi S, Ota Y, Harada Y, Ebita A, Moriya M, Onoda H, Onogi K, Kamahori H, Kobayashi C, Endo H, Miyaoka K, Takahashi K. 2015. The JRA-55 reanalysis: general specifications and basic characteristics. *Journal of the Meteorological Society of Japan* **93**: 5–48, doi: 10.2151/jmsj.2015-001.

Limpasuvan V, Thompson DWJ, Hartmann DL. 2004. The life cycle of the Northern Hemisphere sudden stratospheric warmings. *Journal of Climate* **17**: 2584–2596, doi: 10.1175/1520-0442(2004)017<2584:TLCOTN>2.0.CO;2.

Limpasuvan V, Hartmann DL, Thompson DWJ, Jeev K, Yung YL. 2005. Stratosphere-troposphere evolution during polar vortex intensification. *Journal of Geophysical Research* **110**: D24101, doi: 10.1029/2005JD006302.

Maycock AC, Hitchcock P. 2015. Do split and displacement sudden stratospheric warmings have different annular mode signatures? *Geophysical Research Letters* **42**: 10943–10951, doi: 10.1002/2015GL066754.

Mitchell DM, Charlton-Perez AJ, Gray LJ. 2011. Characterizing the variability and extremes of the stratospheric polar vortices using 2D moment analysis. *Journal of the Atmospheric Sciences* **68**: 1194–1213, doi: 10.1175/2010JAS3555.1.

Mitchell DM, Gray LJ, Anstey J, Baldwin MP, Charlton-Perez AJ. 2013. The influence of stratospheric vortex displacements and splits on surface climate. *Journal of Climate* **26**: 2668–2682, doi: 10.1175/JCLI-D-12-00030.1.

Mukougawa H, Hirooka T. 2004. Predictability of stratospheric sudden warming: a case study for 1998/99 winter. *Monthly Weather Review* **132**: 1764–1776, doi: 10.1175/1520-0493(2004)132<1764:POSSWA>2.0.CO;2.

Nishii K, Nakamura H, Miyasaka T. 2009. Modulations in the planetary wave field induced by upward-propagating Rossby wave packets prior to stratospheric sudden warming events: a case-study. *Quarterly Journal of the Royal Meteorology Society* **135**: 39–52, doi: 10.1002/qj.359.

O'Callaghan A, Joshi M, Stevens D, Mitchell D. 2014. The effects of different sudden stratospheric warming type on the ocean. *Geophysical Research Letters* **41**: 7739–7745, doi: 10.1002/2014GL062179.

Seviour WJM, Mitchell DM, Gray LJ. 2013. A practical method to identify displaced and split stratospheric polar vortex events. *Geophysical Research Letters* **40**: 5268–5273, doi: 10.1002/grl.50927.

Seviour WJM, Gray LJ, Mitchell DM. 2016. Stratospheric polar vortex splits and displacements in the high-top CMIP5 climate models. *Journal of Geophysical Research* **121**: 1400–1413, doi: 10.1002/2015JD024178.

Smith KL, Kushner PJ. 2012. Linear interference and the initiation of extratropical stratosphere-troposphere interactions. *Journal of Geophysical Research* **117**: D13107, doi: 10.1029/2012JD017587.

Taguchi M. 2015. On the asymmetry of forecast errors in the Northern winter stratosphere between vortex weakening and strengthening conditions. *Journal of the Meteorological Society of Japan* **93**: 443–457, doi: 10.2151/jmsj.2015-029.

Taguchi M. 2016. Connection of predictability of major stratospheric sudden warmings to polar vortex geometry. *Atmospheric Science Letters* **17**: 33–38, doi: 10.1002/asl.595.

Tripathi OP, Baldwin M, Charlton-Perez A, Charron M, Eckermann SD, Gerber E, Harrison RG, Jackson DR, Kim B-M, Kuroda Y, Lang A, Mahmood S, Mizuta R, Roff G, Sigmond M, Son S-W. 2015. The predictability of the extratropical stratosphere on monthly time-scales and its impact on the skill of tropospheric forecasts. *Quarterly Journal of the Royal Meteorology Society* **141**: 987–1003, doi: 10.1002/qj.2432.

Tripathi OP, Baldwin M, Charlton-Perez A, Charron M, Cheung JCH, Eckermann SD, Gerber E, Jackson DR, Kuroda Y, Lang A, McLay J, Mizuta R, Reynolds C, Roff G, Sigmond M, Son S-W, Stockdale T. 2016. Examining the predictability of the stratospheric sudden warming of January 2013 using multiple NWP systems. *Monthly Weather Review* **144**: 1935–1960, doi: 10.1175/MWR-D-15-0010.1.

12

Development of a non-hydrostatic atmospheric model using the Chimera grid method for a steep terrain

Kazushi Takemura,* Keiichi Ishioka and Shoichi Shige

Division of Earth and Planetary Sciences, Graduate School of Science, Kyoto University, Japan

Correspondence to:
Kazushi Takemura, Division of Earth and Planetary Sciences, Graduate School of Science, Kyoto University, Kitashirakawaoiwake-cho, Sakyo-ku, Kyoto-shi 606–8502, Japan.
E-mail:
k_takemura@kugi.kyoto-u.ac.jp

Abstract

In a high-resolution atmospheric model, the terrain is resolved in detail, and thus steeper and more complex terrain can be represented. Thus, we developed an atmospheric model that represents the terrain by using the Chimera grid method, which represents the computational region as a composite of overlapping grids. To test this model, we simulated a lee wave over a semicircular mountain and one over a tall semi-elliptical mountain. The results show that the Chimera grid method can be used to represent a very steep terrain which terrain-following coordinates and numerically generated coordinates cannot represent appropriately.

Keywords: Chimera grid method; overlapping grid method; non-hydrostatic model; high--resolution model; representation of a steep terrain; numerically generated coordinate

1. Introduction

Recent developments in computing have rapidly increased atmospheric models' resolution. In high-resolution models, the terrain is resolved in more detail, and thus steeper and more complex terrain can be resolved. Currently, terrain-following coordinates are usually used to represent the terrain in atmospheric models (Gal-Chen and Somerville, 1975; Clark, 1977). However, over steep and complex terrain, such coordinates are greatly deformed, and thus they induce serious errors (Mesinger and Janjić, 1985). Therefore, a good method for the representation of terrain is one of the challenges of high-resolution atmospheric models.

Previous research has handled this challenge by using one of two approaches. One approach is to use Cartesian coordinates and the other one is to use boundary-fitted coordinates. Cartesian coordinates do not require a transformation. Thus, the coordinates are not deformed, even when the terrain is steep and complex. Moreover, the governing equations can be represented in a simpler form. However, the terrain features are unlikely to coincide with the grid lines of the Cartesian coordinates. Thus, various ways have been proposed for representing the terrain: e.g. a cut-cell method (Adcroft *et al.*, 1997; Yamazaki and Satomura, 2010) and a thin-wall approximation (Steppeler *et al.*, 2002). These methods successfully simulated flow over steep slopes. Conversely, Cartesian coordinates require an additional scheme to efficiently increase the resolution near the ground level (Yamazaki and Satomura, 2012).

On the other hand, boundary-fitted coordinates are transformed following the topography. As the topography boundary coincides with the lower boundary of the model, it is easy to apply boundary conditions. Moreover, it is easy to increase the vertical resolution near the ground. However, with this coordinate system, the truncation error increases due to the deformation of the coordinates over steep terrain (Mesinger and Janjić, 1985). Klemp (2011) proposed smoothed terrain-following coordinates obtained by modifying the transformation to reduce these deformation at higher elevations. These transformed coordinates succeeded in reducing the error, but there are limitations near the ground as it is only a vertical modification. Satomura (1989) used numerically generated coordinates, which have high orthgonality because a vertical transformation and a horizontal transformation are used. This coordinate system successfully simulated flow over a semicircular mountain that was so steep that terrain-following coordinates induced serious errors. However, this system cannot be generated over complex terrain in which the slope angle changes abruptly: e.g. a cliff. More improvements are necessary if such complex terrain is to be represented adequately.

In the field of computational fluid dynamics, complex boundaries such as those of an airplane wing are treated, and the Chimera grid method is commonly used (Benek *et al.*, 1986). This method is also referred to overlapping grids, overset grids or composite overlaid grids. In this method, the computational region is represented by a composite of overlapping grids, and one grid is used to represent the entire computational region. Other grids are used to represent local boundaries that are too complex to be represented by the main grid. In the field of atmospheric sciences, the Chimera grid is used for global models to represent the entire Earth (Dudhia and Bresch, 2002; Peng *et al.*, 2006; Staniforth and Thuburn, 2012). However, this method has not yet been used to represent complex terrain.

In this article, we present an atmospheric model that uses the Chimera grid to represent steep and complex terrain. To assess this method, we simulated flow over a semicircular mountain and over a tall semi-elliptical mountain.

2. Model description

We consider the two-dimensional (x: horizontal, z: vertical) full-compressible atmosphere for a resolution from tens of meters to hundreds of meters. The governing equations of this model were transformed from Cartesian coordinates $(x, z) = (x^1, x^2)$ to general coordinates $(\xi, \eta) = (\xi^1, \xi^2)$. The transformed governing equations are

$$\frac{\partial U^i}{\partial t} = -\sum_{j=1}^{2} v^j \frac{\partial U^i}{\partial \xi^j} - \frac{1}{\rho}\frac{\partial p'}{\partial x^i} + \frac{\rho'}{\rho}g^i + \textit{diff. } U^i \quad (i = 1, 2)$$
$$(1)$$

$$\frac{\partial \rho'}{\partial t} = -\sum_{i=1}^{2} \frac{1}{J}\frac{\partial (J\rho v^i)}{\partial \xi^i}$$
$$(2)$$

$$\frac{\partial \theta'}{\partial t} = -\sum_{i=1}^{2} v^i \frac{\partial \theta}{\partial \xi^i} + \textit{diff.}\theta$$
$$(3)$$

$$p = p' + \overline{p} = \left(\rho R_d \theta p_0^{-\frac{R_d}{C_p}}\right)^{1-\frac{R_d}{C_p}}$$
$$(4)$$

where (U^1, U^2) are the (x^1, x^2) components of the flow velocity and (v^1, v^2) are the contravariant component of the flow velocity in the (ξ^1, ξ^2) coordinate, respectively, p is the total pressure, ρ is the total density, θ is the total potential temperature, $(g^1, g^2) = (0, -g)$ is the gravitational acceleration, R_d is the gas constant, p_0 is a reference pressure of 10^5 Pa, C_p is the specific heat at constant pressure, $J = \frac{\partial x^1}{\partial \xi^1}\frac{\partial x^2}{\partial \xi^2} - \frac{\partial x^1}{\partial \xi^2}\frac{\partial x^2}{\partial \xi^1}$ is the Jacobian. Note that some of these variables can be divided into basic components $\overline{p}(z)$, $\overline{p}(z)$ and $\overline{\theta}(z)$, and perturbation components $p'(x, z, t)$, $\rho'(x, z, t)$ and $\theta'(x, z, t)$, respectively. Basic components $\overline{p}(z)$, $\overline{p}(z)$, and $\overline{\theta}(z)$ fulfill:

$$\overline{p} = \left(\overline{p}R_d \overline{\theta}p_0^{-\frac{R_d}{C_p}}\right)^{1-\frac{R_d}{C_p}}$$
$$(5)$$

$$\frac{\partial \overline{p}}{\partial z} = -\overline{\rho}g$$
$$(6)$$

The last terms in the right-hand side of Equations (1) and (3), $\textit{diff. } U^i$ and $\textit{diff. }\theta$ are source terms due to mixing and diffusion, for which we employed the first-order gradient transport theory as:

$$\textit{diff. } U^i = \sum_{j=1}^{2} \frac{\partial}{\partial x^j}K_m\frac{\partial}{\partial x^j}U^i, \ \textit{diff.}\theta = \sum_{i=1}^{2}\frac{\partial}{\partial x^i}K_h\frac{\partial}{\partial x^i}\theta$$
$$(7)$$

where K_m and K_h are the coefficients of mixing and diffusion for the momentum and the heat, respectively. The value of K_m is constant value of 0.1 m²/s and $K_h = 3K_m$ (Deardorff, 1972). To damp upward propagating waves before they are reflected from the upper boundary, we added a Rayleigh damping term $-\gamma(z)U^i$ to the right-hand side of the Equation (1). The value of γ is determined as following.

$$\gamma(z) = \gamma_0 \left\{1 + \left(\frac{z}{D}\right)^2\right\}^3$$
$$(8)$$

where D is a height scale, which is set to 15 km. In addition, to simulate flows stably by removing high-frequency waves, we also added fourth-order diffusion terms in the ξ^1 and ξ^2 directions to the right hand side of the Equations (1)–(3). At points next to a boundary point, the fourth-order diffusion terms were replaced by the second-order diffusion terms.

We adopted the second-order central difference for the space discretization and the fourth-order Runge–Kutta scheme for time integration. For generating the grid, we used the variational method proposed by Brackbill and Saltzman (1982). As shown in Figure 1(a), scalar variables and velocity components were arranged on the centers of cells and the corners of cells, respectively, which was used in Yamazaki and Satomura (2010).

3. The Chimera grid method

In the Chimera grid, the computational region is represented by a composite of overlapping grids. In this article, we refer to the grid used to represent the entire computational region as the global grid, and we refer to a grid used to represent a detail of the terrain as a local grid. These grids are partially overlapping around each grid boundary, as shown in Figure 1(b). At each time step, the governing equations are solved at each grid point, and then the boundary point u^I is obtained by interpolating between surrounding grid points u_{ij}, using them with a boundary condition:

$$u^I = \sum_{i,j=1}^{N} S_{ij}u_{ij}$$
$$(9)$$

where S_{ij} is interpolation coefficient and N is the order of the interpolation stencil. We used a first-order bilinear interpolation with stencil order $N = 2$ and a third-order Lagrange interpolation with stencil order $N = 4$. However, we used the bilinear interpolation instead of the Lagrange interpolation near the ground where the number of points used for interpolation was not sufficient. The interpolation was not performed in the physical space (x, z) but in the computational space (ξ, η), as the physical space coordinates were deformed, and this caused the interpolation to be complex. In the computational space (ξ, η), the grid points were ordered on a rectangle. The interpolation coefficients S_{ij} were

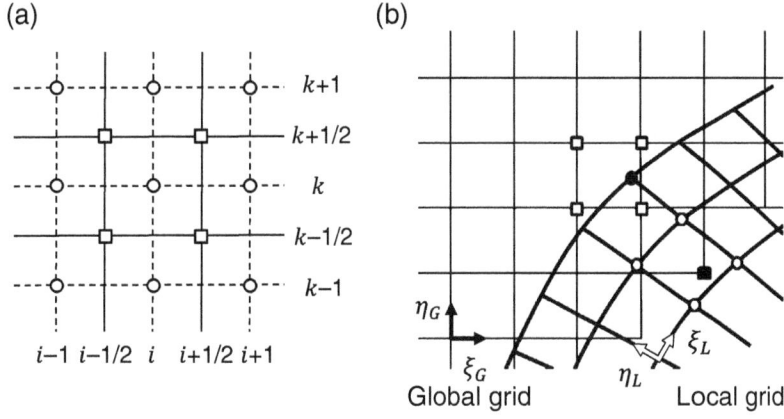

Figure 1. (a) Arrangement of variables. Open circles and open squares represent scalar points and velocity points, respectively. (b) Grid interaction. Overlapping area of the global grid and the local grid. Point closed square and closed circle are the boundary point of the global grid and the local grid, respectively. These points are interpolated from around point of the other gird open circle and open square.

represented as a product of the interpolation coefficients s_i, s_j for each direction:

$$S_{ij} = s_i s_j = \prod_{l=1, l \neq i}^{N} \frac{\xi^I - \xi_l}{\xi_i - \xi_l} \prod_{k=1, k \neq j}^{N} \frac{\eta^I - \eta_k}{\eta_j - \eta_k} \quad (10)$$

where (ξ^I, η^I), (ξ_i, η_i) are the coordinates of the interpolation point u^I and the surrounding grid points u_{ij}, respectively. The coordinates (ξ^I, η^I) are needed to evaluate the interpolation coefficients S_{ij}, but, in general, they are unknown. Sherer and Scott (2005) proposed high-order offsets, and with these, (ξ^I, η^I) are evaluated by using the two-dimensional Newton method:

$$\begin{cases} F_x = \sum S_{i+k,j+l} x_{i+k,j+l} - x^I = 0 \\ F_z = \sum S_{i+k,j+l} z_{i+k,j+l} - z^I = 0 \end{cases} \quad (11)$$

$$\begin{pmatrix} \xi^I \\ \eta^I \end{pmatrix}^{n+1} = \begin{pmatrix} \xi^I \\ \eta^I \end{pmatrix}^{n} - \left[J^{-1} \times \begin{pmatrix} F_x \\ F_z \end{pmatrix} \right]^{n} \quad (12)$$

$$J^{-1} = \frac{1}{\frac{\partial F_x}{\partial \xi} \frac{\partial F_z}{\partial \eta} - \frac{\partial F_x}{\partial \eta} \frac{\partial F_z}{\partial \xi}} \begin{pmatrix} \frac{\partial F_z}{\partial \eta} & -\frac{\partial F_x}{\partial \eta} \\ -\frac{\partial F_z}{\partial \xi} & \frac{\partial F_x}{\partial \xi} \end{pmatrix} \quad (13)$$

In Equation (11), (F_x, F_z) is the error of the interpolated physical space coordinates, defined as the difference between $(\sum S_{i+k,j+l} x_{i+k,j+l}, \sum S_{i+k,j+l} z_{i+k,j+l})$ and the true physical space coordinates (x^I, z^I). Iteration continues until the error (F_x, F_z) is smaller than a predetermined tolerance. The grid interface was set to fulfill the following condition. The ratio of the size of an interpolation point cell to that of the surrounding grid points should be close to unity. The area of the overlapping region must be large enough for each interpolation point to have N stencil points but should be as small as possible.

When introducing the Chimera grid to an atmospheric model, it is necessary to consider the stratification of the atmosphere. As thermodynamic variables change vertically in a stratified atmosphere, vertical interpolation may induce errors. Therefore, only the perturbation components were interpolated.

4. Simulation and results

To test the Chimera grid, we simulated a lee wave over a semicircular mountain and one over a tall semi-elliptical mountain. At the foot of each of these mountains, the slope angle changes rapidly and it becomes very steep. Thus, terrain-following coordinates induce serious errors (Satomura, 1989). We compared the numerical solutions with the analytical solution for Bousinessq system by Miles (1968). To mimic Bousinessq system within our fully compressible system model, we set the value of surface potential temperature θ_0 at $4000\,\mathrm{K}$, where the sound speed increased about four times of that at $300\,\mathrm{K}$.

4.1. Semicircular mountain

We simulated a lee wave over a semicircular mountain of height $H = 1000$ m. For the initial condition, we assumed that the atmosphere had a constant buoyant frequency of $N = 0.01\,\mathrm{s}^{-1}$, and a constant horizontal velocity of $U = 10\,\mathrm{m\,s}^{-1}$. The horizontal and vertical grid intervals were $\Delta x = \Delta z = 250$ m, respectively. The domain consisted of 2000 horizontal and 120 vertical cells, respectively. We performed the simulation using the terrain-following coordinates, the numerically generated coordinate, and the Chimera grid. Figure 2 shows the grid for each coordinate system. In the Chimera grid, the global grid uses Cartesian coordinates, and around the mountain, the local grid uses polar coordinates. The terrain-following coordinates were greatly deformed above the mountain, but there was less deformation with the numerically generated coordinates and the Chimera grid. Figure 3 shows the analytical solution of the vertical velocity over the semicircular mountain and the differences of the numerical solutions at $t = 10$

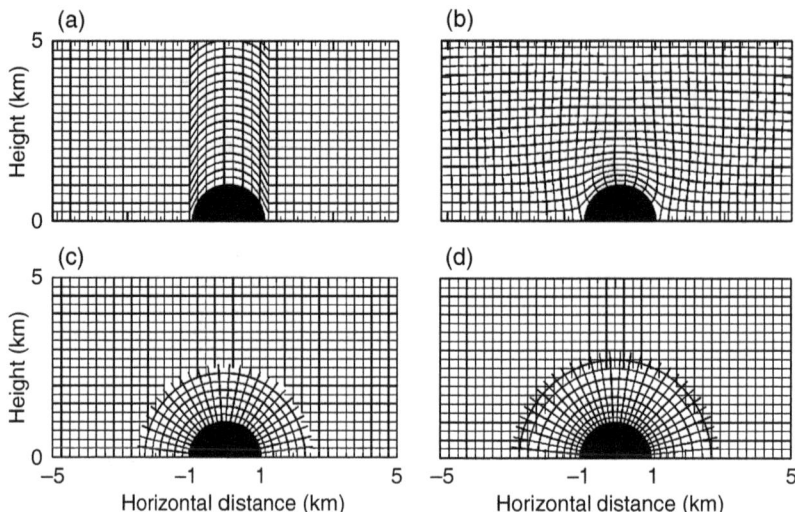

Figure 2. Representation of a semicircular mountain of height $H = 1000$ m: (a) the terrain-following coordinates; (b) the numerically generated coordinates; (c) Chimera grid with bilinear interpolation; and (d) Chimera grid with Lagrange interpolation.

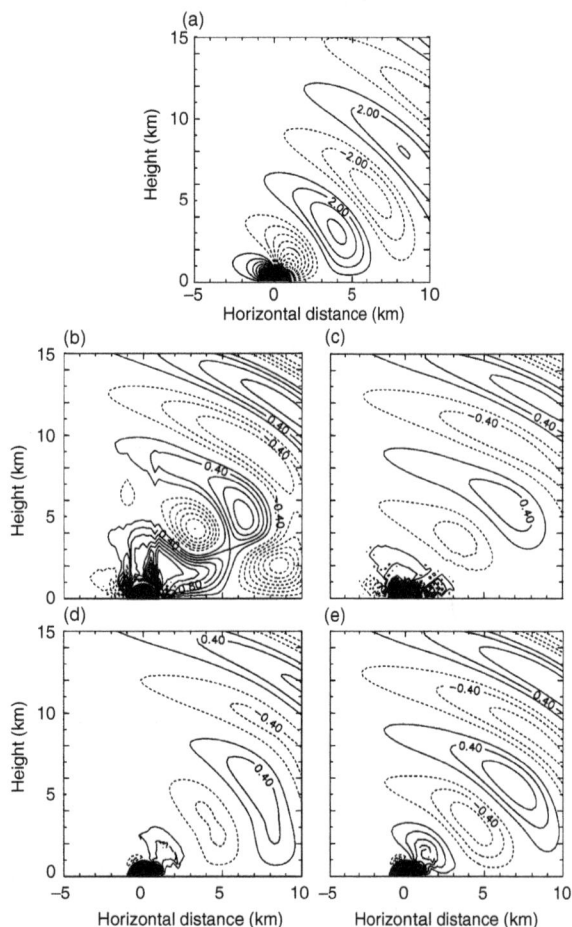

Figure 3. Vertical velocity field and the difference from the analytical solution after 10 h over a semicircular mountain. (a) analytical solution of Miles (1968) (Courtesy of S. Masuda) with contour interval of 1.0 m s^{-1}; (b)–(e) the differences of numerical solution from the analytical solution with contour interval of 0.2 m s^{-1}. (b) terrain-following coordinates; (c) numerically generated coordinates; (d) Chimera grid with bilinear interpolation; and (e) Chimera grid with third-order Lagrange interpolation. The discontinuity of contour near the mountain in panel (d)–(e) is due to the interface of grids.

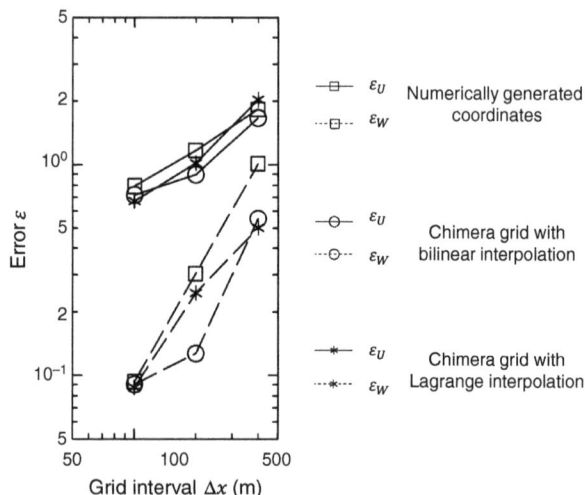

Figure 4. Horizontal velocity error ε_U and vertical velocity error ε_W, with grid intervals Δx.

h from the analytical solution. The difference is clearly smaller with the Chimera grid ((d) and (e)) than with the terrain-following coordinates (b) near the mountain. This superiority of the Chimera grid is likely due to less deformation of the grid near the mountain. The solution with the numerical generated coordinates (c) shows as small difference as the solutions with the Chimera grid. Comparing the bilinear interpolation with the Lagrange interpolation, the bilinear interpolation shows smaller error than the Lagrange interpolation. The increase of overlapping region with the Lagrange interpolation may have induced larger difference. The increase of the difference leeward (b)–(e) is due to the numerical damping.

We performed an quantitative comparison of the Chimera grid and the numerically generated coordinates, and we evaluated the accuracy of our proposed model over the semicircular mountain. We used Equations (14) and (15) to compute the horizontal

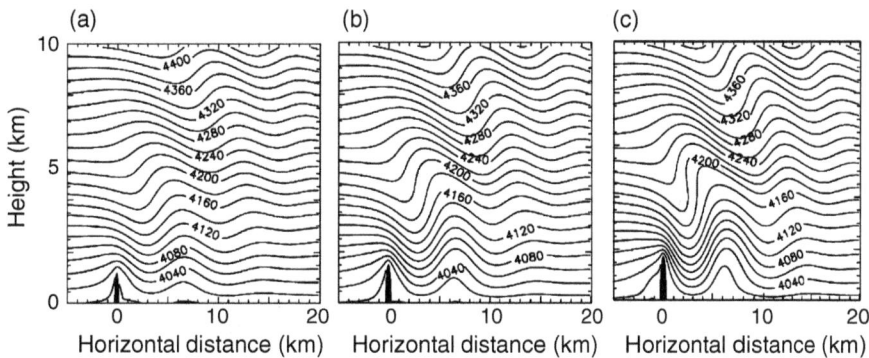

Figure 5. Potential temperature field after 60 min over the tall semi-elliptical mountain for three cases of the mountain height H. Contour interval is 20 K. (a) $H = 1200$ m, $\kappa = 1.2 < \kappa_c$; (b) $H = 1500$ m, $\kappa = 1.5 \approx \kappa_c$; (c) $H = 1730$ m, $\kappa = 1.73 > \kappa_c$.

velocity error ε_U and the vertical velocity error ε_W.

$$\varepsilon_U = \frac{1}{N_x N_z} \sum_{i,k=1}^{N_x, N_z} \left| U_{i,k}^c - U_{i,k} \right| \qquad (14)$$

$$\varepsilon_W = \frac{1}{N_x N_z} \sum_{i,k=1}^{N_x, N_z} \left| W_{i,k}^c - W_{i,k} \right| \qquad (15)$$

where N_x and N_z are the numbers of grid points in the x and z directions, respectively; and $U_{i,k}^c$ and $W_{i,k}^c$ are the analytical solution of the horizontal and vertical velocities, respectively. The domain for the error evaluation was a rectangular domain of -3 km $\leq x \leq 3$ km and 0 km $\leq z \leq 3$ km. Three values of the grid intervals $(\Delta x, \Delta z)$ were tested, 400 m, 200 m and 100 m. Figure 4 shows how the error changed with changes in the grid interval. For each coordinate system, the order of the convergence is less than second-order although we adopted the second-order finite difference scheme. This may be partially due to the difference between the compressible system and the Bousinessq system. However, the error seems to be converging for each coordinate system. Moreover, for each grid interval, the error for the Chimera grid is almost equal to or smaller than that for the numerically generated coordinates. Thus, it can be concluded that the Chimera grid yields an equally or more accurate numerical solution than the numerically generated coordinates in the semicircular mountain case. This is likely due to the fact that the grid of the Chimera grid was less deformed than that of the numerically generated coordinates at the foot of the mountain. Moreover, a comparison of the results of the bilinear interpolation with those of the Lagrange interpolation shows that the error of the bilinear interpolation is almost equal to or smaller than that of the Lagrange interpolation. Thus, it can also be concluded that the decrease in accuracy is negligibly small when using the first-order bilinear interpolation.

4.2. Tall semi-elliptical mountain

Finally, we simulated a lee wave over a tall semi-elliptical mountain. As the slope changes sharply at the foot of the mountain, numerically generated

coordinates cannot properly simulate this topography. Huppert and Miles (1969) obtained the analytical solution of a lee wave over a tall semi-elliptical mountain. They concluded that the behavior of such a wave depends on the ratio of the mountain's height H to its width B, on $\varepsilon = H/B$, and on the nondimensional height $\kappa = NH/U$, which is the reciprocal Froude number. When $\kappa > \kappa_c$, the amplitude of a lee wave develops so large that the θ field becomes convectively unstable. When $\kappa \leq \kappa_c$, a stable lee wave develops and converges to a stationary solution. Here, the critical non-dimensional height κ_c monotonically increases with ε. We simulated a lee wave for this topography with a fixed ratio of $\varepsilon = 5$, and with an average slope of 80°. According to Huppert and Miles (1969), when $\varepsilon = 5$, κ_c is around 1.6. For the initial conditions, we assumed the atmosphere had a horizontal velocity of $U = 10$ ms^{-1}, and a buoyant frequency of $N = 0.01$ s^{-1}. The horizontal and vertical grid intervals were $\Delta x = \Delta z = 150$ m. We performed the simulation for three types of mountain: (a) height $H = 1200$ m, width $B = 240$ m, and $\kappa = 1.2 < \kappa_c$; (b) height $H = 1500$ m, width $B = 300$ m, and $\kappa = 1.5 \approx \kappa_c$; and (c) height $H = 1730$ m, width $B = 346$ m, and $\kappa = 1.73 > \kappa_c$. Figure 5 shows the potential temperature field after 60 min of simulation for each mountain type in using the Chimera grid with the bilinear interpolation. In Figure 5(a) and (b), we can see that stable lee wave simulations are simulated when $\kappa \leq \kappa_c$, and in Figure 5(c), we can see that a lee wave develops too largely and becomes unstable around 4 km above the mountain when $\kappa > \kappa_c$. These results are consistent with the analysis of a lee wave over similar topography by Huppert and Miles (1969). It is evident that the Chimera grid can appropriately simulate time evolutions of flow fields complex terrain.

5. Summary

We have developed an atmospheric model using the Chimera grid method and have performed a high-resolution simulation of air flow over steep and complex terrain. The results show that the Chimera grid can reduce the errors that are produced by using terrain-following coordinates for steep terrain, and can simulate the flow appropriately over a very steep mountain for which the numerical coordinates cannot be generated. The Chimera grid would be useful for the simulation over such a very steep terrain, e.g. a cliff. However, the criterion for deciding which coordinate system should be used is not clear now. Moreover, the accuracy of the results highly depends on the grid configuration and shape of terrain. As future work, it is necessary to clarify the dependence of the accuracy of the numerical solutions on the grid configuration and the shape of topography in more detail.

The interpolation method used in this article did not consider the global conservation of physical quantities. To satisfy the global conservation, we should introduce a conservative interpolation, which interpolates fluxes of physical quantities around interpolation points (Chesshire and Henshaw, 1994). A conservative interpolation method was also developed for a global model (Peng et al., 2006), and would be able to be introduced into our model without extra difficulty. As for the extension of our model to three dimension, it will not be difficult as three dimensional grid generation was developed in the field of the computational fluid dynamics (Petersson, 1999).

Acknowledgements

The authors thank the late T. Satomura for motivate us to start this research. This study was supported by JSPS KAKENHI Grant Number 26282109. We also thank two anonymous reviewers for helpful comments. Figures were produced by GFD-DENNOU Library.

Reference

Adcroft A, Hill C, Marshall J. 1997. Representation of topography by shaved cells in a height coordinate ocean model. *Monthly Weather Review* **125**: 2293–2315, doi: 10.1175/1520-0493(1997)125<2293:ROTBSC>2.0.CO;2.

Benek JA, Steger JL, Dougherty FC, Buning PG. 1986. *Chimera: A Grid-Embedding Technique*. AEDC-TR-85-64, 129 pp.

Brackbill JU, Saltzman JS. 1982. Adaptive zoning for singular problems in two dimensions. *Journal of Computational Physics* **46**: 342–368, doi: 10.1016/0021-9991(82)90020-1.

Chesshire G, Henshaw WD. 1994. A scheme for conservative interpolation on overlapping grids. *SIAM Journal on Scientific Computing* **15**: 819–845, doi: 10.1137/0915051.

Clark TL. 1977. A small-scale dynamic model using a terrain-following coordinate transformation. *Journal of Computational Physics* **24**: 186–215, doi: 10.1016/0021-9991(77)90057-2.

Deardorff JW. 1972. Numerical investigation of neutral and unstable planetary boundary layers. *Journal of the Atmospheric Sciences* **29**: 91–115, doi: 10.1175/1520-0469(1972)029<0091:NIONAU>2.0.CO;2.

Dudhia J, Bresch JF. 2002. A global version of the PSU–NCAR mesoscale model. *Monthly Weather Review* **130**: 2989–3007, doi: 10.1175/1520-0493(2002)130<2989:AGVOTP>2.0.CO;2.

Gal-Chen T, Somerville RCJ. 1975. On the use of a coordinate transformation for the solution of the Navier–Stokes equations. *Journal of Computational Physics* **17**: 209–228, doi: 10.1016/0021-9991(75)90037-6.

Huppert HE, Miles JW. 1969. Lee waves in a stratified flow Part 3. Semi-elliptical obstacle. *Journal of Fluid Mechanics* **35**: 481–496, doi: 10.1017/S0022112069001236.

Klemp JB. 2011. A terrain-following coordinate with smoothed coordinate surfaces. *Monthly Weather Review* **139**: 2163–2169, doi: 10.1175/MWR-D-10-05046.1.

Mesinger F, Janjić ZI. 1985. Problems and numerical methods of the incorporation of mountains in atmospheric models. *Lectures in Applied Mathematics* **22**: 81–121.

Miles JW. 1968. Lee waves in a stratified flow. Part 2. Semi-circular obstacle. *Journal of Fluid Mechanics* **33**: 803, doi: 10.1017/S0022112068001680.

Peng X, Xiao F, Takahashi K. 2006. Conservative constraint for a quasi-uniform overset grid on the sphere. *Quarterly Journal of the Royal Meteorological Society* **132**: 979–996, doi: 10.1256/qj.05.18.

Petersson NA. 1999. Hole-cutting for three-dimensional overlapping grids. *SIAM Journal on Scientific Computing* **21**: 646–665, doi: 10.1137/S1064827597329102.

Satomura T. 1989. Compressible flow simulations on numerically generated grids. *Journal of the Meteorological Society of Japan* **67**: 473–482.

Sherer SE, Scott JN. 2005. High-order compact finite-difference methods on general overset grids. *Journal of Computational Physics* **210**: 459–496, doi: 10.1016/j.jcp.2005.04.017.

Staniforth A, Thuburn J. 2012. Horizontal grids for global weather and climate prediction models: a review. *Quarterly Journal of the Royal Meteorological Society* **138**: 1–26, doi: 10.1002/qj.958.

Steppeler J, Bitzer H-W, Minotte M, Bonaventura L. 2002. Nonhydrostatic atmospheric modeling using a z-coordinate representation. *Monthly Weather Review* **130**: 2143–2149, doi: 10.1175/1520-0493(2002)130<2143:NAMUAZ>2.0.CO;2.

Yamazaki H, Satomura T. 2010. Nonhydrostatic atmospheric modeling using a combined Cartesian grid. *Monthly Weather Review* **138**: 3932–3945, doi: 10.1175/2010MWR3252.1.

Yamazaki H, Satomura T. 2012. Non-hydrostatic atmospheric cut cell model on a block-structured mesh. *Atmospheric Science Letters* **13**: 29–35, doi: 10.1002/asl.358.

13

Tracing the source of ENSO simulation differences to the atmospheric component of two CGCMs

Yanli Tang,[1] Lijuan Li,[2] Wenjie Dong[1]* and Bin Wang[2,3]

[1] State Key Laboratory of Earth Surface Processes and Resource Ecology, Future Earth Research Institute, Beijing Normal University, China
[2] State Key Laboratory of Numerical Modeling for Atmospheric Sciences and Geophysical Fluid Dynamics (LASG), Institute of Atmospheric Physics, Chinese Academy of Sciences, Beijing, China
[3] Ministry of Education Key Laboratory for Earth System Modeling, Center of Earth System Science (CESS), Tsinghua University, Beijing, China

*Correspondence to:
W. Dong, ESPRE, Beijing
Normal University, No. 19,
Xinjiekouwai Street, Beijing
100875, China.
E-mail: dongwj@bnu.edu.cn

Abstract

To explore why the Community Earth System Model (CESM) exhibits too strong El Niño–Southern Oscillation (ENSO), its atmospheric component is replaced by another Atmospheric General Circulation Model (AGCM). Differences among the two simulations and another 'parent' model are then analyzed with reference to their underlying mechanisms. The results indicate that too large ENSO amplitude in the CESM is reduced to half by the new AGCM, mainly due to shortwave radiation feedback. Weaker shortwave radiation feedback in the CESM is found to be closely related to the too negative feedbacks of the cloud fraction and cloud liquid amount in the lower layers.

Keywords: ENSO amplitude; CESM; CESM-GAMIL2; shortwave radiation feedback

1. Introduction

The El Niño–Southern Oscillation [ENSO; i.e., the large-scale sea surface temperature (SST) anomalies that occur every 2–7 years in the central and eastern tropical Pacific] is the dominant mode of climate variability over seasonal to interannual timescales (Wang et al., 2012). ENSO fluctuations have a large impact on the ecology of the tropical Pacific as well as global climate through atmospheric teleconnections (McPhaden et al., 2006). Therefore, understanding and predicting ENSO is critical for scientists and governments worldwide. In recent decades, climate/earth system models have made steady advances in simulating ENSO characteristics (Guilyardi et al., 2012a, 2012b). For example, the ENSO amplitudes in the Coupled Model Intercomparison Project Phase 5 (CMIP5) models better converge around observations than those in the CMIP3 models (Guilyardi, 2006; Yu and Kim, 2010; Kim and Yu, 2012; Bellenger et al., 2014).

However, two important ENSO-related feedbacks remain underestimated in CMIP5 models: the wind–SST feedback and the heat flux–SST feedback. The wind–SST feedback, also known as the Bjerknes feedback, is measured as the regression coefficient between zonal wind stress in the Niño4 region (5°N–5°S, 160°E–210°E) and SST anomalies averaged over the Niño3 region (5°N–5°S, 150°–90°W), and is underestimated by 20–50% in CMIP5 models. The heat flux–SST feedback, measured as the regression coefficient between net heat flux at the surface and SST anomalies in the Niño3 region, is underestimated by a factor of two (Bellenger et al., 2014; Kim et al., 2014). The heat flux–SST feedback comprises of four components: latent heat (LH), shortwave radiation (SW), longwave radiation (LW) and sensible heat (SH), of which LH and SW are the dominant components (Lloyd et al., 2009, 2012). The wind–SST feedback is a positive feedback and can enhance the ENSO amplitude, whereas the heat flux–SST feedback is a negative feedback that can act to dampen SST anomalies (Lloyd et al., 2009, 2012). Guilyardi et al. (2009) showed that the L'Institut Pierre-Simon Laplace Coupled Model (Version 4; IPSL CM4) correctly simulated ENSO amplitude when using the Kerry–Emanuel (KE) convection scheme; however, this outcome resulted from error compensation between the too weak Bjerknes feedback and the too weak heat flux feedback. These feedback errors likely originated from cloud-related processes, such as the convective parameterization scheme (Neale et al., 2008; Guilyardi et al., 2009), non-convective condensation processes (Li et al., 2014) and their uncertain parameters (Watanabe et al., 2011), SST biases (Sun et al., 2009) and the atmosphere–ocean coupling process (Lloyd et al., 2012). As these factors are highly complex and general circulation model (GCM)-dependent, it is difficult to determine how a single atmospheric general circulation model (AGCM) contributes to the feedback errors and ENSO amplitude. In this study, the community earth system model (CESM), known to simulate a strong ENSO, is selected to be coupled with the atmospheric component of the Flexible Global Ocean-Atmosphere-Land System Model (Grid-point Version 2; FGOALS-g2), which has an ENSO simulation close to observation (Bellenger et al., 2014; Kim et al., 2014). The ENSO amplitudes simulated using CESM, the new constructed CGCM and another 'parent' model FGOALS-g2 are then analyzed to further reveal the mechanisms that underlie the atmospheric feedbacks that affect the ENSO, with an emphasis on factors that contribute to weak SW feedback in the CESM.

2. Model, experimental setup and data

2.1. Model and experimental setup

The CESM (version 1.2.0) developed by the National Center for Atmospheric Research (NCAR) is used to simulate ENSO events in this study. For simplification, the ecosystem and chemistry components of the CESM are not included in the model. The atmospheric model used in the CESM is the Community Atmosphere Model (Version 4.0; CAM4; Neale *et al.*, 2013), with an approximate 2° finite volume grid in the horizontal direction and a 30-layer hybrid pressure sigma coordinate in the vertical direction. The land surface model employed by the CESM is the Community Land Surface Model (Version 4.0; CLM4; Oleson *et al.*, 2010), which shares the same horizontal grid as CAM4. The CESM ocean model is the Parallel Ocean Program (Version 2; POP2; Smith *et al.*, 2010), and the CESM use the Los Alamos sea ice model (Version 4; CICE4; Hunke and Lipscomb, 2008), which uses a displaced pole grid with a horizontal resolution of 1° that is compatible with POP2.

To investigate how different AGCMs simulate ENSO amplitudes, the Grid-point Atmospheric Model of IAP LASG (Version 2; GAMIL2) is integrated with the CESM (herein referred to as CESM-GAMIL2). The only difference between the CESM and the CESM-GAMIL2 is the atmospheric component. GAMIL2 employs a dynamical core that includes a finite difference scheme and a two-step shape-preserving advection scheme (TSPAS). In addition, the GAMIL2 uses a hybrid horizontal grid consisting of a 2.8° Gaussian grid between 65.58°S and 65.58°N and a weighted equal-area grid poleward from 65.58°, and a 26-layer sigma coordinate in the vertical direction (Li *et al.*, 2013; Wang *et al.*, 2004). Furthermore, GAMIL2 is the atmospheric component of FGOALS-g2, which produces an ENSO simulation (e.g. the amplitude and the ENSO-related feedbacks) close to observations as a participant in the CMIP5 (Bellenger *et al.*, 2014; Kim *et al.*, 2014).

To eliminate the influence of any external forcing, the CESM and CESM-GAMIL2 are run under pre-industrial (PI) conditions and integrated over 500 years. The first 100 years are model spin-up and monthly simulations from the following 200 years (i.e. years 101–300 in the simulation) are used to conduct ENSO analyses.

2.2. Validation datasets

The following datasets are used to verify model performance: The SST is from the merged products of HADISST1 [Met. Office Hadley Centre sea ice and SST dataset (1870 onward)] and OI.v2 [National Oceanic and Atmospheric Administration (NOAA) Optimum Interpolation SST Version 2 dataset (November 1981 onward)] (Hurrell *et al.*, 2008). The cloud cover and LWP for the period from 1984–2009 are obtained from the International Satellite Cloud Climatology Project (ISCCP; Rossow and Schiffer, 1999). The LWP from the Special Sensor Microwave Imager (SSM/I; 1992-2009) is also used for comparison (Weng *et al.*, 1997). The precipitation dataset for the period from 1984–2009 is obtained from the Climate Prediction Center (CPC) Merged Analysis of Precipitation (CMAP; Xie and Arkin, 1997) and from the Global Precipitation Climatology Project (GPCP; Adler *et al.*, 2003). The Vertical velocities at 500-hPa and surface wind stress data are supplied by the 40 year European Centre for Medium-Range Weather Forecasts Re-Analysis (ERA-40), which covered the period 1958–2001 (Uppala *et al.*, 2005). Finally, objectively Analyzed Air–Sea Fluxes (OAFlux) combined with ISCCP values of short and longwave radiation are used for the period 1984–2009 (Yu and Weller, 2007).

3. Results

3.1. ENSO variability

Area-averaged monthly SST anomalies (SSTA) over the Niño3 region (5°N–5°S, 150°–90°W) provide an index typically used to represent ENSO variability. Time series of the Niño3 index from HadISST observations and the CESM-GAMIL2 and CESM PI-control runs are compared in Figure 1. The PI-control simulation (years 101–300) from another 'parent' model (i.e. FGOALS-g2 from CMIP5) is also included for comparison. Compared with observations, the CESM simulates a much stronger SSTA variation, whereas the CESM-GAMIL2 shows relatively weak variability, and the FGOALS-g2 exhibits a variability most closely matching observations. When quantifying the ENSO amplitude using the Niño3 index standard deviation, the HadISST is approximately 0.81 K, the CESM-GAMIL2 is 0.64 K, the FGOALS-g2 is 0.779 K and the CESM is 1.25 K. Analogous standard deviations based on the Niño3.4 (5°N–5°S, 160°E–150°W) index are 0.77, 0.543, 0.778 and 1.22 K, while the Niño4 (5°N–5°S, 160°E–210°E) index gives 0.56, 0.40, 0.495 and 0.97 K for HadISST, CESM-GAMIL2, FGOALS-g2 and CESM, respectively. These results show that the ENSO amplitude in CESM is twice as strong as that in CESM-GAMIL2 (the large ENSO amplitudes in CESM were also shown in other studies, e.g. Kang *et al.*, 2014; Bellenger *et al.*, 2014), independent of index selection, and the observation value lies between the two models, while FGOALS-g2 is much closer to observations. Based on this result, ENSO variability is herein represented by the Niño3 index.

Previous studies have suggested that ENSO amplitude shows a negative correlation with the strength

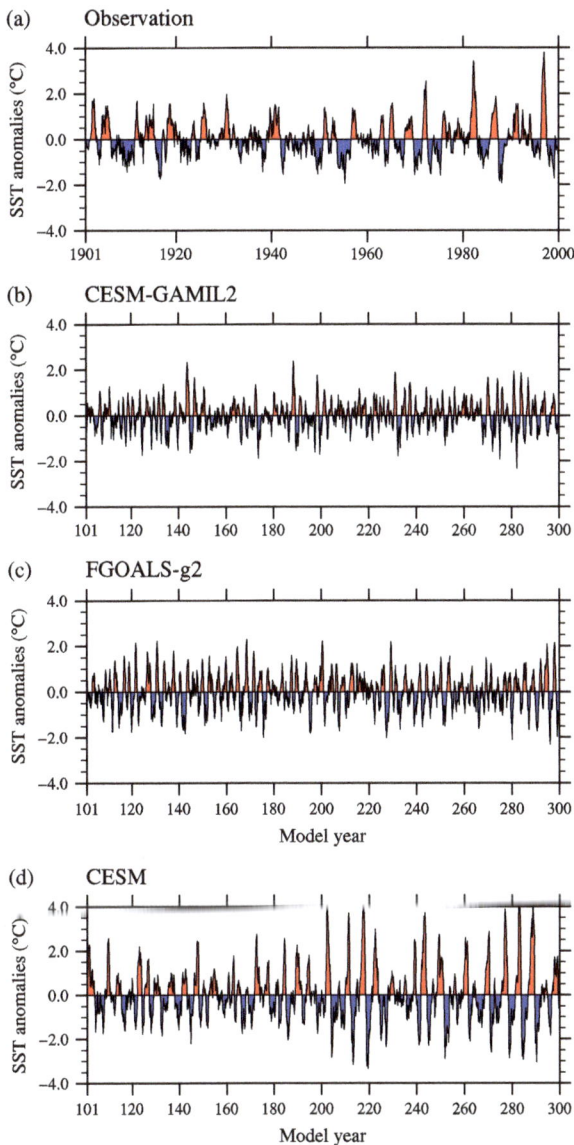

Figure 1. Time series of the Niño3 (5°N–5°S, 150°–90°W) index for (a) HadISST observations during 1901–2000, (b) CESM-GAMIL2, (c) FGOALS-g2 and (d) CESM during years 101–300 in the PI-control run.

of the annual cycle in the eastern equatorial Pacific because variation in SST for the Niño3 region is generally dominated by both the annual cycle and the ENSO signal (Guilyardi, 2006; An and Choi, 2013). To further understand the ENSO differences, normalized spectra of Niño3 full monthly SST for observation and simulations are calculated, and their corresponding percentages of total spectral energy in annual and semi-annual cycles are presented in Figure 2. Based on observations, the cycle within one year accounts for 35.6% of the total energy. The corresponding values for CESM-GAMIL2 and FGOALS-g2 are 30.7% and 30.1%, respectively; i.e., close to the observed value. In the CESM, however, this value is only 14.4%; i.e. 85.6% of the total energy is available for the interannual signals. These energy distributions indicate that if El Niño is viewed as a disruption of the seasonal cycle (Guilyardi, 2006), then it follows that the weak

seasonal cycle in the CESM is more easily disrupted than in CESM-GAMIL2. A dynamical interpretation of the inverse relationship between the annual cycle strength and the ENSO amplitude in the eastern equatorial Pacific, which requires further study using the CESM, is presented by An and Choi (2013) using the Geophysical Fluid Dynamics Laboratory Couple Climate Model (Version 2.0; GFDL-CM2.0).

3.2. Atmospheric feedbacks

On the basis that the only difference between the CESM-GAMIL2 and the CESM is the atmospheric model, atmospheric processes and feedbacks that govern the large difference in ENSO amplitude are explored, with FGOALS-g2 results included for comparison. Atmospheric dynamic and thermal feedbacks in the observation and three simulations are listed in Table 1. Bjerknes feedbacks for the three models compare well in spite of μ being too weak with respect to observations, while heat flux feedbacks, α, vary significantly. The value for α in the CESM is -8.71 W m^{-2} K^{-1}, about half of the values of -18.64 and -17.02 W m^{-2} K^{-1} simulated in CESM-GAMIL2 and FGOALS-g2, respectively. The latter two are much closer to -16.70 W m^{-2} K^{-1} provided by the ERA40 and -15.92 W m^{-2} K^{-1} provided by OAFlux. This result may indicate that a weak α value means the CESM cannot effectively suppress El Niño warming and thus produces a stronger ENSO. As shown in Table 1, the SW feedback, α_{SW}, is the main cause of the α difference among the models, although the LH feedback, α_{LH}, also plays a role.

To further understand the impact of α_{SW} on SSTA, a composite analysis of SW flux anomalies and SSTA during El Niño events is performed (Figure 3). Each El Niño event in the models (13 in CESM-GAMIL2, 13 in FGOALS-g2 and 17 in CESM) is defined as an SSTA in Niño3 greater than 1.5 times its standard deviation for at least three consecutive months (Lloyd *et al.*, 2012). Each event in the observation is selected according to the El Niño definition provided by the Climate Prediction Center (Null, 2015), where weak and moderate events are included. In OAFLUX, negative SW anomalies occur at the end of El Niño warming due to the time required to establish a sufficient SST to trigger convection (Guilyardi *et al.*, 2009). In CESM-GAMIL2, negative SW anomalies cover almost the entire warming episode and rapid warming is suppressed, conducive to a weak El Niño. Conversely, for the CESM, large positive SW anomalies occur in the eastern Pacific at the start of an El Niño event, thereby amplifying El Niño development. The evolution of SW anomalies in FGOALS-g2 is closest to that in OAFlux in the eastern Pacific, and correspondingly, the El Niño strength of FGOALS-g2 is much closer to observations. Stronger negative SW anomalies are collocated with weaker positive interannual SSTA in the CESM-GAMIL2 compared with the CESM, consistent with their feedback results obtained by linear regression. In the western and

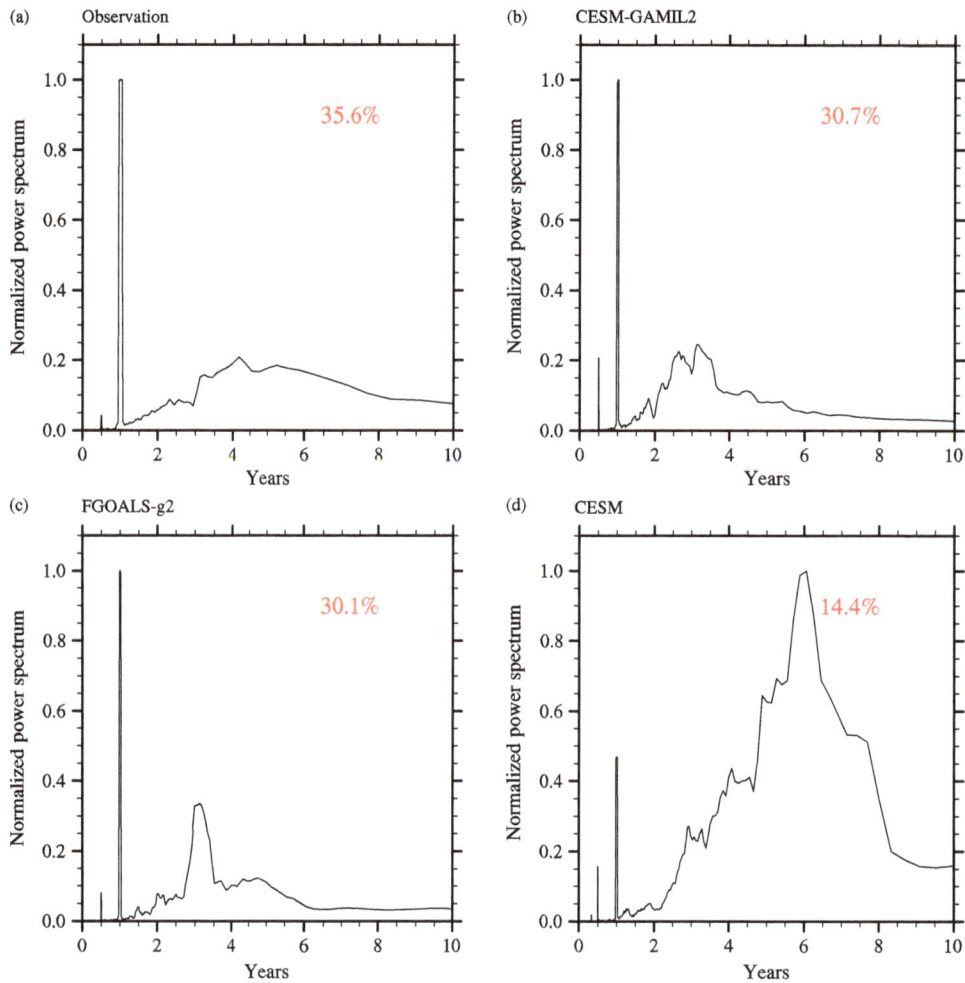

Figure 2. Normalized power spectra of full monthly Niño3 SST for (a) HadISST, (b) CESM-GAMIL2, (c) FGOALS-g2 and (d) CESM. The red values indicate the percentage of total spectral energy in the annual and semi-annual cycles, respectively.

Table I. Coefficients of linear regression against SST of the surface net heat flux, shortwave radiation, and latent heat (W m^{-2} K^{-1}); total, convective, and stratiform precipitation (mm day^{-1} K^{-1}); total liquid water path (g m^{-2} K^{-1}); 500 hPa vertical velocity (hPa day^{-1} K^{-1}); and total-, high-, middle-, and low-cloud fraction (% K^{-1}) over the Niño-3 region and average annual Bjerknes feedback in the Niño4 region (10^{-3} N m^{-2} K^{-1}) from observations, CESM-GAMIL2, FGOALS-g2 and CESM, and their Niño3 amplitudes (K).

	Observation	CESM-GAMIL2	FGOALS-g2	CESM
Niño3	0.81 (HadISST)	0.64	0.78	1.25
α	−16.70/−15.92 (ERA40/OAFlux)	−18.64	−17.02	−8.71
α_{SW}	−11.32/−5.63 (ERA40/OAFlux)	−11.40	−9.05	−2.19
α_{LH}	−6.40/−9.50 (ERA40/OAFlux)	−9.19	−9.45	−6.20
μ	11.06 (ERA40)	8.24	8.06	9.05
α_{pr}	1.11/1.03 (GPCP/CMAP)	1.40	1.14	1.08
α_{prc}	—	0.90	0.74	0.95
α_{prl}	—	0.50	0.40	0.13
α_{lwp}	19.9/4.42 (SSM/I/ISCCP)	13.47	10.55	10.22
α_{w500}	−9.42 (ERA-40)	−11.40	−9.48	−7.79
α_{cldtot}	4.52 (ISCCP)	4.78	3.67	3.11
α_{cldhgh}	3.45 (ISCCP)	4.18	4.20	9.00
α_{cldmid}	2.76 (ISCCP)	3.51	2.97	4.52
α_{cldlow}	−0.42 (ISCCP)	4.16	2.41	−3.34

central Pacific, CESM reproduces the observed evolutions of negative SW anomalies during El Niño development period, while the other two models produce the negative SW anomalies only during mature period.

As reported previously, SW flux is associated mainly with the extent of cloud cover, the cloud liquid water path and dynamical circulation. Thus, the SW feedback is decomposed into cloud fraction feedback, LWP feedback and dynamics (vertical velocity at 500 hPa) feedback (Lloyd *et al.*, 2012; Li *et al.*, 2014). The Niño3 averaged feedbacks of the total cloud fraction (α_{cldtot}), total LWP (α_{lwp}), and dynamics (α_{w500}) in CESM-GAMIL2 are 4.78% K^{-1}, 13.47 g m^{-2} K^{-1} and −11.40 hPa day^{-1} K^{-1}, respectively, while in CESM they are 3.11% K^{-1}, 10.22 g m^{-2} K^{-1} and −7.79 hPa day^{-1} K^{-1}, respectively. All three component feedbacks contribute to weak SW feedback in the CESM. For FGOALS-g2, both the α_{SW} and its three component feedbacks are intermediate between those of the CESM and the CESM-GAMIL2, and are much closer to the latter, suggesting the dominate role of α_{SW} in the atmospheric model. Furthermore, the vertical distributions of the cloud fraction and the cloud liquid amount (CLDLIQ) feedbacks (Figure 4) show that in CESM, negative cloud fraction and CLDLIQ feedbacks

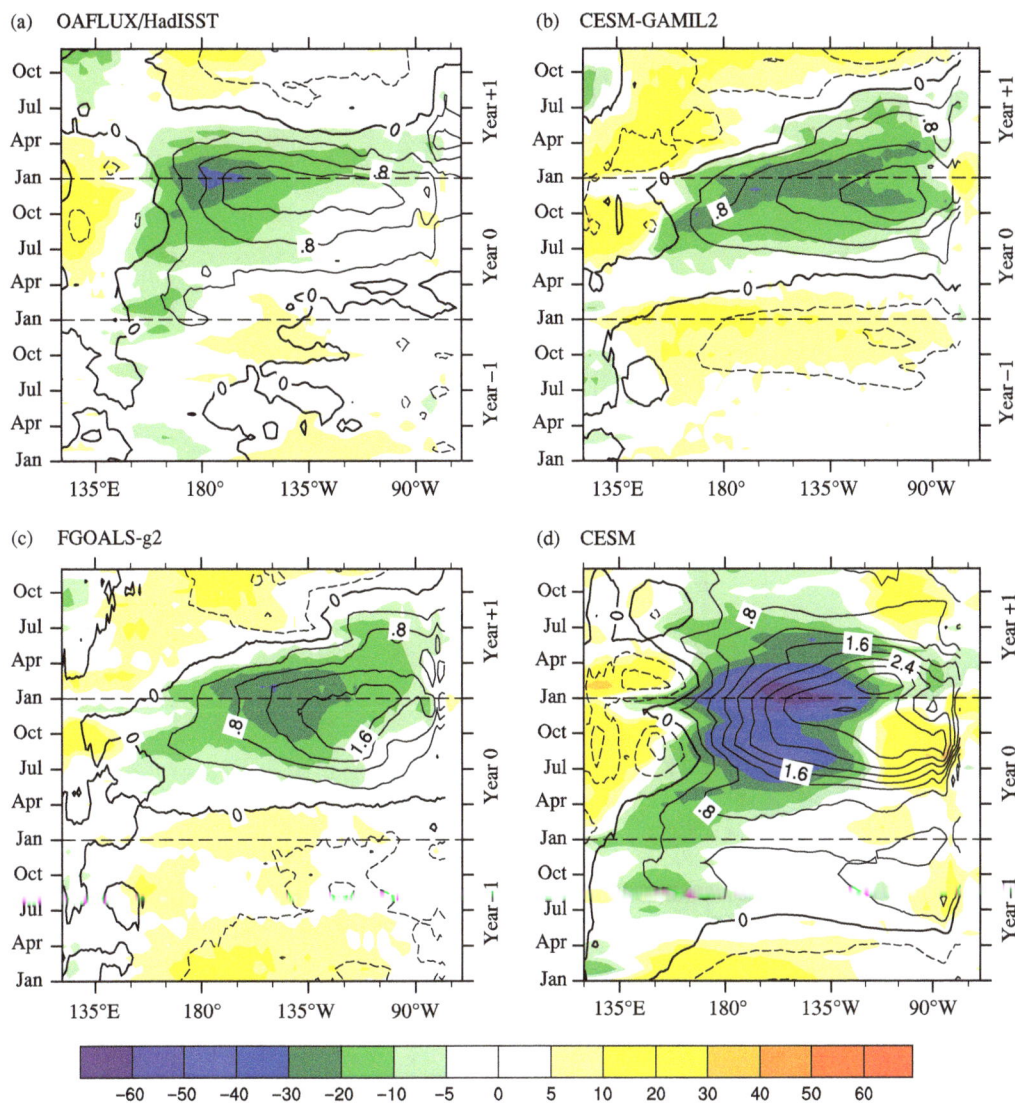

Figure 3. Composite El Niño evolution along the equator for the shortwave flux anomaly (shading) and SST anomaly. The contour interval for SST is 0.4 K. (a) OAFLUX (which includes ISCCP radiative fluxes; 1984–2009), (b) CESM-GAMIL2, (c) FGOALS-g2 and (d) CESM.

below 700 hPa are the main causes of weaker α_{cldtot} and α_{lwp} as well as positive SW anomalies in the eastern Pacific. The too negative cloud fraction feedback and LWP feedback in the lower layers are the common problem in most CMIP5 models, possibly arising from the same root as the 'too few too bright' low-cloud problems (Li *et al.*, 2015, pers. comm.). The small stratiform rainfall feedback in models may be one important factor (Li *et al.*, 2014).

4. Discussion and conclusions

The ENSO simulations of this study are conducted using two CESM-based GCMs (i.e. CESM and CESM-GAMIL2) that differ only in their atmospheric components. The ENSO amplitudes simulated by the two models are different; e.g. the standard deviation of the Niño3 index in the CESM is up to twice that in CESM-GAMIL2. The strong amplitude in the CESM

is consistent with its weak seasonal cycle strength, which accounts for only 14.4% of the total energy compared with 30.7% in CESM-GAMIL and 35.6% in observations. Less energy within the seasonal cycle indicates that more is available for interannual signals (Guilyardi, 2006).

For comparison, another 'parent' model (i.e. FGOALS-g2 from CMIP5) is also included, and the differences in the simulations of ENSO amplitude are attributed to differences in two atmospheric feedbacks, i.e., the positive Bjerknes feedback and negative heat flux feedback, with a main contribution from the heat flux feedback. In the CESM, the weak negative heat flux feedback, approximately half the size of the corresponding feedback calculated from observations and the CESM-GAMIL2 simulation, is found to be incapable of effectively dampening El Niño warming, thereby favoring a strong ENSO. Among the four components of heat flux feedback, the SW component is found to be the main cause of weak

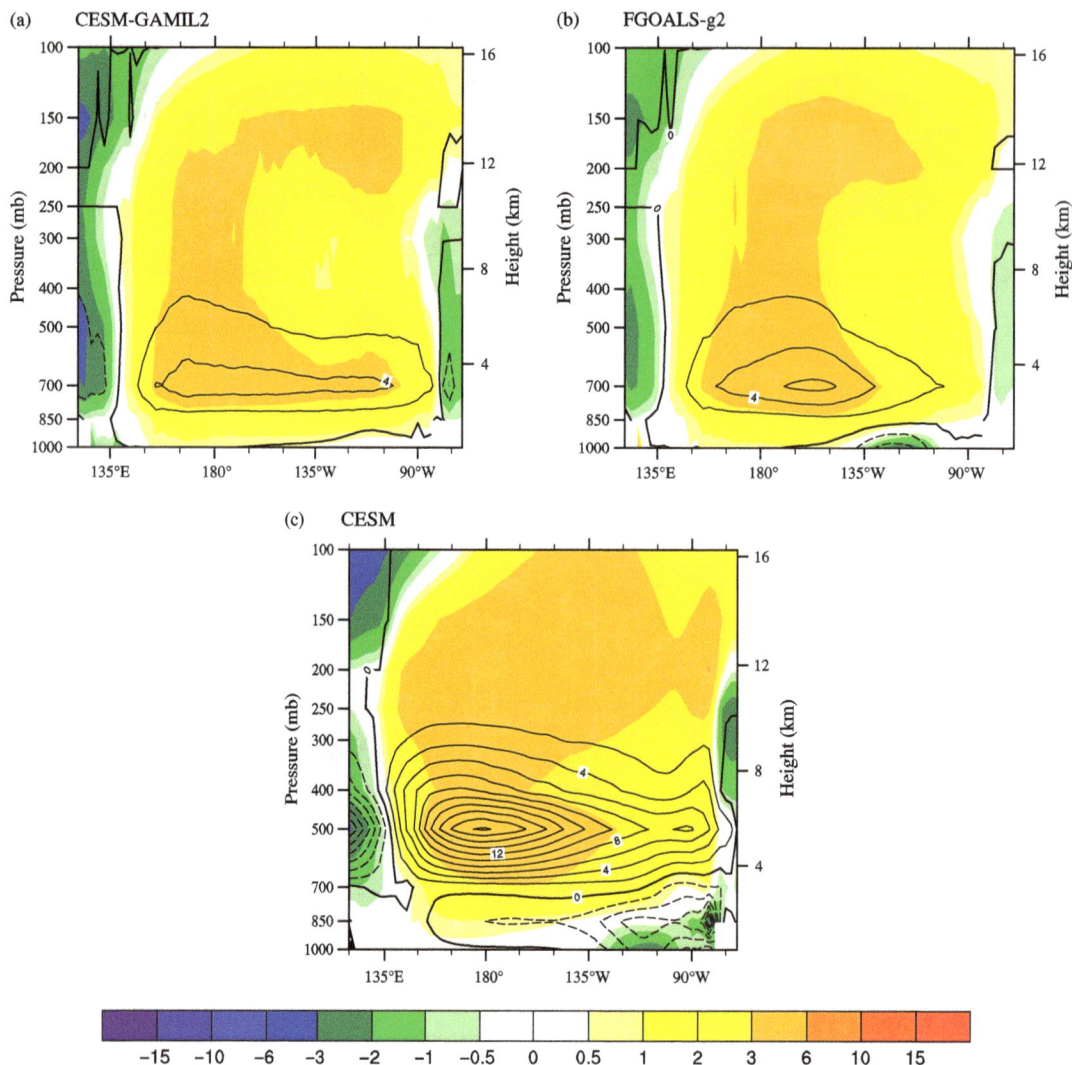

Figure 4. Vertical cross-section of cloud fraction feedback (color shading) and cloud liquid amount (CLDLIQ) feedback (black contours) for (a) CESM-GAMIL2, (b) FGOALS-g2 and (c) CESM averaged over the equatorial Pacific (5°N–5°S) (units: percent for cloud fraction and 10^{-6} Kg Kg^{-1} for CLDLIQ).

heat flux feedback in the CESM, although the LH component also played a role. Further examination indicates that the negative low-cloud fraction feedback and the negative low-cloud liquid amount feedback are the main contributors to weaker α_{SW} in the CESM in the Niño3 region; this result may be associated with its small stratiform rainfall feedback (Li *et al.*, 2014).

ENSO amplitudes in FGOALS-g2 and CESM-GAMIL are moderately different. Therefore, the role of ocean feedbacks in ENSO simulations should be further investigated. In addition, as ENSO simulations are related to their climatological mean state (Guilyardi, 2006), their relationship between the ENSO amplitude and climatological mean state in the CESM and CESM-GAMIL, as well as their uncoupled models, should be explored in future studies. Moreover, the inverse relationship between the mechanism of ENSO amplitude and season cycle merits further investigation. The significance of differences in ENSO amplitude

as well as error bars in longer time scales (such as millennial scale) will also be researched in future.

Acknowledgements

We thank Nan Ding and Wei Xue for help in establishing the CESM-GAMIL2. This work was funded by the National Natural Science Foundation of China (No. 41330527) and the Chinese Academy of Sciences Strategic Priority Research Program (Grant No. XDA05110304).

References

Adler RF, Huffman GJ, Chang A, Ferraro R, Xie PP, Janowiak J, Rudolf B, Schneider U, Curtis S, Bolvin D, Gruber A, Susskind J, Arkin P, Nelkin E. 2003. The version-2 global precipitation climatology project (GPCP) monthly precipitation analysis (1979–present). *Journal of Hydrometeorology* **4**: 1147–1167.

An SI, Choi J. 2013. Inverse relationship between the equatorial eastern Pacific annual-cycle and ENSO amplitudes in a coupled general circulation model. *Climate Dynamics* **40**: 663–675.

Bellenger H, Guilyardi E, Leloup J, Lengaigne M, Vialard J. 2014. ENSO representation in climate models: from CMIP3 to CMIP5. *Climate Dynamics* **42**: 1999–2018.

Guilyardi E. 2006. El Niño–mean state–seasonal cycle interactions in a multi-model ensemble. *Climate Dynamics* **26**: 329–348.

Guilyardi E, Braconnot P, Jin FF, Kim ST, Kolasinski M, Li T, Musat I. 2009. Atmosphere feedbacks during ENSO in a coupled GCM with a modified atmospheric convection scheme. *Journal of Climate* **22**: 5698–5718.

Guilyardi E, Cai W, Collins M, Fedorov A, Jin FF, Kumar A, Sun DZ, Wittenberg A. 2012a. New strategies for evaluating ENSO processes in climate models. *Bulletin of the American Meteorological Society* **93**: 235–238.

Guilyardi E, Bellenger H, Collins M, Ferrett S, Cai W, Wittenberg A. 2012b. A first look at ENSO in CMIP5. *CLIVAR Exchanges* **17**: 29–32.

Hunke EC Lipscomb WH. 2008. CICE: The Los Alamos sea ice model user's manual, Version 4, Los Alamos National Laboratory Tech. Rep. LA-CC-06-012, 76 pp.

Hurrell JW, Hack JJ, Shea D, Caron JM, Rosinski J. 2008. A new sea surface temperature and sea ice boundary dataset for the community atmosphere model. *Journal of Climate* **21**: 5145–5153.

Kang X, Huang R, Wang Z, Zhang RH. 2014. Sensitivity of ENSO variability to Pacific freshwater flux adjustment in the community earth system model. *Advances in Atmospheric Sciences* **31**: 1009–1021.

Kim ST, Cai W, Jin FF, Yu JY. 2014. ENSO stability in coupled climate models and its association with mean state. *Climate Dynamics* **42**: 3313–3321.

Kim ST, Yu JY. 2012. The two types of ENSO in CMIP5 models. *Geophysical Research Letters* **39**: L11704.

Li LJ, Wang B, Zhang GJ. 2014. The role of nonconvective condensation processes in response of surface shortwave cloud radiative forcing to El Niño warming. *Journal of Climate* **27**: 6721–6735.

Li LJ, Wang B, Dong L, Liu L, Shen S, Hu N, Sun WQ, Wang Y, Huang WY, Shi XJ, Fu Y, Yang GW. 2013. Evaluation of grid-point atmospheric model of IAP LASG Version 2 (GAMIL2). *Advances in Atmospheric Sciences* **30**: 855–867.

Lloyd J, Guilyardi E, Weller H. 2009. The role of atmosphere feedbacks during ENSO in the CMIP3 models. *Atmospheric Science Letters* **10**: 170–176.

Lloyd J, Guilyardi E, Weller H. 2012. The role of atmosphere feedbacks during ENSO in the CMIP3 models. Part III: the shortwave flux feedback. *Journal of Climate* **25**: 4275–4293.

McPhaden MJ, Zebiak SE, Glantz MH. 2006. ENSO as an integrating concept in earth science. *Science* **314**: 1740–1745.

Neale RB, Richter JH, Jochum M. 2008. The impact of convection on ENSO: From a delayed oscillator to a series of events. *Journal of Climate* **21**: 5904–5924.

Neale RB, Richter J, Park S, Lauritzen PH, Vavrus SJ, Rasch PJ, Zhang M. 2013. The mean climate of the community atmosphere model (CAM4) in forced SST and fully coupled experiments. *Journal of Climate* **26**: 5150–5168.

Null J. 20157-11-04. El Niño and La Niña Years and Intensities. http://ggweather.com/enso/oni.htm (accessed 10 March 2015).

Oleson KW, Lawrence DM, Bonan GB, Flanner MG, Kluzek E, Lawrence PJ, Levis S, Swenson SC, Thornton PE, Dai A, Decker M, Dickinson R, Feddema J, Heald CL, Hoffman F, Lamarque JF, Mahowald N, Niu GY, Qian T, Randerson J, Running S, Sakaguchi K, Slater A, Stöckli R, Wang A, Yang ZL, Zeng X, Zeng X. 2010. Technical description of version 4.0 of the Community Land Model (CLM), NCAR Tech. Note NCAR/TN-478+STR, 257 pp.

Rossow WB, Schiffer RA. 1999. Advances in understanding clouds from ISCCP. *Bulletin of the American Meteorological Society* **80**: 2261–2288.

Smith RD, Jones P, Briegleb B, Bryan F, Danabasoglu G, Dennis J, Dukowicz J, Eden C, Fox-Kemper B, Gent P, Hecht M, Jayne S, Jochum M, Large W, Lindsay K, Maltrud M, Norton N, Peacock S, Vertenstein M, Yeager S. 2010. The Parallel Ocean Program (POP) reference manual, Los Alamos National Laboratory Tech. Rep. LAUR-10-01853, 140 pp.

Sun DZ, Yu Y, Zhang T. 2009. Tropical water vapor and cloud feedbacks in climate models: a further assessment using coupled simulations. *Journal of Climate* **22**: 1287–1304.

Uppala SM, Kållberg PW, Simmons AJ, Andrae U, Bechtold VDC, Fiorino M, Gibson JK, Haseler J, Hernandez A, Kelly GA, Li X, Onogi K, Saarinen S, Sokka N, Allan RP, Andersson E, Arpe K, Balmaseda MA, Beljaars ACM, Van De Berg L, Bidlot J, Bormann N, Caires S, Chevallier F, Dethof A, Dragosavac M, Fisher M, Fuentes M, Hagemann S, Hólm E, Hoskins BJ, Isaksen L, Janssen PAEM, Jenne R, Mcnally AP, Mahfouf JF, Morcrette JJ, Rayner NA, Saunders RW, Simon P, Sterl A, Trenberth KE, Untch A, Vasiljevic D, Viterbo P, Woollen J. 2005. The ERA-40 Re-Analysis. *Quarterly Journal of the Royal Meteorological Society* **131**: 2961–3012.

Wang B, Wang H, Ji ZZ, Zhang X, Yu RC, Yu YQ, Liu HT. 2004. Design of a new dynamical core for global atmospheric models based on some efficient numerical methods. *Science China. Mathematics* **47**: 4–21.

Wang C, Deser C, Yu JY, DiNezio P, Clement A. 2012. *El Nino and Southern Oscillation (ENSO): A Review, A Chapter for Springer Book: Coral Reefs of the Eastern Pacific.* Springer-Verlag.

Watanabe M, Chikira M, Imada Y, Kimoto M. 2011. Convective control of ENSO simulated in MIROC. *Journal of Climate* **24**: 543–562.

Weng F, Grody N, Ferraro R, Zhao Q, Chen C. 1997. Global cloud water distribution derived from special sensor microwave imager/sounder and its comparison with GCM simulation. *Advances in Space Research* **19**: 407–411.

Xie P, Arkin PA. 1997. Global precipitation: a 17-year monthly analysis based on gauge observations, satellite estimates, and numerical model outputs. *Bulletin of the American Meteorological Society* **78**: 2539–2558.

Yu L, Weller RA. 2007. Objectively analyzed air–sea heat fluxes for the global ice-free oceans (1981–2005). *Bulletin of the American Meteorological Society* **88**: 527–539.

Yu JY, Kim ST. 2010. Identification of Central-Pacific and Eastern-Pacific types of ENSO in CMIP3 models. *Geophysical Research Letters* **37**: L15705.

14

The improved Noah land surface model based on storage capacity curve and Muskingum method and application in GRAPES model

Lili Wang, Dehui Chen and Hongjun Bao*

National Meteorological Centre, China Meteorological Administration, Beijing, P. R. China

Correspondence to:
Hongjun Bao, National
Meteorological Centre, China
Meteorological Administration,
46 Zhongguancun Nandajie,
Beijing, 100081, P. R. China.
E-mail: baohongjun@cma.gov.cn

Abstract

Noah-LSM, applied in GRAPES model, cannot express the changes of runoff-yield area effectively and describe hydrological cycle completely. It is improved as follows: (1) Add saturation excess runoff generation module, which considers the change in runoff-yielding area based on a storage capacity curve for the better grid runoff simulation; (2) introduce routing module based on Muskingum method, to consider the redistribution of soil moisture in two-dimensional level. The improved model application results indicated that the meteorological elements in the surface layer simulated are influenced by land surface water cycle process, such as soil moisture, 2 m temperature, is closer to the observed.

Keywords: improvement; Noah-LSM; storage capacity curve; Muskingum; GRAPES model

1. Introduction

Land surface can remember weather events through variations in its heat and water storages. In turn, land heat and water storage can affect atmospheric predictability through their effects on surface energy and water fluxes (Roesch *et al.*, 2001; Jiang *et al.*, 2009). As the water exchange in the different forms and the different scales, the relationship between atmosphere and hydrology is mutually related to each other and mutually restricts each other. Hydrological cycling between the atmosphere, vegetation, and soil plays an important role in land surface processes. The changes of elements in hydrological cycle are feedback to the atmosphere, through the latent heat flux and sensible heat flux. Therefore, hydrological process in land surface model is a critical link between land and atmosphere. It is crucial to understand interactions between hydrological cycle and atmospheric system, which includes controls of atmospheric conditions to regional hydrologic system and feedbacks of hydrological cycle to atmospheric system. The simulation of water balance is important in atmospheric system which is directly influence the evapotranspiration, surface runoff, interflow, and underground runoff in hydrological processes (Cao and Liu, 2005). However, in most land surface model, the implemented hydrological process simulation is too simple to predict soil moisture accurately, even the water cycle. Several important problems exist in most land surface models. First, heterogeneity of the runoff-yielding area is not fully considered. For example, infiltration capacity is usually assumed to be homogeneous within a grid cell, but it is likely highly heterogeneous within a cell. Second, land surface models usually lack sophisticated runoff concentration. It is true that physical mechanisms

and theories of surface hydrologic processes have been well studied for a long time; hydrologists are able to model these processes adequately – even though the systems and processes are spatially heterogeneous –if many important topographic, soil, vegetation, and climatic parameters are known. In China, the Xin'anjiang model is widely applied in flood forecasting and runoff simulation (Zhao, 1983; Li *et al.*, 2006). It is a well-tested hydrological model. Therefore, it can be used to improve the modelling of hydrological processes in land surface models in terms of providing better simulation runoff. As a result, it should be able to largely improve the simulation of land surface water cycle in land surface models.

Noah-LSM is a well-known land surface model, which couples the diurnally dependent Penman potential evaporation approach of Mahrt and Ek (1984), the multilayer soil model of Mahrt and Pan (1984), and the primitive canopy model of Pan and Mahrt (1987). It has been extended by Chen *et al.* (1996) and Chen and Jimy (2001) to include the modestly complex canopy resistance approach of Noilhan and Planton (1989) and Jacquemin and Noilhan (1990). During the 1990s, National Centers for Environmental Prediction (NCEP) greatly expanded its land surface modelling collaborations via several components of the Global Energy and Water Cycle Experiment (GEWEX), most notably, the GEWEX Continental-Scale International Project (GCIP) and the Project for Intercomparison of Land-surface Parameterization Schemes (PILPS). These collaborations included the Office of Hydrological Development (OHD) of the National Weather Service, National Environmental Satellite Data and Information Service (NESDIS), NASA, National Center for Atmospheric Research (NCAR), the US Air

Force, and OSU and other university partners. As an outgrowth of these collaborations and their broad scope of LSM testing in both uncoupled and coupled mode over a wide range of space scales and timescales, NCEP substantially enhanced the OSU LSM, now renamed the Noah-LSM in recognition of the broad partnership above (Ek *et al.*, 2003).

The Global/Regional Assimilation and Prediction System (GRAPES, Chen and Shen, 2006; Chen *et al.*, 2008; Xue and Chen, 2008) is an important operational numerical weather prediction system in the China Meteorological Administration (CMA). This model adopts a structure of standardized and module-based software in accordance with the strict requirements of software engineering.A preliminary study (Wu *et al.*, 2005) shows that the application of GRAPES model meets the requirement for sustainable development of the numerical prediction system of China. Nowadays, GRAPES model has been expanded to applications in various fields, such as GRAPES_Meso for mesoscale weather prediction, GRAPES_TCM for typhoon prediction, GRAPES_SDM for sandstorm forecast, and GRAPES_SWIFT for short-time weather forecasting (Zhao and Li, 2006; Zhu *et al.*, 2007). Further development of GRAPES is undergoing. So far, however, the operational GRAPES model in the CMA has not yet been able to directly predict runoff and flood events. Therefore, it is necessary to improve the Noah-LSM model that accommodates the whole process of the closed land surface water cycle, and is able to predict the runoff or discharge. But there is obviously insufficient that is the Noah-LSM land surface model for hydrological processes especially runoff description. It cannot simulate the complete description of the hydrological cycle, on the basis of the land surface hydrological using parameterization scheme is a simple water balance model SWB (Simple Water Balance, Entekhabi *et al.*, 1999).

The aims of this article, therefore, are to (1) improve Noah-LSM with introducing saturation excess runoff generation to descript the inhomogeneity of the yield runoff area and routing module, which can represent better land surface water cycle processes and (2) apply the newly improved Noah-LSM to describe hydrological process of GRAPES_MESO model and compare with the forecast results of the near-surface meteorological elements of the original model.

2. Improvement on hydrological process of Noah-LSM

2.1. Hydrological process on Noah-LSM

In Noah-LSM, the Simple Water Balance (SWB) model is used to calculate the surface runoff (R). The SWB model (Schaake *et al.*, 1996) is a two-reservoir hydrological model. A thin upper layer consists of the vegetation canopy and the soil surface. A lower layer includes both the root zone of the vegetation and the groundwater system. The surface runoff, Rs, is difference between

precipitation (P_d) and maximum infiltration (I_{max}):

$$R_s = P_d - I_{max} \quad (1)$$

I_{max} is formulated as

$$I_{max} = P_d D_x \left[1 - \exp\left(-kdt \; \delta_t\right)\right] /$$
$$\left[P_d + \left[1 - \exp\left(-kdt \; \delta_t\right)\right]\right] \quad (2)$$

$$D_x = \sum_{i=1}^{4} \Delta Z_i \left(\Theta_s - \Theta_i\right) \quad (3)$$

$$kdt = kdt_{ref} K_s / K_{ref} \quad (4)$$

where D_x is the soil water diffusivity; Θ_s is the maximum soil moisture (porosity); Θ_i is the soil moisture content of the *i*th layer; ΔZ_i is the thickness of the *i*th soil layer; $kdt = 3.0$ and $K_{ref} = 2 \times 10^{-6}$ m s^{-1}; δ_i is the conversion of the current model time step δ_t (in terms of seconds) into daily values.

The underground runoff is as computed in Noah-LSM model as:

$$R_g = \begin{cases} Q_{max}\left(1 - D_b/S_{max}\right), & D_b < S_{max} \\ 0, & \text{Others} \end{cases}$$
$$(5)$$

where Q_{max} is the maximum underground runoff, S_{max} is the critical value of soil water moisture deficit, D_b is the soil water shortage in the lower soil layer, the maximum value is the lower soil water content.

2.2. Problems in hydrological process modelling of NOAH-LSM and their improvements

2.2.1. Problems in hydrological process modelling of Noah-LSM

In the rainfall-runoff process, runoff yielding area is an area that contributes to the runoff flow in the basin. It changes during the precipitation process. According to Equations (1) and (4), only vertical movement of water is considered in the runoff module in Noah-LSM model. It cannot effectively simulate spatial changes in runoff-producing area during the rainfall (Abbott *et al.*, 1986). Therefore, runoff module of NOAH-LSM should be improved in the runoff generation and routing.

2.2.2. Improvement on hydrological process modelling of Noah-LSM

The concept of runoff formation on repletion of storage holds true so far as one single point, in the sense of a very small area, of a basin is concerned. For an entire watershed, things are more complicated. The moisture deficit often varies from place to place. This non-uniform distribution effects the runoff production of a whole basin significantly. To solve the problem, the tension water storage capacity curve (Zhao, 1983; Zhao and Liu, 1995), which represents the concept of runoff formation on repletion of Storage, is suggested to improve the runoff module in Noah-LSM.

$$1 - f/F = \left(1 - W'/W_{MM}\right)^B \qquad (6)$$

where f/F represents the proportion of the pervious area of the basin whose tension water capacity is less than or equal to the value of the ordinate; W' is point tension water capacity, varies from zero to a maximum W_{MM} according to the relationship (mm); W_{MM} is the maximum tension water capacity (mm); B is an exponent of the tension water capacity distribution curve.

Based on the relationship between storage capacity curve and exchange between rainfall and runoff, the improved runoff generation in Noah-LSM is computed as follows:

When $P - E + A < W_{MM}$, only part of the grid cell is producing runoff:

$$R = P - E - \left(W_M - W_0\right) + W_M$$
$$\times \left(1 - (P - E + A)/W_{MM}\right)^{(1+B)} \qquad (7)$$

When $P - E + A \geq W_{MM}$, the whole grid cell is producing runoff:

$$R = P - E - \left(W_M - W_0\right) \qquad (8)$$

where P is areal mean rainfall in the grid (mm); E is areal mean evapotranspiration in the grid; W_0 is the initial areal mean tension water storage in the grid (mm); W_M is the areal mean tension water capacity in a grid cell (mm), W_M and W_{MM} are related through the parameter B; R is the runoff generation in the grid cell (mm); A is determined by $A = W_{MM} \times \left[1 - \left(1 - \frac{W}{W_M}\right)^{\frac{1}{1+B}}\right]$.

In the tension water storage capacity curve and runoff generation model, P is the input, E and W_0 can be calculated with according to measured pan evaporation and soil moisture, the relation between W_M and W_{MM} is easy to be shown by $W_{MM} = W_M(1 + B)$, and B can be calibrated by basin area and underlying surface factors (Zhao and Liu, 1995). Therefore, the parameter estimate of W_M is very important and necessary.

W_M corresponds to available soil water defined as the difference between field capacity and wilting point:

$$W_M = \left(\theta_{fc} - \theta_{wp}\right) \times L_a \qquad (9)$$

where θ_{fc} is field capacity; θ_{wp} is wilting point; L_a is thickness of aeration zone (mm).

The values of θ_{fc} and θ_{wp} were determined from soil texture following Williams et al. (1998) and Anderson et al. (2002). To derive the spatial distribution of W_M, thicknesses of aeration zone and humus soil layer should be given beforehand. Soil thickness, which is an important factor in hydrologic processes, can vary spatially because of complex interactions of many different factors; practical measurement of soil thickness is laborious and time consuming (Tesfa et al., 2009). Thus, we developed a simple method to predict roughly the spatial pattern of soil thickness from topographic, soil texture, and land cover attributes.

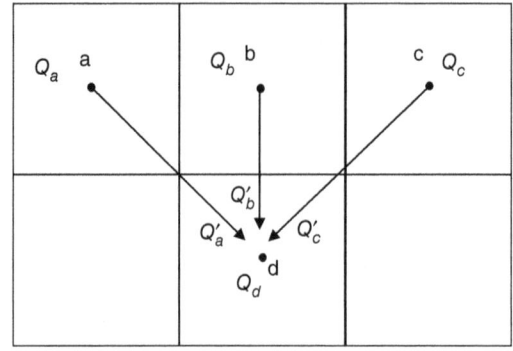

Figure 1. Schematic representation of flow routing between adjacent grid cells.

Table 1. Locations of the soil moisture stations.

Station name	Latitude (°N)	Longitude (°E)	Basin
Nanchong	30.80	106.08	Yangtze River basin
Linfen	36.06	111.50	Yellow River basin

Assuming that topography is the dominant source of heterogeneity in thickness of aeration zone L_a, a simple scheme based on topographic index (Beven and Kirkby, 1979) was developed:

$$L_a = \xi_a \times TI + \xi_b \qquad (10)$$

where TI is topographic index, ξ_a and ξ_b are two coefficients.

The TI reflects the tendency of flow to accumulate in the watershed and the tendency for gravitational forces to move that flow downslope (Quinn et al., 1991; Yao et al., 2012). A location with high TI value usually has a high water table, i.e. a thin aeration zone. It is reasonable to expect relatively low moisture capacity W_M in such a location. It is thus valid to assume that hydrologic unit with the minimum or maximum value of TI correspond to the unit with the minimum or maximum value of W_M (Zhao and Liu, 1995; Yao et al., 2012).

In terms of the soil textures, ξ_a and ξ_b are calculated by (Yao et al., 2012):

$$\begin{cases} \xi_a \times TI_{min} + \xi_b = L_{a\,max} = \dfrac{W_{M\,max}}{\theta_{fc,TI_{min}} - \theta_{wp,TI_{min}}} \\ \xi_a \times TI_{max} + \xi_b = L_{a\,min} = \dfrac{W_{M\,min}}{\theta_{fc,TI_{max}} - \theta_{wp,TI_{max}}} \end{cases} \qquad (11)$$

where $W_{M\,max}$ and $W_{M\,min}$ could be easily pre-set by referring to the literature (Zhao, 1992; Zhao and Liu, 1995) or the experience with implementation of the Xinanjiang model.

2.3. Adding routing module to hydrological process modelling of Noah-LSM

The runoff flow produced by the rainfall could flow with the terrain. If there is no routing module, runoff redistribution in the horizontal direction cannot be accounted

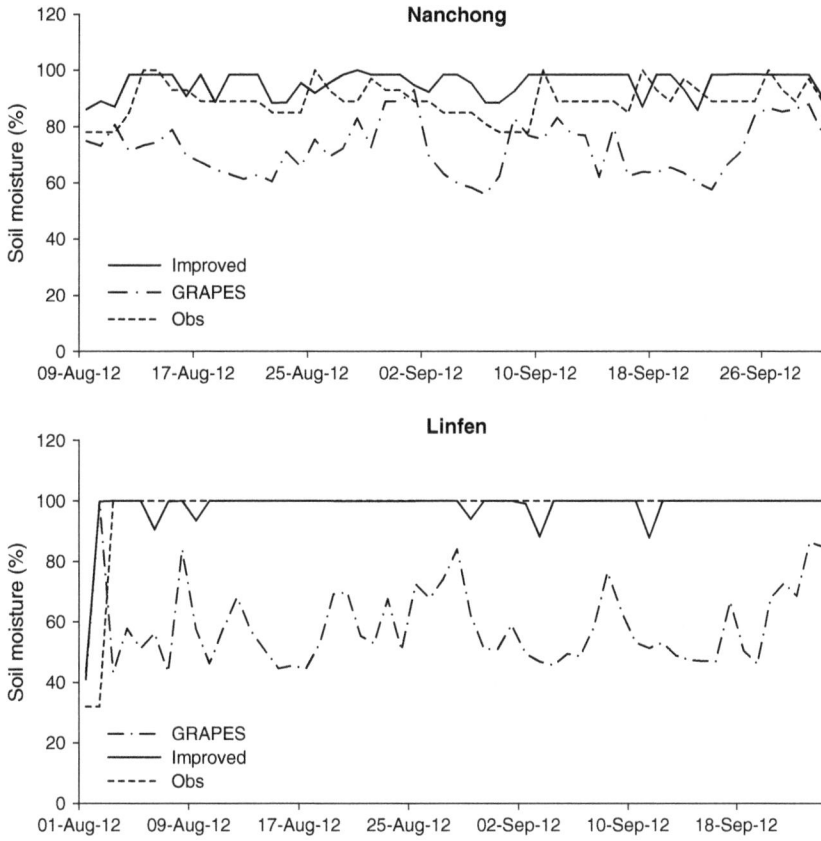

Figure 2. Comparison of observed and modelled soil water content at the depth of 10 cm.

for; it will cause the runoff to accumulate in the grid cell without flowing to the downstream areas, and an incomplete simulation of water cycle. In land surface model, the runoff directly affects the sensible heat flux and the latent heat flux, which have important feedbacks to the meteorological model. By adding the routing module, it consider the movement of the runoff in the horizontal direction, and making the simulation closer the reality and potentially more accurate.

The Muskingum routing method (Bates and De Roo, 2000) is adopted into Noah-LSM in this study. As shown in Figure 1, the flow in the grid cells a, b, and c can all flow into the grid d; the outflow are Q_a, Q_b, and Q_c, respectively. Q_a', Q_b', and Q_c' are obtained by the Muskingum flow routing method,

$$Q_{i+1}^{t+1} = C_1 Q_i^t + C_2 Q_i^{t+1} + C_3 Q_{i+1}^t \qquad (12)$$

where

$$C_1 = \left(0.5\Delta t - x_e k_e\right) / \left(\left(1 - x_e\right) k_e + 0.5\Delta t\right) \qquad (13)$$

$$C_2 = \left(0.5\Delta t + x_e k_e\right) / \left(\left(1 - x_e\right) k_e + 0.5\Delta\right) \qquad (14)$$

$$C_3 = \left(\left(1 - x_e\right) k_e - 0.5\Delta t\right) / \left(\left(1 - x_e\right) k_e + 0.5\Delta t\right) \qquad (15)$$

where x_e and k_e are the two parameters of the Muskingum flow routing method.

As a result, the outflow Q_d of the grid d is the sum of Q_a', Q_b', Q_c' and the runoff Q_{d0} produced at the grid d:

$$Q_d^t = Q_a^{tt} + Q_b^{tt} + Q_c^{tt} + Q_d^{tt} \qquad (16)$$

The depth of flow and discharge in reach can be derived from empirical relationships recommended by the US Reclamation Service (Chow, 1959), hence, the Muskingum k_e and x_e parameters can be estimated with simplifying overland routing for wide rectangular channel routing, channel routing for parabolic channel routing (Tewolde and Smithers, 2006; Bao, 2009; Li et al., 2008).

The Muskingum k_e is estimated from Equation (17) as follows (Chow, 1959; Fread, 1993):

$$K_e = \frac{\Delta L}{V_w} \qquad (17)$$

where ΔL is reach length (m), V_w is celerity (m s^{-1}).

For a wide rectangular channel, the celerity may be estimated by $V_w = \frac{5}{3} V_{av}$. For a parabolic channel, the celerity may be estimated by $V_w = \frac{11}{9} V_{av}$, where V_{av} is the average velocity, which can be calculated from the Manning equation (Chow, 1959; Tewolde and Smithers, 2006; Bao, 2009).

The Muskingum x_e is estimated by (Fread, 1993; Tewolde and Smithers, 2006):

$$x_e = \frac{1}{2} - \frac{Q_0}{2SPV_W\Delta L} \qquad (18)$$

Table 2. Locations of the air-temperature observation stations.

Station name	Latitude (°N)	Longitude (°E)	Basin
Shouxian	32.55	116.78	Huai River basin
Macheng	31.18	114.97	Yangtze River basin
Shangqiu	34.45	115.67	Huai River basin

Table 3. Root mean square error of temperature in 2 m.

	Temperature in 2 m (°C)	
Name	IM	G
ShouXian	3.15	3.27
Macheng	3.10	3.29
Shangqiu	3.81	3.86

'IM' means improved hydrological process on GRAPES NOAH-LSM; 'G' means GRAPES_MESO.

where Q_0 is reference flow, which can be estimated by $Q_0 = I_b + 0.5(I_P - I_b)$. I_b is minimum discharge and I_P is peak discharge. S is slope, P is wetted perimeter (m), which can be calculted by $P = c\sqrt{Q_0}$. c is coefficient (between 4.71 and 4.81) (Bao, 2009).

3. Experiments and analysis

In order to compare the performance of the improved Noah-LSM model with the original one, an experiment

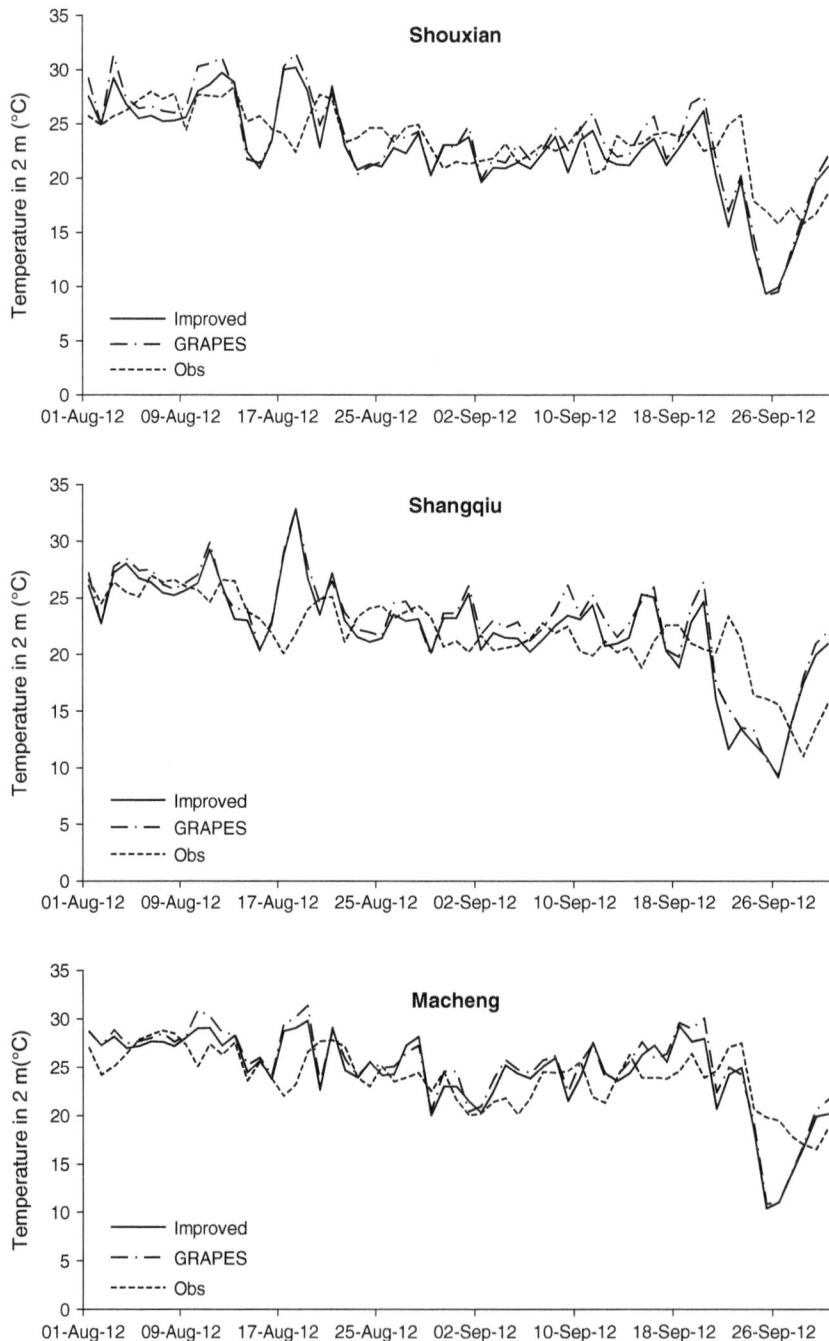

Figure 3. Comparison of observed and modelled air temperature at 2 m above the ground.

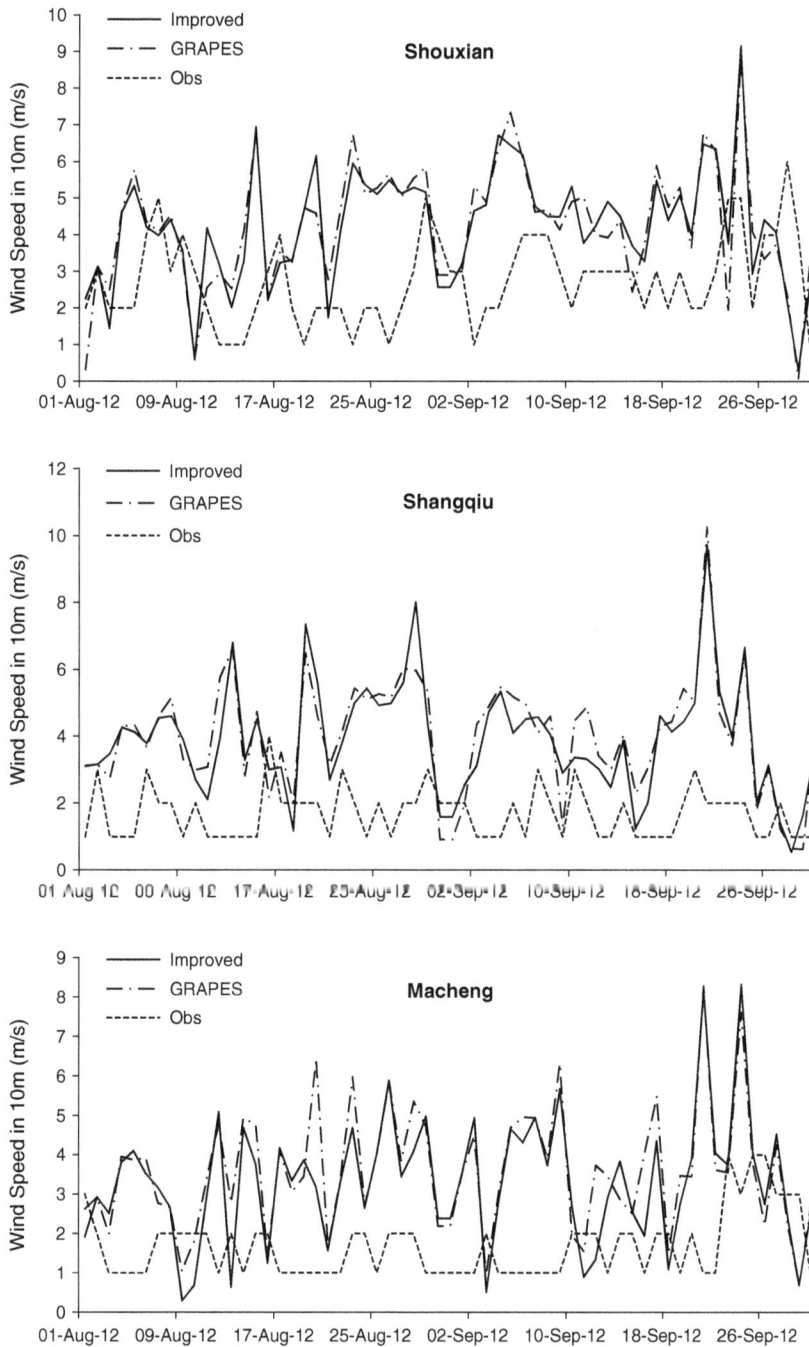

Figure 4. Comparison of observed and modelled wind speed at 10 m above the ground.

is conducted for simulating the precipitation from 1 August 08 BJT (Beijing Time), corresponded for 1 August 00 UTC, to 1 October 08 BJT. The prediction time is 48 h for GRAPES_MESO model within the domain (15°–64.5°N, 70°–145.3°E), on 08 BJT daily.

The experiment are conducted with the initial fields and lateral boundaries provided by NCEP forecast fields with 1° × 1° resolution, while the GRAPES_MESO model has a 0.15° × 0.15° horizontal resolution. Because the improvement is on the hydrological processes of Noah-LSM, soil moisture, especially the soil moisture of the first level (in 10 cm below the ground), is directly affected, and 2 m temperature is a

Table 4. Root mean square error of wind speed in 10 m.

Station name	Wind speed (m s⁻¹)	
	IM	G
Shouxian	2.46	2.56
Macheng	2.50	2.66
Shangqiu	2.80	2.92

'IM' means improved hydrological process on GRAPES NOAH-LSM; 'G' means GRAPES_MESO.

key element for meteorological model. Precipitation is the final forecasting result for GRAPES_MESO model. The runoff flow should show the biggest changes

Table 5. Threat score and bias error.

Model	Light rain		Moderate rain		Heavy rain	
	TS	**B**	**TS**	**B**	**TS**	**B**
IM	0.222	1.235	0.090	1.319	0.030	0.272
G	0.223	1.229	0.090	1.212	0.028	0.242

'IM' means improved hydrological process on GRAPES NOAH-LSM; 'G' means GRAPES_MESO. TS, threat score; B, bias error.

Table 6. Latitude and longitude of the stations.

Station name	Latitude (°N)	Longitude (°E)	Basin
Macheng	31.18	114.97	Yangtze River Basin
Wuyi	37.80	115.88	Hai River Basin

after the improvement. Therefore, these variables are selected for comparison. More importantly, all these variables were also measured. It is worth to note that the distribution of the soil moisture stations is different from that of the 2 m temperature stations.

3.1. Soil moisture

Comparing to the original Noah-LSM, not only the vertical variation of water but also its horizontal redistributionhas been changed in the improved Noah-LSM. The overland flow affects the horizontal distribution of the soil moisture, especially the soil moisture of the first level. Land surface water cycle caused the soil moisture change at first. The observed soil moisture records at 10 cm of Nanchong station and Linfen station

were chosen for the comparison. Table 1 shows the latitude and longitude of the stations. The observed soil moisture at Nanchong station starts to rise at 08 BJT 1 August as shown in Figure 2(a). It becomes saturated from 3 August to 30 September (Figure 2(b)) in Linfen station. Comparing to the observation, the GRAPES_MESO results are generally low, but the improved results are much closer to the observations. This showing that improving the new runoff module and adding the routing module help improve the simulation of soil moisture indeed. In land surface model, change in soil moisture directly affects the simulation of soil temperature. Furtherly, this effect can provide feedback to the atmospheric model, to impact the simulation of the temperature in 2 m, wind speed in 10 m, and other variables (Chen *et al.*, 1996; Peters-Lidard *et al.*, 1998; Chen and Jimy, 2001).

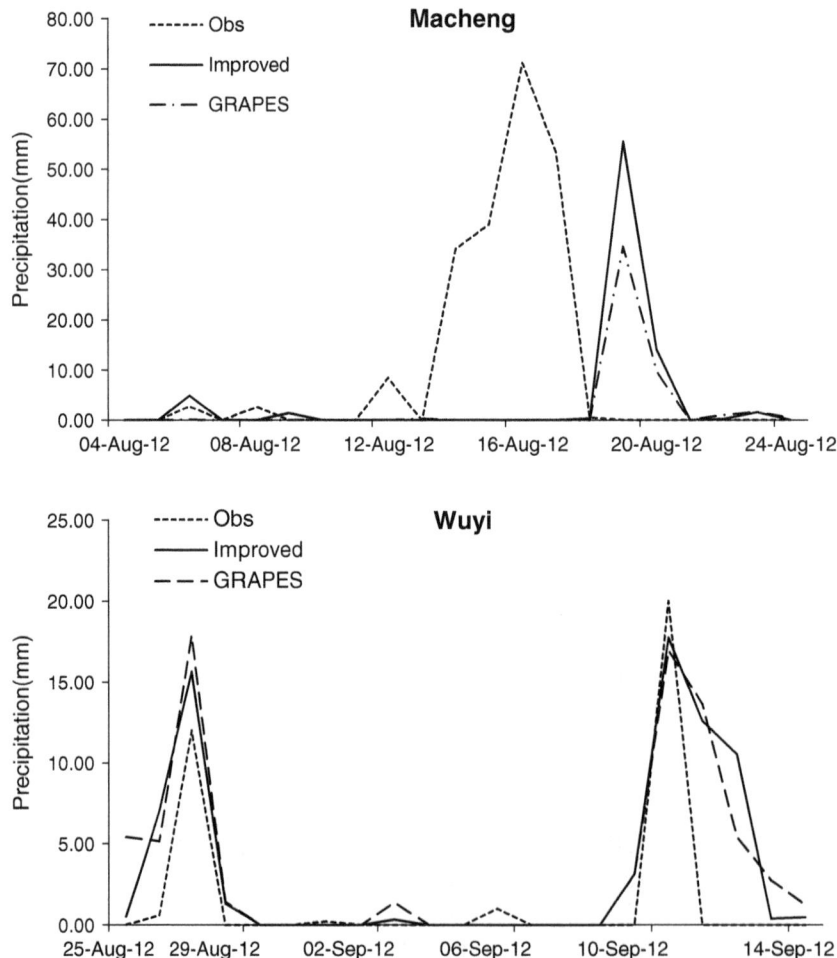

Figure 5. Comparison of observed and simulated precipitation at Macheng and Wuyi stations.

3.2. Temperature in 2 m

The change of the soil moisture can affected the 2 m temperature, which is another key variable for GRAPES_MESO model in our continuous experiments. In this study, the 2-m observation stations include Shouxian, Shangqiu, and Macheng stations (Table 2). It is clear that the 2-m temperature series in GRAPES model are different from those in the revised model in Figure 3. The root mean square error of temperature in 2 m simulated by the improved one is lower than the GRAPES_MESO model (Table 3). This indicates that the improved hydrological process has also help to improve the simulation of temperature in 2 m.

3.3. Wind speed in 10 m

The soil moisture can also affect the simulated on the wind speed in 10 m. Observations at three stations were selected to compare with the simulations (Table 1). In Figure 4, the 10-m wind speed modelled by the improved model is slightly different to the results of the original model. According to the root mean square error on wind speed in 10 m, the improved one is slightly better than the original one (Table 4). The better results in 10-m wind speed simulation highlight the important consequence of the improvement on the hydrological processed conducted in this study.

3.4. Precipitation

Threat score (TS) as the point-based precipitation verification method is commonly used in weather forecast operation and model research (Huang, 2001). In this article, 2513 stations are chosen for precipitation verification. The improved and original models have close TS scores and bias errors as shown in Table 5. We further selected two stations for precipitation verification (in Table 6). It is clear that the simulated precipitation by the improved model is different from this by the original model (Figure 5), which proves that the changes in the hydrological process has a direct consequence on the simulation of the rainfall regime.

3.5. Discharge

In this study, due to the limited stream flow observations, the observed flow at Wangjiaba station from 13 to 16 August area chosen for verification in this experiment. The contributing area of Wangjiaba stationis about $30\,672\,km^2$ and is the upper stream of Huai River, where channel bed slope and flow velocity are generally large. The first key flood control gate of the catchment is located at Wangjiaba station. Behind this gate is the Mengwa flood retention zone, with a design capacity of 750 million m^3 and a design maximum discharge of $1626\,m^3\,s^{-1}$. The area, during drier periods, usually serves as farmland of approximately $12\,000\,ha$ for a local population of about 157 800. The water level at Wangjiaba station is a key flooding

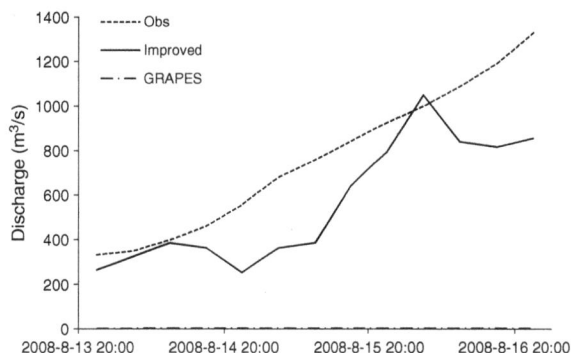

Figure 6. Observed and model flow discharges from 13 to 16 August 2008 at Wangjiaba Station.

indicator for the entire catchment. It is therefore important to obtain reliable forecasting of discharge at Wangjiaba station. Seen from Figure 6, the accuracy of runoff flow simulation has been greatly improved. This is shown that the improved hydrological process in GRAPES_MESO model has the ability of the simulation on the runoff flow.

4. Conclusions

The Noah-LSM in GRAPES_MESO model does not consider the heterogeneity distribution within grid cell. For example, infiltration capacity is usually assumed to be homogeneous within a grid cell, but it is likely highly heterogeneous within a cell. Land surface models usually lack sophisticated runoff concentration, because hydrological module is too simple. In this study, we improved the hydrological process modelling in the Noah-LSM model. First, we incorporated saturation-excess runoff generation module, which considers the change of runoff-yielding area based on a storage capacity curve. Second, we introduced a routing module based on the Muskingum method into the model to consider the redistribution of soil moisture in two-dimensional level and to improve the simulation accuracy of the surface runoff. To evaluate the performance of the improved model, we applied both original and improved models to conduct simulations from August to September 2009 and further investigated the feedback of changes in the land surface water modelling to meteorological simulation. The results show that the revised land surface water modelling has affected the results of simulated meteorological variables such as soil moisture, air temperature, and wind speed and generally led to better results. This suggests that the model improvement conducted in this study is valid and effective. It also highlights the important connection between surface hydrology and meteorology.

Acknowledgements

This work was funded by the National Natural Science Foundation of China (grants no. 41105068, 51509043, 91537211) and the 1st Young Talents Program project of China Meteorological

Administration (2014-2017). The meteorological data were accessed through China Meteorological Administration. The hydrological data were provided by Ministry of Water Resources.

References

Abbott MB, Bathurst JC, Cunge JA, O'Connell PE, Rasmussen J. 1986. An introduction to the European Hydrological System-Systeme Hydrologique Europeen, "SHE", 1: history and philosophy of a physically-based, distributed modelling system. *Journal of Hydrology* **87**(1–2): 45–59.

Andersen J, Dybkjaer G, Jensen KH, Refsgaard JC, Rasmussen K. 2002. Use of remotely sensed precipitation and leaf area index in a distributed hydrological model. *Journal of Hydrology* **264**(1): 34–50.

Bao HJ. 2009. Coupling EPS-hydrologic-hydraulic model for flood forecasting. PhD thesis, Department of Hydrology and water resources, Hohai University, Nanjing, China (in Chinese).

Bates PD, De Roo A. 2000. A simple raster-based model for flood inundation simulation. *Journal of Hydrology* **236**: 54–77.

Beven KJ, Kirkby MJ. 1979. A physically based variable contributing area model of basin hydrology. *Hydrological Sciences Bulletin* **24**: 43–69.

Cao LJ, Liu JM. 2005. Review of land-surface hydrological process studies. *Meteorological Science and Technology* **33**(2): 97–103.

Chen F, Jimy D. 2001. Coupling an advanced land surface-hydrology model with the Penn State–NCAR MM5 modeling system. Part I: model implementation and sensitivity. *Monthly Weather Review* **129**: 569–585.

Chen DH, Shen XS. 2006. Recent progress on GRAPES research and application. *Journal of Applied Meteorological Science* **17**(6): 773–777 (in Chinese).

Chen F, Kenneth M, John S, Xue Y, Pan H, Victor K. 1996. Modeling of land-surface evaporation by four schemes and comparison with FIFE observations. *Journal of Geophysical Research* **101**(D3): 7251–7268.

Chen DH, Xue JS, Yang XS, Zhang HL, Shen XS, Hu JL, Wang Y, Ji LR, Chen JB. 2008. New generation of multi scale NWP system (GRAPES): scientific design. *Chinese Science Bulletin* **53**(20): 2396–2407 (in Chinese).

Chow VT. 1959. *Open Channel Hydraulics.* McGraw-Hill: New York, NY.

Ek MB, Mitchell KE, Lin Y, Rogers E, Grunmann P, Koren V, Gayno G, Tarpley JD. 2003. Implementation of Noah land surface model advances in the National Centers for Environmental Prediction operationalmesoscale Eta model. *Journal of Geophysical Research* **108**: 1–16.

Entekhabi D, Asrar GR, Betts AK, Beven KJ, Rafael L, Duffy CJ. 1999. An agenda for land surface hydrology research and a call for the second international hydrological decade. *Bulletin of the American Meteorological Society* **80**: 2043–2058.

Fread DL. 1993. Chapter 10: Flow routing. In *Hand Book of Hydrology*, Maidment DR (ed). McGraw-Hill: New York, NY; 1–36.

Huang Z. 2001. *Weather Forecasting Product Quality Rating System.* Department of Prediction and Disaster Mitigation, China Meteorological Administration: Beijing; 9–11.

Jacquemin B, Noilhan J. 1990. Sensitivity study and validation of a land surface parameterization using the HAPEX-MOBILHY data set. *Boundary-Layer Meteorology* **52**: 93–134.

Jiang X, Niu G-Y, Yang Z-L. 2009. Impacts of vegetation and groundwater dynamics on warm season precipitation over the Central United States. *Journal of Geophysical Research* **114**: D06109, doi: 10.1029/2008JD010756.

Li ZJ, Yao C, Xu, PZ. 2006. Application of GIS-based hydrological models in humid watersheds. Water for life: surface and ground water resources, In *Proceedings of the 15th APD-IAHR & ISMH*, Madras, India, 685–690.

Li ZJ, Bao HJ, Xue CS, Hu YZ, Fang H. 2008. Real-time flood forecasting of Huai River with flood diversion and retarding areas. *Water Science and Engineering* **1**(2): 10–24.

Mahrt L, Ek M. 1984. The influence of atmospheric stability on potential evaporation. *Journal of Climate and Applied Meteorology* **23**: 222–234.

Mahrt L, Pan HL. 1984. A two-layer model of soil hydrology. *Boundary-Layer Meteorology* **29**: 1–20.

Noilhan J, Planton S. 1989. A simple parameterization of land surface processes for meteorological models. *Monthly Weather Review* **117**: 536–549.

Pan H-L, Mahrt L. 1987. Interaction between soil hydrology and boundary-layer development. *Boundary-Layer Meteorology* **38**: 185–202.

Peters-Lidard CD, Blackburn E, Liang X, Wood EF. 1998. The Effect of soil thermal conductivity parameterization on surface energy fluxes and temperatures. *Journal of the Atmospheric Sciences* **55**: 1209–1224.

Quinn P, Beven K, Planchon O. 1991. The prediction of hillslope flow paths fordistributed hydrological modeling using digital terrain models. *Hydrological Processes* **5**(1): 59–79.

Roesch A, Wild M, Gilgen H, Ohmura A. 2001. A new snow cover fraction parameterization for ECHAM4 GCM. *Climate Dynamics* **17**: 933–946, doi: 10.1007/s003820100153.

Schaake JC, Koren VI, Duan QY, Mitchell K, Chen F. 1996. A simple water balance model (SWB) for estimating runoff at different spatial and temporal scales. *Journal of Geophysical Research* **101**: 7461–7475.

Tesfa TK, Tarboton DG, Chler DG, Mcnamara JP. 2009. Modeling soil depth from topographic and land cover attributes. *Water Resources Research* **45**(10): 1277–1278.

Tewolde MH, Smithers JC. 2006. Flood routing in ungauged catchments using Muskingum methods. *Water SA* **32**(3): 379–388.

Williams JR, Ouyang Y, Chen J, Ravi V. 1998. Estimation of Infiltration Rate in Vadose Zone: Application of Selected Mathematical Models. ALM (METI) Contract No. 68-W5-0011 DO 022. Dynamac Corp. 68-C4-0031.

Wu XJ, Jin ZY, Huang LP, Chen DH. 2005. The software framework and application of GRAPES model. *Journal of Applied Meteorological Science* **16**(4): 539–546.

Xue JS, Chen DH. 2008. *Scientific Design and Application of Numerical Prediction System GRAPES.* Science Press: Beijing; 334–335.

Yao C, Li ZJ, Yu ZB, Zhang K. 2012. A priori parameter estimates for a distributed, grid-based Xinanjiang model using geographically based information. *Journal of Hydrology* **10**: 47–62.

Zhao RJ. 1983. *Watershed Hydrological Model-Xinanjiang Model and Shanbei Model.* Water Power Press: Beijing; 23–35 (in Chinese).

Zhao RJ. 1992. The Xinanjiang model applied in China. *Journal of Hydrology* **135**: 371–38.

Zhao JH, Li YH. 2006. Summarization of sand-dust prediction based on application of GRAPES – SDM. *Arid Meteorology* **24**(1): 7–13.

Zhao RJ, Liu XR. 1995. The Xinanjiang model. In *Computer Models of Watershed Hydrology*, Singh VP (ed). Water Resources Publications: Highlands Ranch, CO; 215–232.

Zhu ZD, Duan YH, Chen DH. 2007. Ananalysis of GRAPES-TCM'S operational experiment results. *Meteorological Monthly* **33**(7): 44–54.

15

A counterexample of aerosol suppressing light rain in Southwest China during 1951–2011

Jian Wu,[1]* Caiyun Ling,[1] Deming Zhao[2] and Bin Zhao[3]

[1] Department of Atmospheric Science, Yunnan University, Kunming, China
[2] Key Laboratory of Regional Climate-Environment for Temperate East Asia, Institute of Atmospheric Physics, Chinese Academy of Sciences, Beijing, China
[3] Meteorological Bureau of Tengchong, China

*Correspondence to:
J. Wu, Department of
Atmospheric Science, Yunnan
University, Kunming 650091,
China.
E-mail: wujian@ynu.edu.cn

Abstract

Surface meteorological observation data and aerosol optical depth (AOD) data from 1951–2011 were analyzed for Tengchong city, which is a clean city in Southwest China. A significant reduction of light rain, accompanied by an increase of visibility and a decrease of AOD, was observed. Thus, the observed light rain reduction in Tengchong was not associated with an increase of aerosols. The main cause of the reduction in light rain was the decrease of relative humidity in the layer between 850 to 500 hPa, which was induced mainly by the increase of temperature. This counterexample indicates that there are some evidences on the short-term scale, but the depression of light rain by aerosols on the long-term scale remains controversial.

Keywords: light rain; aerosol; warming; water vapor; Tengchong

1. Introduction

In recent years, increases in precipitation were found primarily for heavy and extreme precipitation events in the United States (Karl and Knight, 1998), Europe (Klein Tank and Können, 2003), Southeast Asia and the South Pacific (Manton et al., 2001), and China (Zhai et al., 2005). At the same time, reduction of light rain is an aspect of climate change, which has important effects on drought and agriculture. Light rains decreased in China (Gong et al., 2004; Fu et al., 2008), Europe, North America, and Asia (Qian et al., 2010; Huang and Wen, 2013). Moreover, the most distinct reductions of light rain, in terms of amount and days, were for lower intensity light rains (Qian et al., 2007; Liu et al., 2011).

A decrease of cloud droplet radius accompanied by an increase of cloud droplet numbers was observed in cases of aerosol pollution (Warner and Twomey, 1967), and observed aerosol depression of precipitation has been reported (Zhao, et al., 2006; Rosenfeld et al., 2007). In addition, a significant increase of the cloud droplet number concentration and a reduction of droplet sizes, which led to significant reductions in rainfall frequency and amount, were found by simulations (Qian et al., 2009). Short-term simulation from a bin and bulk microphysics model also identified a similar mechanism (Fan et al., 2012). In addition, the water vapor holding capacity of the atmosphere increases in warmer environment compared to that in colder environment according to the Clausius–Clapeyron equation (Trenberth et al., 2003), which means that the dew-point temperature is harder to achieve in a warming environment with stationary water vapor content. Light rain is the transition rainfall grade between stronger rainfall and no rainfall and should be more sensitive to the changes of temperature and water vapor than other stronger rainfall grades. Recently, the warming and the change of water vapor content were identified as two important factors for the decrease of light rain in Eastern China, but the influence of aerosols on the long-term reduction of light rain remained uncertain in this region due to the simultaneous occurrence of severe air pollution, warming, and changes of water vapor content (Wu et al., 2015).

While light rain reduction has been found in many regions of the world (Qian et al., 2010; Huang and Wen, 2013), aerosol optical depth (AOD) was not always high in regions with light rain decreases. Other research has shown distinct spatial variations in the AOD in China and that the significant increase of aerosol occurred mainly after 1996 (Guo et al., 2011), but the light rain reduction in China began in the 1960s (Fu et al., 2008). These observations have raised questions about the role of aerosols in suppressing light rain on the long-term scale.

In this paper, we present a long-term diagnostic analysis of aerosols and light rain reduction in Tengchong city in Southwest China to examine this proposition. The data and methods are presented in Section 2 and the results and analysis in Section 3, followed by the conclusions.

2. Data and methods

Tengchong is a small city located in the western part of Yunnan province, which is a plateau in the low-latitude belt in Southwest China, and the

Tengchong meteorological station (98°30′E, 25°01′N, at the surface height of 1654 m above sea level) belongs to the China Meteorological Administration (CMA). The earliest meteorological records in Tengchong date from the 1930s, and standard daily observations date back to 1951. In recent decades, the ecological environment in Tengchong city has been protected very well, and no significant anthropogenic air pollutant emissions or wide-scale land use and cover changes have occurred in the city and adjacent regions. While the Tengchong meteorological station was relocated twice, in 1956 and 1987, it moved only 600 and 150 m, respectively, which should not result in significant differences in the long-term rainfall and temperature data. These characteristics of the Tengchong station permit examination of the background climate changes.

The daily surface observation data for 1951–2011, including rainfall amount, visibility, cloud fraction, relative humidity (RH) and temperature, and the daily radiosonde data for 1984–2013, including temperature and specific humidity from 850 to 500 hPa, were used in this study. The dataset observed by the CMA observation network including the Tengchong station passed the homogeneity and quality tests and was therefore regarded as the most credible dataset in China (Song *et al.* 2004). The CMA rainfall grade standards were used in our analysis, in which measured daily rain rates were classified into five grades of intensity: light ($0.1–10$ mm d^{-1}); moderate ($10–25$ mm d^{-1}); heavy ($25–50$ mm d^{-1}); storm ($50–100$ mm d^{-1}); and downpour (>100 mm d^{-1}).

Visibility is often used as an indicator of atmospheric purity and can be influenced by the extinction effects of humid air, aerosols, and some types of air pollutants. Daily visibility before 1980 in China was recorded according to 10 ranks based on distance, and

the recorded data have been supplanted by the distance of visibility since 1980. The method introduced by Qin *et al.* (2010) was used to reconcile the two sets of visibility data by converting the daily visibility data during 1980–2010 to the 10 ranks used before 1980. The mean distance of each visibility rank was then calculated, and these mean distances were subsequently used to replace each visibility rank record before 1980. To identify the extinction by aerosols and its effects on light rain well, the sunny visibility data were selected according to the following three conditions: (1) observation time must be at 0600 UTC; (2) data with near-surface RH more than 70% were excluded; and (3) data with total cloud cover more than 40% were rejected (Wu *et al.*, 2012). In addition, the daily AOD data with $1° \times 1°$ resolution observed by the moderate resolution imaging spectro-radiometer (MODIS) over Tengchong during 2001–2014 were used to analyze the influences of aerosols on the light rain changes.

Composite analysis was used to determine the impacts of aerosols, temperature, and water vapor content on light rain amount. Years of abnormal meteorological parameters were determined in the composite analysis based on the principle that the time series of the meteorological parameters should be normalized; then, if the value was more (less) than 1 (-1), its corresponding year was deemed as a positive (negative) abnormal year (Wu *et al.*, 2015). Additionally, Student's *t*-test was used to determine the significance of the data. The linear trend coefficient was calculated using the least-squares method.

3. Results and analysis

The temporal changes of different rainfall grades in Tengchong during 1951–2011 are shown in Figure 1.

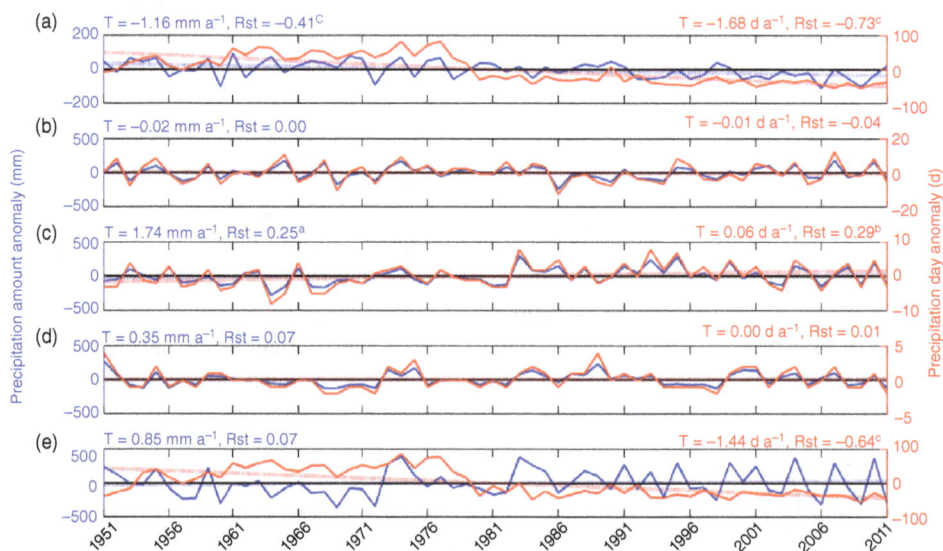

Figure 1. Temporal changes of anomalies for annual four-grade rainfall amounts and rainfall days in Tengchong city (from top to bottom: (a) light rain, (b) moderate rain, (c) heavy rain, (d) storm, (e) total rainfall). T and Rst stand for the climatic tendency and the linear trend coefficient, respectively. Superscripts [a], [b], and [c] refer to the correlation coefficients statistically significant at the 90, 95, and 99% levels, respectively.

There was a significant decrease in the amount and days of light rain during the recent 50 years (Figure 1(a)), and the linear trend coefficients were significant according to the t-test at the 99% level. The decreases of light rainfall amounts and light rain days were more significant after the middle of the 1980s than before. However, there were no statistically significant trends in both the moderate and the storm rainfall amounts and days (Figure 1(b) and (d)), but the heavy rainfall amounts and days increased significantly during 1951–2011 (Figure 1(c)). In addition, the total rainfall amount had a statistically insignificant increasing trend of 0.85 mm a^{-1} (Figure 1(e)), which was due to the increases of heavy and storm rainfall amounts. However, the total rainfall days decreased at the rate of -1.44 d a^{-1}, which was statistically significant at the 99% level, and the distinct decrease of the total rainfall days could be attributed to the significant reduction of light rain days at the rate of -1.68 d a^{-1}. Similar reductions of light rain in the background of the total rainfall increase were also found in many other regions (Gong et al., 2004; Qian et al., 2007; Fu et al., 2008).

The temporal changes of visibility, sunny visibility, and AOD are shown in Figure 2. The visibility and the sunny visibility showed increasing trends of 0.07 and 0.05 km a^{-1}, respectively, during 1951–2011 and were statistically significant according to the t-test at the 99% level. The MODIS AOD data were available only after June 2000 and showed a weakly decreasing trend during 2001–2014. The increases of visibility and sunny visibility and the decrease of AOD indicated that the air quality of Tengchong was well protected in the last 50 years. The correlation coefficients and the composite analyses between light rain amount and visibility, sunny visibility, visibility on light rain days, and AOD are shown in Tables 1 and 2, respectively. A statistically significant correlation coefficient was found between light rain amount and visibility only, and other correlation coefficients between light rain amount and sunny visibility, visibility on light rain days, and AOD were statistically indistinctive. The results of composite analysis revealed some insignificant visibility and AOD differences between the higher and lower light rainfall amount years. Therefore, the pronounced decreases of light rainfall amount and light rainfall days in the long-term observation data, especially after 1980, could not be explained by the increase of aerosol, as reported in other polluted regions (Zhao et al., 2006; Qian et al., 2009; Fan et al., 2012; Fu and Dan, 2014).

In earlier research, another mechanism of light rain reduction was reported (Wu et al. 2015), in which the light rain in many regions of Eastern China was distinctly affected by the low-level RH, which in turn was dominated by the warming and the change of water

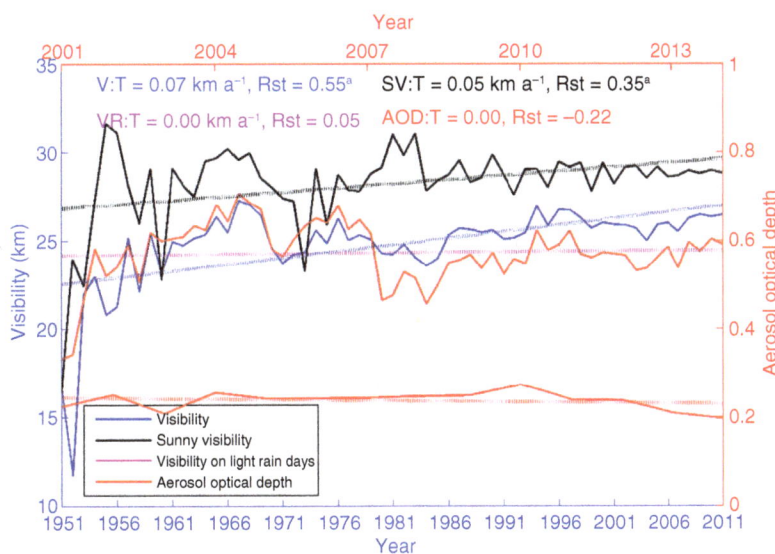

Figure 2. Temporal changes of annual visibility, sunny visibility, AOD, and their annual means on light rain days in Tengchong city. T and Rst stand for the climatic tendency and the linear trend coefficient, respectively. Superscript [a] refers to the correlation coefficients statistically significant at the 99% levels.

Table 1. Correlation coefficients between light rainfall amount and visibility, AOD, and RH in Tengchong city.

Light rainfall amount	Visibility	Sunny visibility	Visibility on light rain days	MODIS AOD	MODIS AOD on light rain days	RH	RH on light rain days
Correlation coefficient	-0.34^{b}	0.02	0.00	-0.24	-0.22	0.37^{a}	0.29

The period of available daily visibility data is 1951–2011, RH refers to the air mass-weight RH from 850 to 500 hPa during 1984–2011, and AOD data are available only for 2001–2011.
Superscripts [a] and [b] refer to the correlation coefficients and the differences of composite analysis statistically significant at the 90% and 99% levels, respectively.

Table 2. Composite analysis results between light rainfall amount and visibility, AOD, and RH in Tengchong city.

Light rainfall amount	Visibility	Sunny visibility	Visibility on light rain days	MODIS AOD	MODIS AOD on light rain days	RH	RH on light rain days
Correlation coefficient	−0.34[b]	0.02	0.00	−0.24	−0.22	0.37[a]	0.29
Results of composite analysis-1	−0.72 km	0.35 km	0.73 km	0.00	−0.01	6.02%[a]	2.20%
Results of composite analysis-2	30.93 mm	23.75 mm	2.16 mm	−10.60 mm	−4.42 mm	47.50 mm	52.11 mm

The period of available daily visibility data is 1951–2011, RH refers to the air mass-weight RH from 850 to 500 hPa during 1984–2011, and AOD data are available only for 2001–2011.

Results of composite analysis-1 indicate the differences of each factor in the first line between the positive and negative years of abnormal amounts of light rain, and results of composite analysis-2 indicate the differences in light rainfall amounts between the positive and negative abnormal years for each factor in the first line.

Superscripts [a] and [b] refer to the correlation coefficients and the differences of composite analysis statistically significant at the 90% and 99% levels, respectively.

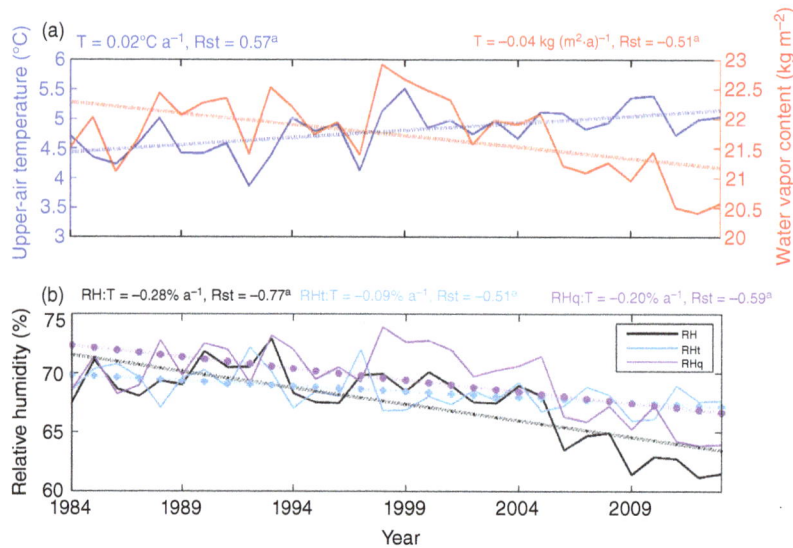

Figure 3. Temporal changes of air mass-weighted temperature and column water vapor (a) and air mass-weighted RH between 850 to 500 hPa, as well as relative contributions from changes in temperature (RH_t) and water vapor (RH_q) to the long-term change of RH in Tengchong city (b). T and Rst stand for the climatic tendency and the linear trend coefficient, respectively. Superscript [a] refers to the correlation coefficients statistically significant at the 99% levels.

vapor according to the Clausius–Clapeyron equation. The long-term decrease of the light rain amount in Tengchong showed different temporal characteristics from the air quality status. Thus the effects of warming and the change of water vapor on the light rain reduction were analyzed. The air mass-weighted RH between 850 to 500 hPa had a statistically significant correlation coefficient at the 90% level with the light rainfall amount during 1984–2011 (Table 1), and the difference of the annual average mass-weighted RH between years with high and low light rain amounts reached 6.02%, which was statistically significant at the 90% confidence level. Meanwhile, the difference of the annual light rain amount between years with high and low mass-weighted RH reached 47.50 mm, which was 12.5% of the annual mean light rain amount and 3.1% of the annual mean total rainfall amount during 1984–2011 (Table 2). The above-mentioned characteristics indicated that the changes of air mass-weighted RH between 850 to 500 hPa had important effects on the long-term reduction of the light rain amount in Tengchong.

The effects of warming and the change of water vapor content on the long-term changes of mass-weighted RH are shown in Figure 3. The air mass-weighted temperature between 850 to 500 hPa increased from 4.5 to 5.0 °C during 1984–2013, which included a rapid increase after the middle of the 1990s. Meanwhile, the column water vapor content in the same vertical range decreased from 22 to 21.5 kg m^{-2}, which included a rapid decrease after 2000 (Figure 3(a)). These two changes could affect the mass-weighted RH distinctly. Figure 3(b) shows the long-term changes of mass-weighted RH and the individual contributions from the warming (RH_t) and from the changes of water vapor content (RH_q), calculated using the Clausius–Clapeyron equation by constraining the column water vapor content and the temperature at their 1984 values, respectively. RH decreased from 68% in 1984 to 62% in 2013 with a decreasing rate of −0.28% a^{-1}, which was statistically significant at the 99% level. A fluctuation in RH was observed during 1984–1993, and the most distinct decrease was found after 2004. RH_t showed a constant decreasing trend during 1984–2013 with a linear trend of $-0.15\%\,a^{-1}$. At the same time, RH_q fluctuated during 1984–1998, followed by a distinct decrease after 1999, especially after 2004, and the linear trend of RH_q reached $-0.2\%\,a^{-1}$ during 1984–2013. The contribution from the linear trend of RH_q to RH was more significant than that of RH_t, and the decrease of RH during 1984–2013 was 5.97%, for which RH_t and RH_q accounted for 0.9 and 4.06%, respectively. In addition, the large decrease of RH after 2004 was mainly caused by the corresponding decrease of RH_q, and the constant decreasing trend of RH, especially during 1984–1999, can be explained by the steady reduction of RH_t because RH_q fluctuated during the same period. Thus, the warming and the reduction of column water vapor content are two factors inducing the light rain decrease in Tengchong during 1984–2013.

4. Summary

We examined the light rain reduction in Tengchong city in Southwest China during the period 1951–2011, when the air was consistently clean, as indicated by the weak increase of visibility over 60 years and a steady fluctuation of satellite AOD after 2001. It was obvious that the significant reductions of both the amounts and days of light rain in Tengchong could not be explained by an increase of aerosols, as was reported in other regions with high concentrations of air pollutants. At the same time, the air mass-weighted temperature from 850 to 500 hPa in the city was increasing markedly, whereas the column water vapor content in the same layers was decreasing, so the two factors decreased the air mass-weighted RH, which in turn improved conditions for condensation and thus decreased the occurrences of light rain. In conclusion, the effects of the changes in temperature and water vapor on the changes in light rain should be stressed, especially in clean air regions.

Acknowledgments

The surface meteorological data are available from the China Meteorological Data Sharing Service System, and the radiosonde data were arranged by the Meteorological Bureau of Tengchong. MODIS AOD data are available from NASA (http://modis.gsfc.nasa.gov/), and we thank all the data sources. This study was sponsored by the Chinese Natural Science Foundation (41275162,41305103) and the Chinese Academy of Sciences Strategic Priority Program under grant No. XDA05090206.

References

Fan J, Leung LR, Li Z, Morrison H, Chen H, Zhou Y, Qian Y, Wang Y. 2012. Aerosol impacts on clouds and precipitation in eastern China: Results from bin and bulk microphysics. *Journal of Geophysical Research* 117: D00K36, doi: 10.1029/2011JD016537.

Fu CB, Dan L. 2014. Trends in the different grades of precipitation over South China during 1960–2010 and the possible link with anthropogenic aerosols. *Advances in Atmospheric Sciences* 31(2): 480–491, doi: 10.1007/s00376-013-2102-7.

Fu J, Qian W, Lin X, Chen D. 2008. Trends of graded precipitation days in China from 1961 to 2000. *Advances in Atmospheric Sciences* 25(2): 267–278, doi: 10.1007/s00376-008-0267-2.

Gong D-Y, Shi P-J, Wang J-A. 2004. Daily precipitation changes in semiarid region over northern China. *Journal of Arid Environments* 59: 771–784, doi: 10.1016/j.jaridenv.2004.02.006.

Guo JP, Zhang XY, Wu YR, Zhaxi YZ, Che HZ, La B, Wang W, Li XW. 2011. Spatio-temporal variation trends of satellite-based aerosol optical depth in China during 1980–2008. *Atmospheric Environment* 45: 6802–6811, doi: 10.1016/j.atmosenv.2011.03.068.

Huang G, Wen GH. 2013. Spatial and temporal variations of light rain days over China and the mid-high latitudes of the northern hemisphere. *Chinese Science Bulletin* 58(12): 1402–1411.

Karl TR, Knight RW. 1998. secular trends of precipitation amount, frequency, and intensity in the United States. *Bulletin of the American Meteorological Society* 79(2): 231–241.

Klein Tank AMG, Können GP. 2003. Trends in indices of daily temperature and precipitation extremes in Europe, 1946–99. *Journal of Climate* 16: 3665–3680.

Liu B, Xu M, Henderson M. 2011. Where have all the showers gone? Regional declines in light precipitation events in china, 1960–2000. *International Journal of Climatology* 31(8): 1177–1191.

Manton MJ, Della-Marta PM, Haylock MR, Hennessy KJ, Nicholls N, Chambers LE, Collins DA, Daw G, Finet A, Gunawan D, Inape K, Isobe H, Kestin TS, Lefale P, Leyu CH, Lwin T, Maitrepierre L, Ouprasitwong N, Page CM, Pahalad J, Plummer N, Salinger MJ, Suppiah R, Tran VL, Trewin B, Tibig I, Yee D. 2001. Trends in extreme daily rainfall and temperature in Southeast Asia and the South Pacific: 1961–1998. *International Journal of Climatology* 21: 269–284, doi: 10.1002/joc.610.

Qian WH, Fu J, Yan Z. 2007. Decrease of light rain events in summer associated with a warming environment in China during 1961–2005. *Geophysical Research Letters* 34: L11705, doi: 10.1029/2007GL029631.

Qian Y, Gong D, Fan J, Leung LR, Bennartz R, Chen D, Wang W. 2009. Heavy pollution suppresses light rain in China: observations and modeling. *Journal of Geophysical Research* 114: D00K02, doi: 10.1029/2008JD011575.

Qian Y, Gong D, Leung LR. 2010. Light rain events change over North America, Europe, and Asia for 1973–2009. *Atmospheric Science Letters* 11: 301–306.

Qin SG, Shi GY, Chen L. 2010. Long-term variation of aerosol optical depth in China based on meteorological horizontal visibility observation. *Chinese Journal of Atmospheric Sciences* 34(2): 449–456, (in Chinese).

Rosenfeld D, Dai J, Yu X, Yao ZY, Xu XH, Yang X, Du CL. 2007. Inverse relations between amounts of air pollution and orographic precipitation. *Science* 315: 1396–1398.

Song F, Hu Q, Qian WH. 2004. Quality control of daily meteorological data in China, 1951–2000: a new dataset. *International Journal of Climatology* 24: 853–870.

Trenberth KE, Dai A, Rasmussen RM, Parsons DB. 2003. The changing character of precipitation. *Bulletin of the American Meteorological Society* 84(9): 1205–1217.

Warner J, Twomey S. 1967. The production of cloud nuclei by cane fires and the effect on cloud droplet concentration. *Journal of the Atmospheric Sciences* 24: 704–706.

Wu J, Fu CB, Zhang LY. 2012. Trends of visibility on sunny days in China in the recent 50 years. *Atmospheric Environment* 55: 339–346, doi: 10.1016/j.atmosenv.2012.03.037.

Wu J, Zhang L, Zhao D, Tang J. 2015. Impacts of warming and water vapor content on the decrease in light rain days during the warm season over eastern China. *Climate Dynamics* 45: 1841–1857, doi: 10.1007/s00382-014-2438-4.

Zhai P, Zhang X, Wan H, Pan X. 2005. Trends in total precipitation and frequency of daily precipitation extremes over China. *Journal of Climate* 18(7): 1096–1108.

Zhao C, Tie X, Lin Y. 2006. A possible positive feedback of reduction of precipitation and increase in aerosols over eastern central China. *Geophysical Research Letters* 33: L11814, doi: 10.1029/2006GL025959.

16

Interdecadal circumglobal teleconnection pattern during boreal summer

Bo Wu,[1,2]* Jianshe Lin[1,3] and Tianjun Zhou[1,2]

[1] State Key Laboratory of Numerical Modeling for Atmospheric Sciences and Geophysical Fluid Dynamics, Institute of Atmospheric Physics, Chinese Academy of Sciences, Beijing, China
[2] Joint Center for Global Change Studies (JCGCS), Beijing, China
[3] University of Chinese Academy of Sciences, Beijing, China

*Correspondence to:
B. Wu, State Key Laboratory of Numerical Modeling for Atmospheric Sciences and Geophysical Fluid Dynamics, Institute of Atmospheric Physics, Chinese Academy of Sciences, No. 40 Huayanli, Chaoyang District, Beijing, China.
E-mail: wubo@mail.iap.ac.cn

Abstract

A new atmospheric teleconnection pattern on the interdecadal time scale, interdecadal circumglobal teleconnection pattern (ID-CGT pattern), is identified. The ID-CGT pattern dominates the interdecadal variability of the Northern Hemisphere (NH) extratropical circulation during boreal summer. The ID-CGT pattern has a great climate effect and can cause an alternatively positive and negative pattern of the NH mid-latitude land surface temperature anomalies. Compared with its counterpart on the interannual time scale (IA-CGT pattern), the ID-CGT pattern shows following three distinctive features. Firstly, although both ID-CGT and IA-CGT patterns show zonal wavenumber-5 structures, the former nodes shift westward relative to the latter by about 1/4 wavelength. Secondly, all the five nodes of the ID-CGT pattern possess barotropic structures, unlike the IA-CGT, which has a baroclinic node. Thirdly, the ID-CGT pattern is associated with the Atlantic multi-decadal oscillation, whereas the IA-CGT pattern is with the Indian summer monsoon precipitation anomalies.

Keywords: decadal variability; atmospheric teleconnection; AMO

1. Introduction

Circumglobal teleconnection (CGT) pattern is one of two dominant modes of the Northern Hemisphere (NH) extratropical upper-tropospheric atmospheric circulation anomalies during boreal summer on the interannual time scale (Ding and Wang, 2005). It exhibits a wave-like structure with wavenumber 5 and all the five nodes are confined within the waveguide associated with the NH westerly jet stream (Ding and Wang, 2005; Ding et al., 2011; Wang et al., 2012). The CGT pattern is excited by the Indian summer monsoon rainfall anomalies (Lin, 2009; Ding et al., 2011) and maintained by extratropical atmospheric internal dynamics (Yasui and Watanabe, 2010; Ding et al., 2011). The CGT pattern has been identified as one of the major predictability sources of the NH extratropical atmospheric circulation (Wang et al., 2009; Lee et al., 2010, 2011). However, the studies of the CGT pattern are so far confined to the interannual or shorter time scales.

Recently, a wave-like circulation pattern along the NH mid-latitude westerly jet stream was found to be associated with the interdecadal variability of the East Asian summer monsoon (Wu et al., 2016). The result inspired us to investigate whether the CGT pattern exists on the interdecadal time scale. Furthermore, if the interdecadal CGT (hereafter ID-CGT) pattern exists as hypothesized, what differences are there with the conventional CGT pattern on the interannual time scale (hereafter IA-CGT pattern)? In this study, we

show evidences that the ID-CGT pattern is an intrinsic mode of the interdecadal variability of extratropical atmospheric circulations.

2. Data and method

Datasets used in the study include: (1) circulation fields from the Twentieth Century Reanalysis (20CR) dataset for the period of 1920–2012 (Compo et al., 2010); (2) observational sea surface temperature (SST) from the HadISST 1.1 dataset for the period of 1920–2012 (Rayner et al., 2003); and (3) observational land surface temperature from the University of Delaware's Air Temperature database (v3.01) for the period of 1920–2010 (Willmott and Matsuura, 2012).

We identify the ID-CGT pattern through an objective and consistent way with that for the IA-CGT pattern. Major analysis processes include following three steps. Firstly, the Lanczos filtering method is used for time scale separation. A total of 8-year low-pass (high-pass) filtering is applied to the raw data to obtain low-frequency (LF) [high-frequency (HF)] fields. We also try filtering with window widths of 7 and 9 years. The results are not sensitive to the selection of window width (figure not shown). In addition, variability associated with the nearly global mean SST (60°S–60°N), which is the proxy of the global warming, are removed explicitly in the LF fields through a linear regression method (Ting et al., 2009). Secondly, an EOF analysis is applied to the June–September (JJAS) mean

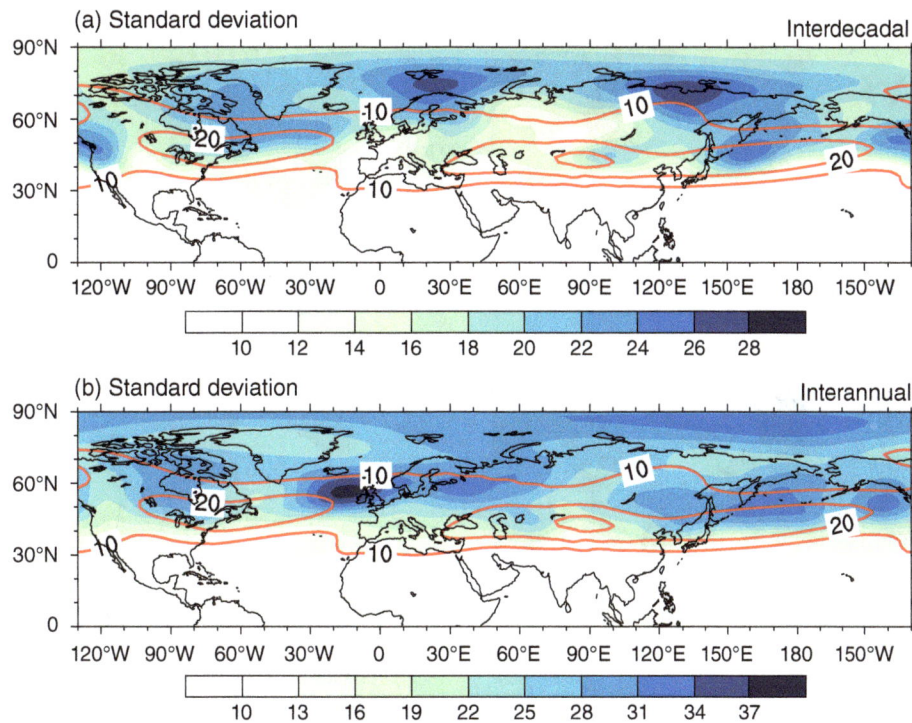

Figure 1. (a) Standard deviation of JJAS-mean 8-year low-pass filtered Z200, in which the signals associated with the global warming are removed through a regression analysis (shading, units: m) and climatological 200 hPa zonal wind (contour, units: m s^{-1}). (b) As in (a), but for 8-year high-pass filtered Z200.

LF and HF 200 hPa geopotential height (Z200) in the latitudinal band of 0°–70°N, respectively. The first mode of LF Z200 and the second mode of HF Z200 represent the ID-CGT and IA-CGT pattern, respectively, as shown below. Thirdly, other variable anomalies associated with the ID-CGT and IA-CGT patterns are obtained through regressing onto the corresponding principal component (PC) time series.

The statistical significance of the regression analyses is estimated using a 'random-phase' test developed by Ebisuzaki (1997), which is constructed based on a Monte Carlo simulation (details are in the Appendix S1, Supporting Information).

3. Results

Figure 1 shows the spatial distributions of the standard deviation of the LF and HF Z200. Both the LF and HF Z200 have a strong wave-like oscillation band along the jet stream, approximate 30°–70°N. However, oscillation centers of the Z200 are different between the two time scales, suggesting that LF and HF mid-latitude circulation anomalies are distinct. The magnitude of the LF standard deviation is about 70% of the HF over the band of 30°–70°N.

We applied EOF analyses to LH and HF Z200 over the NH (0°–70°N), respectively. For the interdecadal (interannual) time scale, the first (second) EOF mode, which account for 48.5% (13.7%) of total LF (HF) variance, is dominated by a zonal wave train along the NH jet stream (Figures 2 and S1). Hence, we take

the first (second) EOF mode of the LF (HF) Z200 as the ID-CGT (IA-CGT) pattern, and corresponding PC time series as the time series of the ID-CGT (IA-CGT) pattern.

Both the ID-CGT and IA-CGT patterns have five nodes. For the ID-CGT pattern, the five nodes are located at the eastern coast of North America, eastern Europe, northern East Asia, western North Pacific and western coast of North America, generally corresponds to the centers of the LF Z200 standard deviation (Figures 1(a) and 2(a)). In contrast, for the IA-CGT pattern, the five nodes are located at eastern North Atlantic, Caspian Sea, northeastern East Asia, central North Pacific and eastern North America. The correspondence between the nodes of the IA-CGT pattern and the centers of the HF Z200 standard deviation is much lower than that on the interdecadal time scale, especially for the standard deviation centers over the North Atlantic and Europe (Figures 1(b) and 2(b)). The spatial distributions of the contributions of the ID-CGT and IA-CGT patterns to the LF and HF variances are investigated (Figure 2(c) and (d)). All the nodes of the ID-CGT pattern account for more than 50% of the local LF variances. In contrast, for the IA-CGT pattern, only nodes over the East Asia and central North Pacific account for more than 40% of the local HF variances. We also estimate the variance contributions of the two modes to the raw (unfiltered) Z200 (Figure 2(e) and (f)). The spatial distributions of the variance contributions for the raw data are generally consistent with those for the filtered data, but magnitudes are much smaller (Figure 2(c)–(f)). The ID-CGT (IA-CGT) accounts for

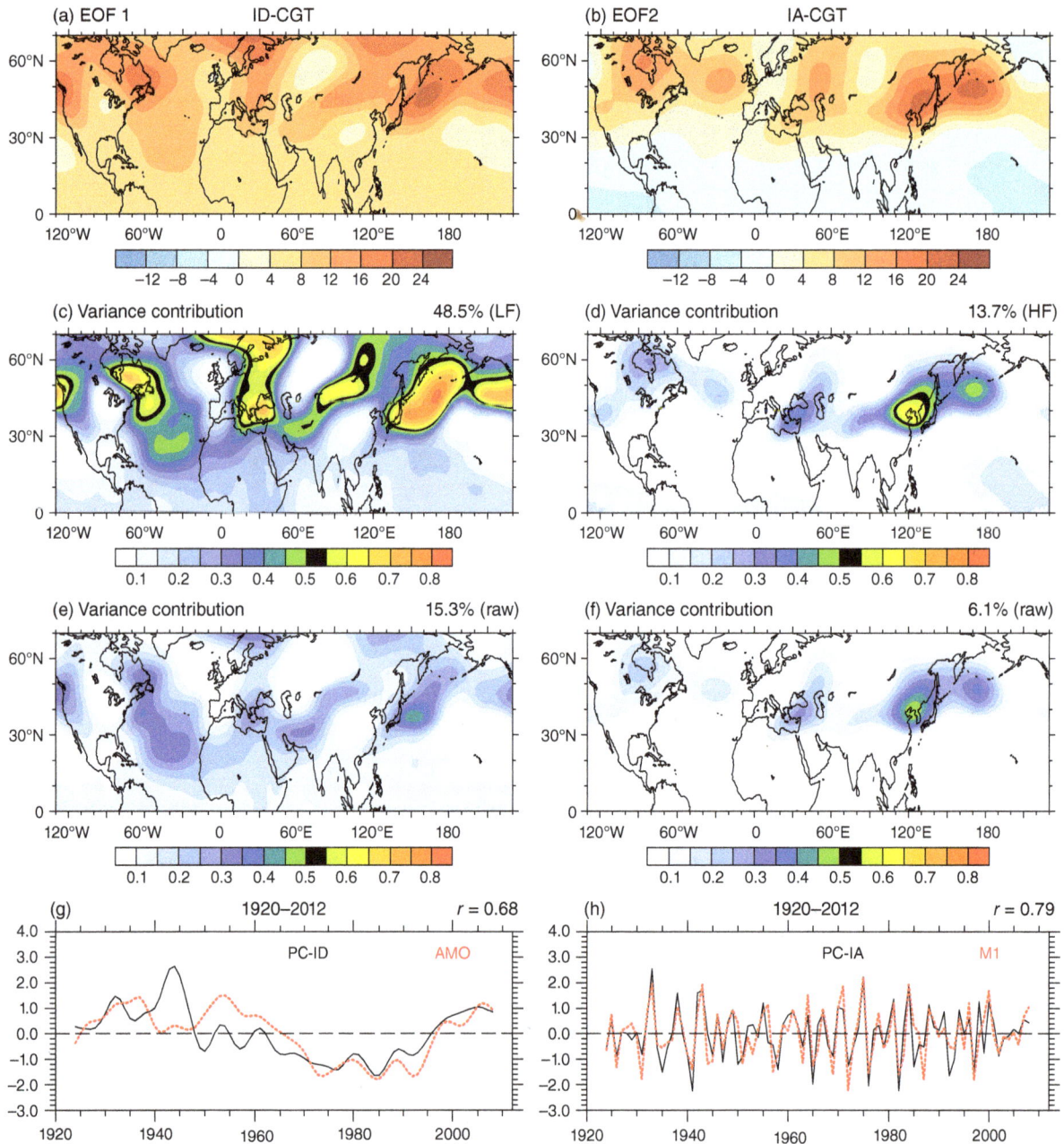

Figure 2. (a) The first EOF mode of the JJAS-mean NH Z200 on the interdecadal time scale. (c) Contribution of the mode to the variance of the Z200 on the interdecadal time scale. (e) Contribution of the mode to the total variance of the raw (unfiltered) Z200. (b, d, f) As in (a, c, e), but for the second EOF mode on the interannual time scale. (g and h) Corresponding PC time series of the two modes (solid black), AMO index and IA-CGT time series derived from the MCA analysis (dotted red lines).

15.3% (6.1%) of the total variances of the NH raw Z200 (0°–70°N).

It is worth noting that the IA-CGT pattern is obtained through a maximum covariance analysis (MCA) in Ding *et al.* (2011). To check the reliability of the results derived from the EOF, we also conducted the MCA using 20CR, which extracts dominant modes explaining the greatest covariance between HF NH (0°–90°N) Z200 and HF tropical (15°S–30°N) precipitation. The obtained first MCA mode corresponds well to the second EOF mode of the HF Z200, with correlation coefficient between their time series reaching 0.79 (Figure 2(h)), which is statistically significant at

the 5% level. In addition, the IA-CGT pattern derived from the 20CR for the period of 1920–2012 by using the two different methods generally resemble the CGT pattern in Ding *et al.* (2011) for the period 1948–2009, indicating that the 20CR is reliable in the extra-tropical atmospheric circulation.

To investigate the differences in the three-dimensional structure between the ID-CGT and IA-CGT patterns, we show the corresponding 200, 500 and 850 hPa geopotential height anomalies (Figure 3). For the IA-CGT pattern, four out the five nodes exhibit barotropic structures, except for the node extending from north of the Indian peninsula northeastward to the

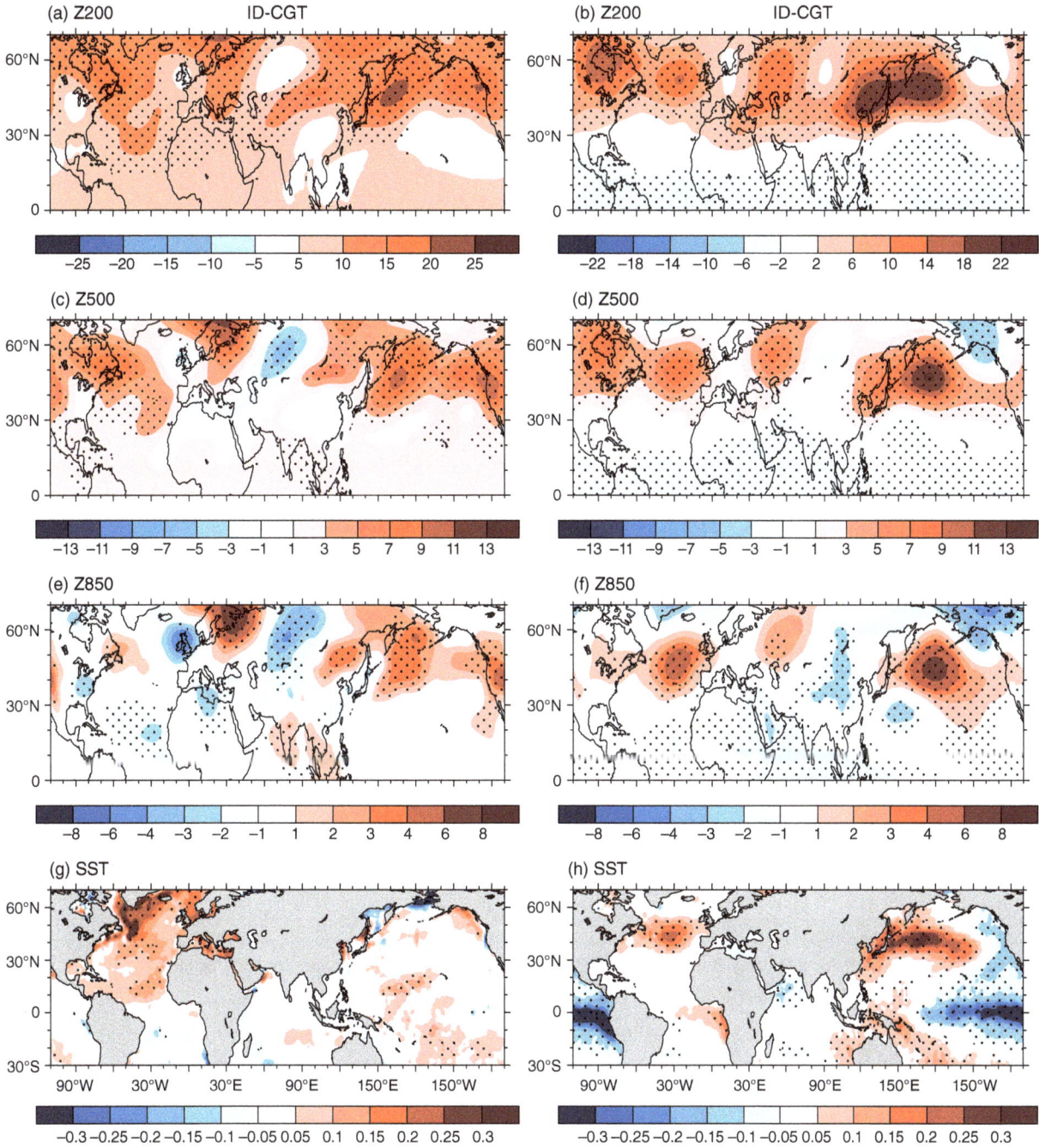

Figure 3. Left panel: (a) 200 hPa, (c) 500 hPa and (e) 850 hPa geopotential height and (g) SST anomalies associated with the ID-CGT pattern. They are obtained through regressing 8-year low-pass filtered variables on the corresponding PC time series. Dots denote values passing 10% significance level. Right panel (b, d, f, h): as in the left panel, but for the IA-CGT pattern.

northern East Asia (right panel of Figures 3 and S1(f)), consistent with the CGT pattern in Ding and Wang (2005) and Ding *et al.* (2011). On the basis of this characteristic, Ding *et al.* (2011) proposed that the CGT pattern is excited by the Indian summer monsoon precipitation anomalies during La Nina developing phase (Figure 3(h)), and then maintained by barotropic instability of the mid-latitude basic flow and propagation of wave energy along the wave guide associated with the westerly jet stream (Figure S1(b) and (d)). In the 20CR, the correlation coefficient of the IA-CGT time series with the contemporary observational All-Indian

Rainfall Index (AIRI) (data from ftp://www.tropmet. res.in/pub/data/rain/iitm-regionrf.txt, Kothawale *et al.*, 2006) is 0.44, reaching the 5% significance level.

The wave energy associated with the ID-CGT pattern also propagates eastward along the wave guide associated with the NH stream jet, consistent with the IA-CGT pattern (Figure S1(c) and (d)). However, the vertical structure of the ID-CGT pattern is different from that of the IA-CGT pattern. All five nodes of the ID-CGT pattern exhibit barotropic structures (left panel of Figures 3 and S1(e)), suggesting that the ID-CGT pattern is excited by different mechanisms from the

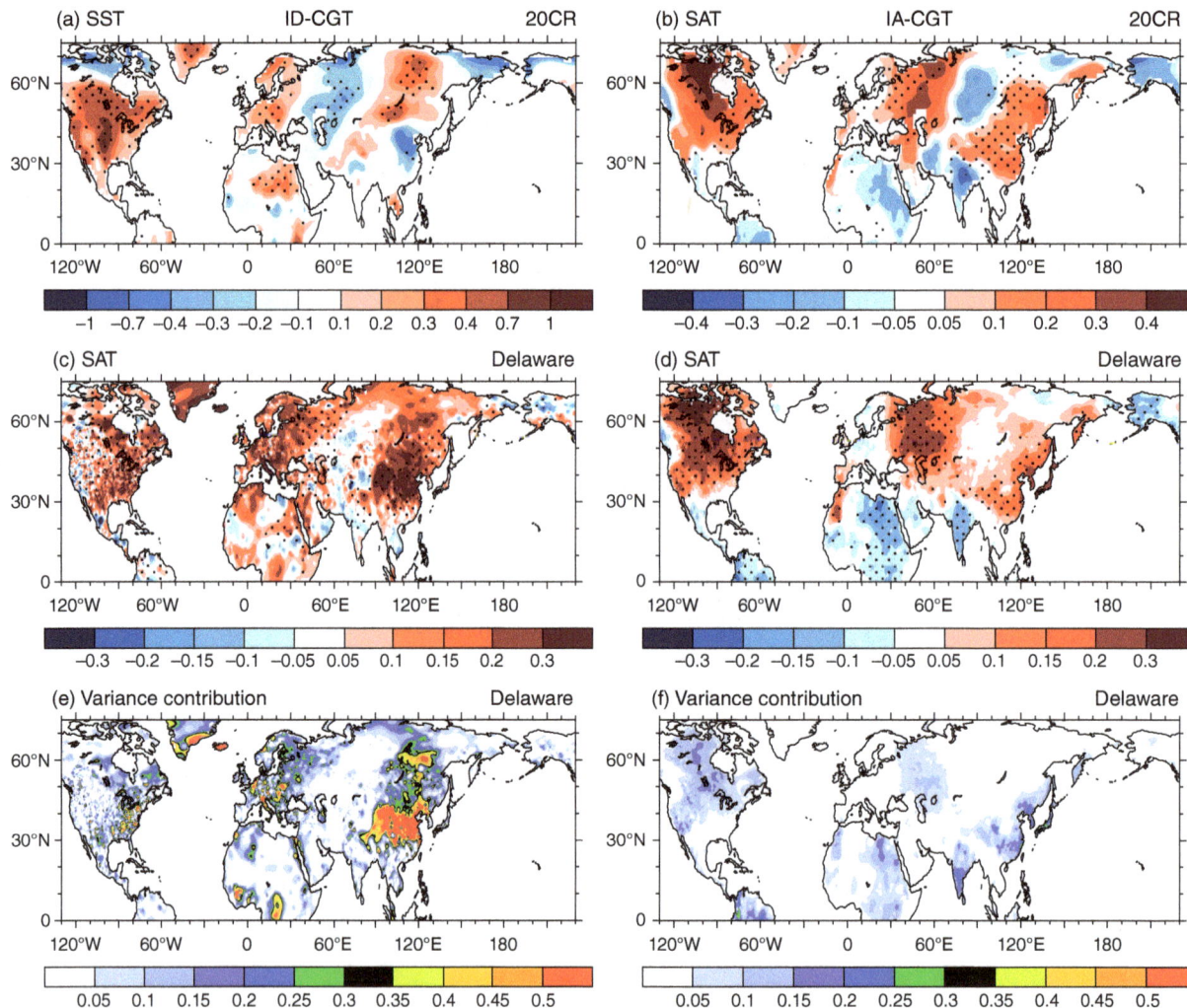

Figure 4. (a, c) Land surface air temperature anomalies associated with the ID-CGT pattern, derived from the 20CR and Delaware datasets, respectively. (e) Percentage of the LF variances of the land SAT associated with the ID-CGT mode, derived from the Delaware dataset. (b, d, f) As in (a, c, e), but for the IA-CGT pattern. Dots denote values passing 10% significance level.

IA-CGT pattern. To investigate what forcing factors responsible for the ID-CGT pattern, we show the associated SST anomalies (SSTAs) (Figure 3(g)). The North Atlantic is covered by basin scale warm SSTAs, which have two centers, located in the Labrador Sea and the tropical North Atlantic, respectively, resembling the spatial pattern of the Atlantic multi-decadal oscillation (AMO) (e.g. Kushnir, 1994; Kerr, 2000; Delworth *et al.*, 2007). Meanwhile, the time series of the ID-CGT pattern is highly correlated with the AMO index, with correlation coefficient reaching 0.68, passing 5% significance test (Figure 2(g)). Here the AMO index is defined as JJAS-mean LF area-averaged SST in the North Atlantic ($0°-60°N$, $0°-80°W$) minus nearly global mean SST ($60°S-60°N$), following Trenberth and Shea (2006).

The IA-CGT pattern tends to influence climate along its path (Ding *et al.* 2011). We compare the land surface air temperature (SAT) anomalies associated with the IA-CGT and ID-CGT patterns (Figure 4). The land SAT anomalies in the NH middle latitudes derived from the 20CR show alternatively positive and negative

anomalies for both the IA-CGT and ID-CGT patterns. The warm (cold) anomalies generally correspond to overlying anticyclone (cyclone) anomalies, although their spatial phases zonally shift slightly. As a result, the wave-like SAT anomalies associated with the ID-CGT pattern shift westward relative to those associated with the IA-CGT pattern by about 1/4 wavelength due to the spatial phase shift between the ID-CGT and IA-CGT patterns. Observational land SAT anomalies derived from the Delaware dataset are generally consistent with those from the 20CR, supporting that the ID-CGT and IA-CGT patterns derived from the 20CR are reliable. The distinctions between 20CR and Delaware are also evident. Firstly, the cold anomalies over the East Asia associated with the ID-CGT pattern derived from the 20CR are contrary to warm anomalies from the Delaware. Secondly, all the negative anomalies in the Delaware are not as strong and significant as those in the 20CR in amplitude for both the time scales. What causes the distinctions deserve further study. On the other hand, the percentage of the LF (HF) variances of the land SAT associated with the ID-CGT

(IA-CGT) mode is calculated (Figure 4(e) and (f)). The ID-CGT-related SAT anomalies account for more than 30% of the LF variances in the eastern coast of North American, Europe and northern East Asia. Comparing with the ID-CGT, the contribution of the IA-CGT-related land SAT to the HF variances is much smaller, consistent with the smaller variance contributions of the IA-CGT mode relative to the ID-CGT mode.

This study is primarily based on the 20CR reanalysis dataset, which just assimilated observational surface pressure and sea level pressure and thus can cover the entire 20th century. It has been demonstrated that the 20CR can realistically reproduce the conventional IA-CGT pattern, including its dynamic properties, implying the reliability of the ID-CGT derived from it. The robustness of the ID-CGT pattern and its insensitivities to the datasets and analysis methods are further verified from following four aspects.

1. We applied an EOF analysis to the 8-year low-pass filtered Z200 (global warming signal is not removed as above), the second EOF mode corresponds to the ID-CGT pattern (Figure S2(a)), with its PC time series highly correlated with the ID-CGT time series ($r = 0.76$, Figure S2(b)).
2. A MCA analysis is applied to the LF fields between NH mid-latitude Z200 and SST in the North Atlantic (Figure S3). The first MCA mode explains 71% of the squared covariance. The mode involves the ID-CGT pattern in the Z200 field and the AMO in the SST field, indicating that the ID-CGT pattern is coupled with the AMO.
3. The EOF analysis was applied to the 8-year low-pass filtered Z200 from the JRA-55 reanalysis dataset (Ebita *et al.*, 2011), which covers the period of 1958–2012. The second mode corresponds to the ID-CGT pattern, with locations of the five positive nodes highly consistent with those derived from the 20CR (Figures 2(a) and S4(a)). The corresponding PC time series is highly correlated with the simultaneous AMO index ($r = 0.82$, Figure S4(b)).
4. A MCA analysis is applied to the LF fields between NH mid-latitude Z200 from the 20CR and underlying observational land SAT from the Delaware (Figure S5). The first MCA mode accounts for 76% of the total squared covariance. The spatial distribution of the Z200 anomalies exhibits an ID-CGT-like pattern (Figure S5(a)). Corresponding expansion coefficient time series is highly correlated with the time series of the ID-CGT derived from the EOF analysis ($r = 0.97$, Figure S5(c)). The results indicate that the interdecadal variability NH summer mid-latitude land SAT is dominated by the ID-CGT.

4. Summary

In the study, we revealed that the interdecadal variability of the summertime NH extratropical atmospheric circulation is dominated by a global-scale stationary barotropic wave-like pattern along the jet stream with zonal wavenumber 5. Because of its resemblances with the conventional CGT pattern on the interannual time scale (IA-CGT), the wave-like pattern is referred to as the ID-CGT pattern. We demonstrated that the ID-CGT and IA-CGT patterns can be extracted through a consistent EOF analysis for Z200 on two different time scales. The contribution of the ID-CGT pattern to the variances of the NH raw (unfiltered) Z200 is far larger than that of the IA-CGT pattern (15.3 vs 6.1%).

Although the ID-CGT and IA-CGT patterns show some similarities, such as zonal wavenumber 5 structure and locations of action centers along the westerly jet stream, they have distinctive characteristics on following aspects. Firstly, the ID-CGT pattern shifts westward relative to the IA-CGT pattern by about 1/4 wavelength, consistent with the phenomenon that the locations of the major variability centers of the NH mid-latitude atmospheric circulation are different between interannual and interdecadal time scales (Figure 1). Secondly, the ID-CGT pattern is tightly associated with the AMO, whereas the IA-CGT pattern is with the tropical forcing from the Indian summer monsoon precipitation anomalies. Thirdly, all the five nodes of the ID-CGT pattern exhibits barotropic structure, while the IA-CGT has a baroclinic node located to the north of the Indian peninsula, suggesting that the ID-CGT pattern should not be driven by the tropical forcing as the IA-CGT, although the AMO-related SSTAs have a tropical component.

The ID-CGT pattern can modulate land SAT along it path. The anticyclonic (cyclonic) anomalies correspond to underlying warm (cold) anomalies. Previous studies have noted that the positive phase of the AMO can increase the NH mean surface temperature (e.g. Zhang *et al.*, 2007; DelSole *et al.*, 2011), and cause warm anomalies over the North America and Europe during boreal summer (Sutton and Hodson, 2005). Our results reveal a complete picture of the impact of the AMO on the NH extratropical land SAT during boreal summer, that is, the impacts of AMO on the NH local SAT rely on the spatial phase of the ID-CGT pattern.

Acknowledgements

This work is jointly supported by a National Program on Key Basic Research project (2012CB955202), the NSFC (grant nos 41661144009, 41005040 and 41023002), and R&D Special Fund for Public Welfare Industry (meteorology) (GYHY201506012).

References

Compo GP, Whitaker JS, Sardeshmukh PD, Matsui N, Allan RJ, Yin X, Gleason BE, Vose R, Rutledge G, Bessemoulin P. 2011. The twentieth century reanalysis project. *Quarterly Journal of the Royal Meteorological Society* **137**: 1–28.

DelSole T, Tippett MK, Shukla J. 2011. A significant component of unforced multidecadal variability in the recent acceleration of global warming. *Journal of Climate* **24**: 909–926.

Delworth TL, Zhang R, Mann ME. 2007. Decadal to centennial variability of the Atlantic from observations and models. In *Ocean Circulation: Mechanisms and Impacts-Past and Future Changes of Meridional Overturning*. American Geophysical Union: Washington, D.C.; 131–148.

Ding QH, Wang B. 2005. Circumglobal teleconnection in the Northern Hemisphere summer. *Journal of Climate* **18**: 3483–3505.

Ding Q, Wang B, Wallace JM, Branstator G. 2011. Tropical-extratropical teleconnections in boreal summer: observed interannual variability. *Journal of Climate* **24**: 1878–1896.

Ebisuzaki W. 1997. A method to estimate the statistical significance of a correlation when the data are serially correlated. *Journal of Climate* **10**: 2147–2153.

Ebita A, Kobayashi S, Ota Y, Moriya M, Kumabe R, Onogi K, Harada Y, Yasui S, Miyaoka K, Takahashi K. 2011. The Japanese 55-year Reanalysis" JRA-55": an interim report. *SOLA* **7**: 149–152.

Kerr RA. 2000. A North Atlantic climate pacemaker for the centuries. *Science* **288**: 1984–1985.

Kothawale R, Rupa Kumar K, Mooley D, Munot A, Parthasarathy B, Sontakke N. 2006. *India Regional/Subdivisional Monthly Rainfall Data Set*. Indian Institute of Tropical Meteorology: Pune, Maharashtra, India.

Kushnir Y. 1994. Interdecadal variations in North Atlantic sea surface temperature and associated atmospheric conditions. *Journal of Climate* **7**: 141–157.

Lee J-Y, Wang B, Kang I-S, Shukla J, Kumar A, Kug J-S, Schemm J, Luo J-J, Yamagata T, Fu X. 2010. How are seasonal prediction skills related to models' performance on mean state and annual cycle? *Climate Dynamics* **35**: 267–283.

Lee J-Y, Wang B, Ding Q, Ha K-J, Ahn J-B, Kumar A, Stern B, Alves O. 2011. How predictable is the northern hemisphere summer upper-tropospheric circulation? *Climate Dynamics* **37**: 1189–1203.

Lin H. 2009. Global extratropical response to diabatic heating variability of the Asian summer monsoon. *Journal of the Atmospheric Sciences* **66**: 2697–2713.

Rayner N, Parker DE, Horton E, Folland C, Alexander L, Rowell D, Kent E, Kaplan A. 2003. Global analyses of sea surface temperature, sea ice, and night marine air temperature since the late nineteenth century. *Journal of Geophysical Research, [Atmospheres]* **108**: 4407.

Sutton RT, Hodson DLR. 2005. Atlantic Ocean forcing of North American and European summer climate. *Science* **309**: 115–118.

Ting M, Kushnir Y, Seager R, Li C. 2009. Forced and internal twentieth-century SST trends in the North Atlantic. *Journal of Climate* **22**: 1469–1481.

Trenberth KE, Shea DJ. 2006. Atlantic hurricanes and natural variability in 2005. *Geophysical Research Letters* **33**: L12704.

Wang B, Lee J-Y, Kang I-S, Shukla J, Park C-K, Kumar A, Schemm J, Cocke S, Kug J-S, Luo J-J. 2009. Advance and prospectus of seasonal prediction: assessment of the APCC/CliPAS 14-model ensemble retrospective seasonal prediction (1980–2004). *Climate Dynamics* **33**: 93–117.

Wang H, Wang B, Huang F, Ding Q, Lee JY. 2012. Interdecadal change of the boreal summer circumglobal teleconnection (1958–2010). *Geophysical Research Letters* **39**: L12704.

Willmott C, Matsuura K. 2012. *Terrestrial Air Temperature: 1900–2010 Gridded Monthly Time Series. Version 3.01*. Center for Climatic Research, Department Of Geography, University of Delaware: Newark, NJ. http://climate. geog. udel. edu/~climate.

Wu B, Zhou T, Li T. 2016. Impacts of the Pacific-Japan and circumglobal teleconnection patterns on interdecadal variability of the East Asian summer monsoon. *Journal of Climate* **29**: 3253–3271.

Yasui S, Watanabe M. 2010. Forcing processes of the summertime circumglobal teleconnection pattern in a dry AGCM. *Journal of Climate* **23**: 2093–2114.

Zhang R, Delworth TL, Held IM. 2007. Can the Atlantic Ocean drive the observed multidecadal variability in Northern Hemisphere mean temperature? *Geophysical Research Letters* **34**: 346–358.

17

Model moist bias in the middle and upper troposphere during DEEPWAVE

Yang Yang,* Stuart Moore, Michael Uddstrom, Richard Turner and Trevor Carey-Smith

NIWA, Wellington, New Zealand

*Correspondence to:
Y. Yang, National Institute of
Water and Atmospheric
Research (NIWA), 301 Evans
Bay Parade, Hataitai, Wellington,
New Zealand.
E-mail: y.yang@niwa.co.nz,
yang.yang816@gmail.com

Abstract

Data from 279 dropsonde profiles collected during the Deep Propagating Gravity Wave Experiment (DEEPWAVE) over New Zealand between 4 June and 20 July 2014 were used to verify the relative humidity (RH) fields simulated by regional configurations of the UK Met Office Unified Model (MetUM) in the troposphere. Significant RH biases (predictions up to 28% too high) were found in the middle and upper troposphere during this period. This RH bias was found to be mainly caused by the errors in the simulated-specific humidity. It is demonstrated here that evaporation from the lower boundary (mainly sea surface) is not a factor leading to the moist bias. A similar magnitude of moist bias was also found in the Global UM (the global configuration of the MetUM) and from a preliminary inspection is also very likely to occur in ERA-interim and NCEP-GFS reanalyses. This study suggests that the moist bias is very likely not a regional or a model specific issue.

Keywords: relative humidity; specific humidity; moist bias; DEEPWAVE; numerical weather prediction

1. Introduction

Relative humidity (RH) is an important field in determining the distribution and occurrence of clouds and precipitation (e.g. Price and Wood, 2002; Derbyshire et al., 2004). Changes in RH also affect (1) the distribution of latent heating, so as to affect the atmospheric circulation (Schneider et al., 2010), and (2) the level for the occurrence of deep convection detrainment (Hartmann and Larson, 2002). It has been shown that the size and intensity of tropical cyclones are controlled/impacted by the environmental RH (Holloway and Neelin, 2009) and atmospheric moisture profiles (Dunion and Velden, 2004). Cumulus parameterization schemes based on moisture convergence are widely used in Numerical Weather Prediction (NWP) models with a grid-length of approximately 8 km or coarser. If these models are too moist or too dry at levels where large-scale convergence occurs, convective precipitation forecasts will be affected.

Water vapour plays a key role in the budget of radiation in the troposphere and affects atmospheric radiative transfer (e.g. Held and Soden, 2000). Biases in forecasted large-scale humidity may result in biases in diagnosed clouds, leading to biases in radiation computations. The radiative effects and phase changes of water vapour play a key role in atmospheric processes, and the land–air and sea–air interactions (e.g. Shine and Sinha, 1991).

The significant impact of RH on precipitation, winds, and atmospheric processes means that a reliable analysis and forecast of RH is desirable. In NWP, initial conditions of moisture are provided through assimilation of direct or indirect measurements of moisture or water vapour from rawinsondes, surface stations and satellites. However, because of the limitations and errors in the measurements and analysis methods, deficiencies in moisture analyses and forecasts have been found both temporally and spatially (e.g. Bock and Nuret, 2009; Newman et al., 2015). Thus, verification of RH forecast and analysis using not only routine RH observations, but also field campaign RH observations are necessary to understand how well RH is analysed (initialised) and forecast within NWP systems.

DEEPWAVE studied the dynamics of gravity waves from the surface of the Earth to the upper level of the atmosphere (Fritts et al., 2015). It was conducted over and around New Zealand from 4 June to 20 July 2014 with extensive in situ observations from research aircraft along with surface, airborne, radar wind profiler, dropsondes and radiosondes. These highly valuable, spatially and temporally dense data can be used to verify model simulations in a region which normally has few in situ observations.

During initial DEEPWAVE investigations, simulations from two NWP models; the New Zealand Limited Area Model (NZLAM) and New Zealand Convective Scale model (NZCSM) were compared with profiles from eight dropsondes over the South Island of New Zealand for a flight on 24 June 2014 (Figure S1, Supporting information). Significant differences – due to errors in the specific humidity – leading to up to 50% moist errors in RH were found over the South Island. This raised three questions: (1) is the moist bias a case sensitive issue or a robust feature during DEEPWAVE not only over land but also over the open sea? (2) does the Global UM, from which NZLAM derives its lateral boundary conditions, and other global models have this issue? and (3) if the moist bias is more generally confirmed, what are the possible factors leading to the bias? This paper reports on our further investigation of these questions.

2. Description of models and data

The MetUM is a non-hydrostatic and fully compressible model system (see Section 5 of Webster *et al.*, 2003 and Davies *et al.*, 2005 for detailed descriptions). The NZLAM is a regional configuration of the MetUM. Its domain has 324×324 horizontal grid points (Figure 1(a)) with a grid-length of approximately 12 km and 70 levels in the vertical, with the model top at 80 km. NZLAM performs a 3 hourly data assimilation cycle using an incremental 3DVAR FGAT (first guess at appropriate time) analysis scheme (Lorenc *et al.*, 2000). The lateral boundary conditions are derived from the operational Global UM run at the Met Office (see Yang *et al.*, 2012 for further details of NZLAM and its data assimilation scheme).

During DEEPWAVE, a total of 282 dropsondes were launched from the flight level (approximately 12 km) of the NSF/NCAR HIAPER Gulfstream-V research aircraft. Among them, 279 dropsondes were within the NZLAM domain (Figure 1(a)). The observed RH ranges from 0 to 100%, with an absolute accuracy of $\pm 5\%$ and resolution of 1.0%. The vertical resolution of the dropsonde data is approximately 10 m. Moist biases in the dropsonde observations during DEEPWAVE were found and corrected recently. In this study, the corrected dropsonde data were used. For more information on the NCAR Dropsonde System one may see: http://www.eol.ucar.edu/instrumentation/sounding/dropsonde. Further, these dropsondes were not incorporated in the data assimilation schemes of the Global UM or NZLAM, so can be regarded as an independent validation dataset.

For the verification, the simulations from the operational Global UM and NZLAM were used. The corresponding profiles (i.e. simulated dropsondes) were created from model level data using linear interpolation allowing for the drift of the sondes. Hourly outputs from the models were used for this interpolation. Two groups of simulated dropsondes were made to investigate the effect of forecast range on model moist bias. For the first group, the forecasts started about 6–13 h before the observation time (Group $T + 6$). In the second group, the forecasts started 30–37 h before the observations (Group $T + 30$). The difference in forecast range is 24 h between the two groups. Nine days of Global UM data were not available for the purposes of this study. This led to only 192 simulated dropsondes in Group $T + 6$ and 193 simulated dropsondes in Group $T + 30$. Between the two groups, 124 dropsondes overlapped. All the dropsondes were launched in a time window of a few hours before midnight to a few hours

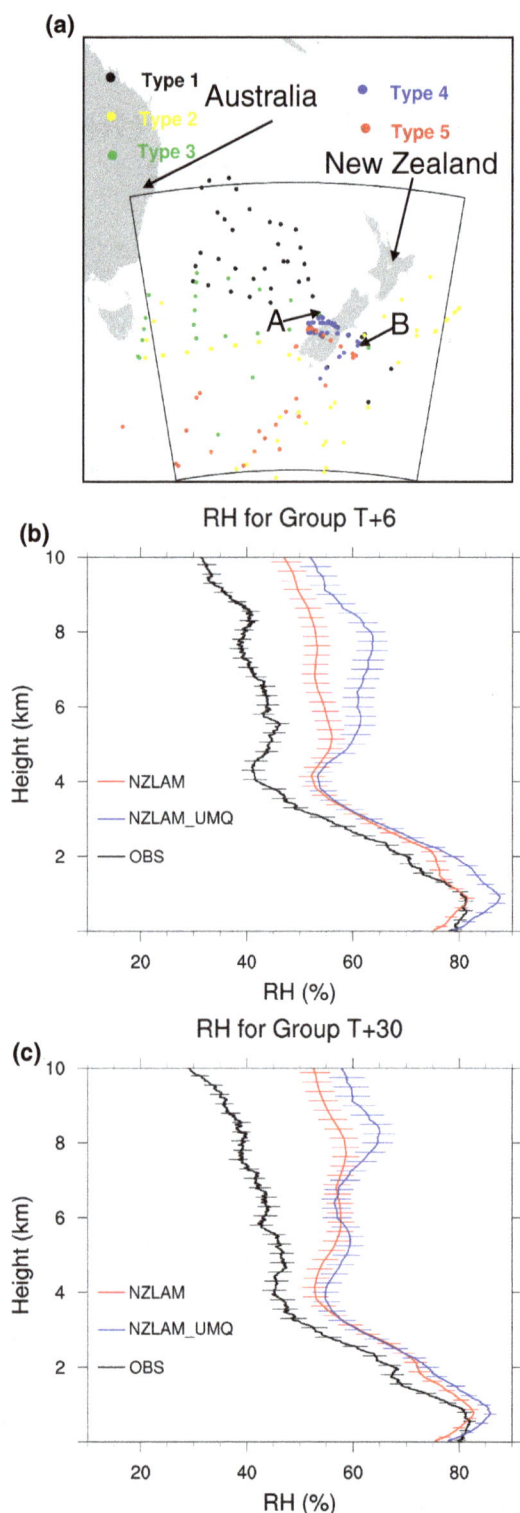

Figure 1. (a) The full NZLAM domain showing locations of the DEEPWAVE dropsondes (dots) used in this study. The five colours correspond to the five weather situations (Type 1 to Type 5 shown in Table S1). The sub-area outlined by the black lines is used for area mean RH profiles shown in Figure 5. The arrows from A and B point to the beginning and end of the cross-sections in Figure S1, respectively. (b) The vertical profile of mean RH for Group $T + 6$ from the dropsondes, the NZLAM simulations and the recalculated RH (NZLAM_UMQ) using the simulated-specific humidity from the global UM. (c) The same as (b) but for Group $T + 30$. Along each of the RH profiles, the length of the horizontal line is twice the standard error of the mean.

after midnight local time (NZST). From the surface simulations and analysis of NZLAM around midnight, five distinct weather situations were classified for the 279 dropsondes (Table S1).

3. Results

3.1. Relative humidity

The mean RH profiles for both the observations and NZLAM simulations are shown in Figures 1(b) and (c). Large differences in RH (10–28%) were found between 4 and 10 km for the two groups. Obvious moist biases (simulations − observations) were found from 4 km upward, with the maximum moist bias of 28% for Group $T+30$ and 22% for Group $T+6$ at 10 km (not shown). These biases in simulated RH by NZLAM from 4 to 10 km were significantly larger than the accuracy of $\pm 5.0\%$ in humidity observations and the standard error of the mean. This bias in RH is referred to as model moist bias in this paper.

Mean profiles of RH for five different synoptic weather situations observed during DEEPWAVE were analysed and a robust moist bias from 4 to 10 km was found in NZLAM for each weather situation (Table S1 and Figure 2(a)). Some differences in the magnitude of the moist bias were found, but these were attributed to the difference in weather situations and circulations, and also to the difference in sample sizes and locations of the dropsondes (Figure 1(a)).

RH can be affected by pressure, temperature and specific humidity. To estimate the relative contribution each of these three fields makes to the moist bias, simulated RH was recalculated by replacing simulated fields with, in turn, observations of pressure (NZLAM_OBSP), air temperature (NZLAM_OBST) and specific humidity (NZLAM_OBSQ) from the dropsondes. Using the observed pressure, the bias of the recalculated RH were almost the same as those of simulated RH (NZLAM, Figures 2(b) and (c)), indicating that errors in the simulated pressure did not contribute to the moist RH bias. Using the observed air temperature to replace the NZLAM simulated air temperature, the 'moist bias' only decreased by approximately 5% (Figures 2(b) and (c)). Using the observed specific humidity, the bias of the recalculated RH (NZLAM_OBSQ) were very close to zero, especially from 4 to 10 km. These results indicate that error in the simulated-specific humidity is the dominant factor leading to the moist bias in the middle and upper troposphere.

The simulated model level data of the Global UM available for this study contains only wind, potential temperature and specific humidity. RH at model levels cannot be calculated. To determine whether the Global UM also has the moist bias found for NZLAM, the simulated-specific humidity by the Global UM was used to replace the specific humidity simulated by NZLAM to recalculate RH (NZLAM_UMQ, Figures 1(b) and (c)). For Group $T+6$ (Figure 1(b)),

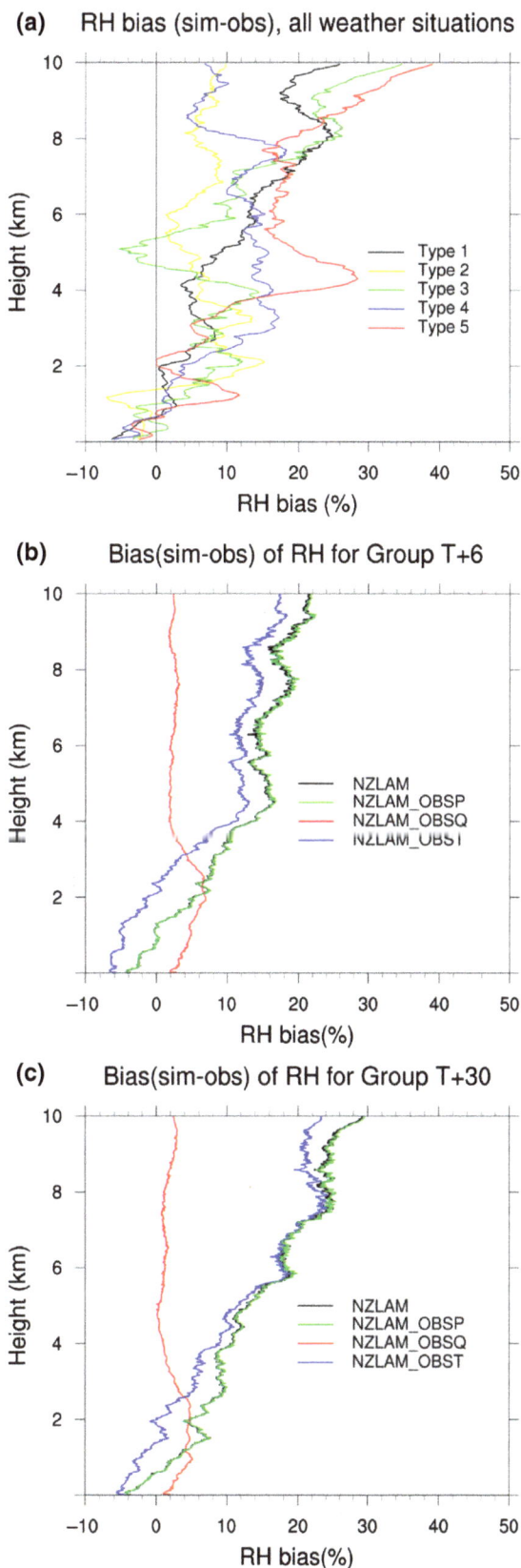

Figure 2. (a) Bias of RH simulated by NZLAM for the five weather situations (Type 1 to Type 5 in Table S1). (b) Bias of RH for Group $T+6$ simulated by NZLAM and bias of the recalculated RH using observed pressure (NZLAM_OBSP), observed mixing ratio (NZLAM_OBSQ) and observed air temperature (NZLAM_OBST) from the dropsondes. (c) is the same as (b) but for Group $T+30$.

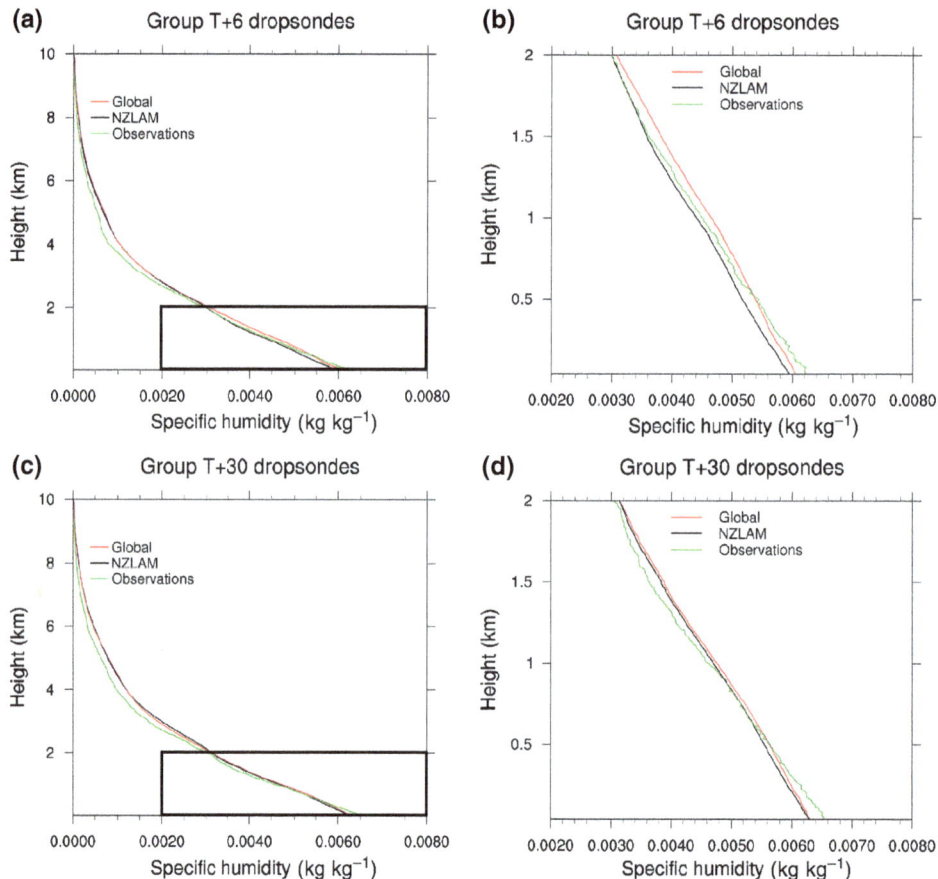

Figure 3. (a) Profiles of the mean specific humidity of Group $T+6$ for observations, NZLAM and the global UM; (b) same as (a) but for the lower boundary layer corresponding to the box in (a). (c) and (d) are the same as (a) and (b) but for Group $T+30$.

below 2 km, the RH for NZLAM_UMQ was 2–7% higher than NZLAM. From 2 to 5 km aloft, both were very close. Above 5 km, the RH for NZLAM_UMQ was 5–10% higher than NZLAM. For Group $T+30$ (Figure 1(c)), below 7 km, the RH for NZLAM_UMQ was either very close or slightly higher (1–2%) than NZLAM. Above 7 km, the RH for NZLAM_UMQ was 3–5% higher than NZLAM. Because the errors in specific humidity are the main factor leading to the moist bias, these facts indicate that a moist bias with similar magnitude is also present in the Global UM in the middle and upper troposphere.

Noh *et al.* (2016) used one-year radiosonde observations from the Global Climate Observing System Reference Upper-Air Network (GRUAN) to verify the Global UM analysis. Moist bias (approximately 3% in RH) was only found in the upper troposphere, much smaller than that in this study. These differences may be due to that fact the radiosonde observations had also been used for the analysis. In addition, the drift of the radiosondes were not considered.

3.2. Mixing ratio

For profiles of the mean specific humidity during DEEPWAVE (Figure 3), the exponential decrease of mixing ratio with altitude was consistent between observations and the simulations. NZLAM was very

close to the Global UM with respect to vertical variations in the specific humidity. Major differences between observations and simulations were found between 3 km and 6 km aloft. In the lower boundary layer for both groups (Figures 3(b) and (d)), observed specific humidity was greater than that forecast by the Global UM and NZLAM below 500 m. The maximum difference between observations and simulations was found at the surface. These features in the lower boundary layer corresponded to the negative bias (simulations − observations) of the simulated-specific humidity (Figures 4(b) and (d)), indicating that evaporation at the lower boundary (mainly sea surface) simulated by the models was generally less than observations. Because the source of the moisture in the atmosphere is from the earth surface, the simulated moist bias in the middle and upper troposphere was not caused by the lower boundary conditions.

For the specific humidity (Figures 4(a)), large errors (MAD, mean absolute differences) were found below 5 km, with NZLAM slightly higher than the Global UM for both groups (Figures 4(a)). Above 5 km, MAD of the specific humidity decreased dramatically with altitude. For the bias (Figure 4(b)), positive values (moist bias) were found above 2 km with the maximum near an altitude of 4 km. Large negative values (dry bias) were found at the surface. Figure 4(c) shows the relative

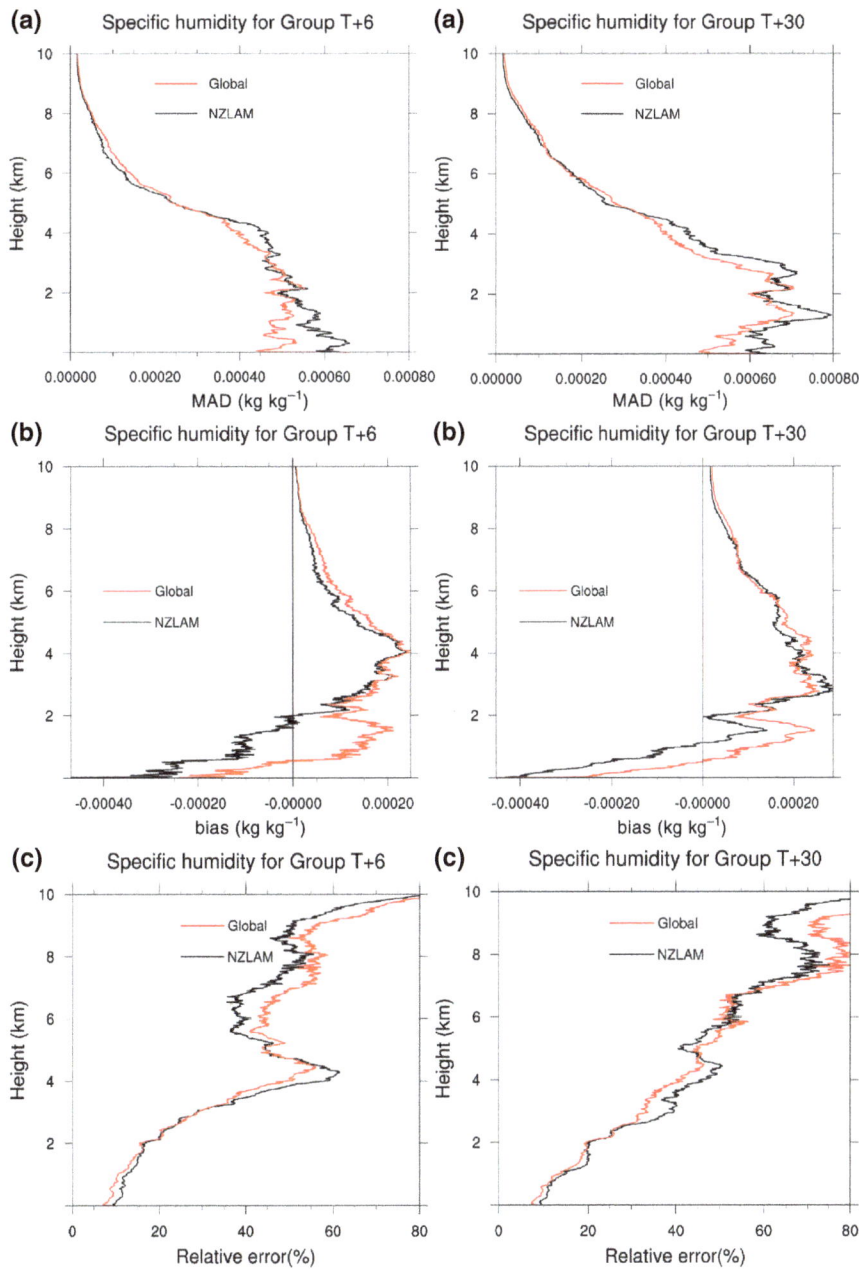

Figure 4. MAD (a), (b) bias (simulation − observation) and relative errors (c) of the simulated-specific humidity by NZLAM and the global UM for Group $T+6$ (left panel) and Group $T+30$ (right panel).

errors of specific humidity $\left(\frac{obs-sim}{obs} \times 100\right)$. Relatively large values (approximately 40% or higher) were found from approximately 4 km upward. Thus, the magnitude of the model moist bias in the middle and upper troposphere is determined by the relative errors of specific humidity, not its absolute errors.

3.3. Moist bias in ERA-Interim and NCEP-GFS analyses

It is of interest to determine whether the moist bias observed in the UM is a feature of this particular model or is also manifest in other NWP systems. Simulated hourly RH on standard pressure levels from both ERA-Interim (Dee *et al.*, 2011) and NCEP-GFS have

been compared with NZLAM over the sub-area from 150.0 to 182.5°E and 32.5 to 55.0°S (Figure 1(a)). This region corresponds with the DEEPWAVE experiment and encompasses 263 DEEPWAVE dropsonde locations. These dropsondes data were not used to create ERA-Interim (0.25° resolution) and NCEP-GFS (0.5° resolution) reanalyses.

Figure 5 shows the mean RH profiles for the sub-area at pressure levels for ERA-Interim and NCEP-GFS analyses at 1200 UTC and the corresponding mean RH profile from the NZLAM analyses from 6 June to 20 June 2014. RH analyses at 1200 UTC was chosen because the dropsondes were launched around 1200 UTC. For ERA-Interim (Figure 5(a)), the mean RH from 650 to 500 hPa was 2–5% lower than that of

(a) Mean of RH at 1200 UT   **(b)** Mean of RH at 1200 UT

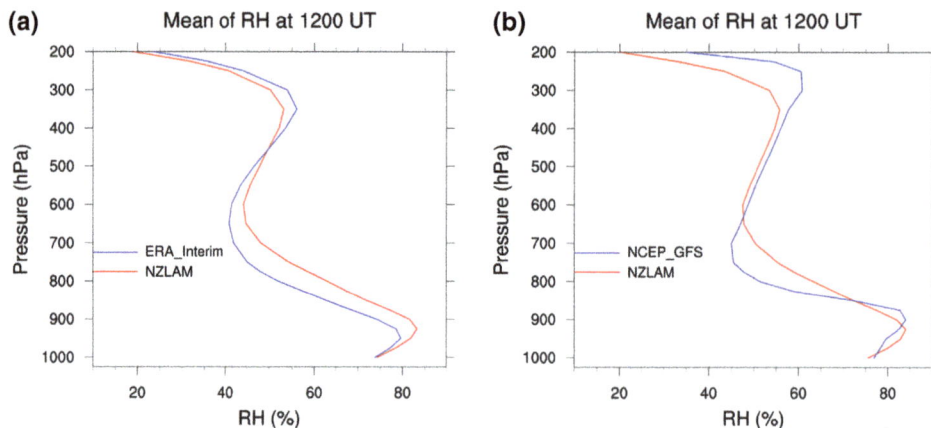

Figure 5. (a) Mean RH profiles for an area of 32.5–55.0°S and 150.0–182.5°E with a resolution of 0.25° from 6 June to 20 July 2014 using analysis data at 1200 UTC. For NZLAM, the RH analysis was regridded to 0.25° grids as ERA-Interim. (b) Same as (a) but for NCEP-GFS 0.5° analysis and NZLAM analysis regridded to a resolution of 0.5°.

NZLAM, but 1–2% higher from 400 to 200 hPa. Overall, the mean RH profile of ERA-Interim was very close to NZLAM from 650 to 200 hPa, indicating that ERA-Interim also had a similar magnitude of moist bias in middle and upper troposphere. These results are consistent with a study made during the Tibetan Plateau Experiment (Bao and Zhang, 2013) which also showed a moist bias (5–25% in RH in their Figures 2 and 3) from 500 to 200 hPa for ERA-Interim. This suggests also that the moist bias may not be specific to the New Zealand region. For NCEP-GFS, from 700 to 200 hPa, the mean RH was 4–10% higher than NZLAM, indicating that NCEP-GFS also had a moist bias of at least the same magnitude as NZLAM, or even larger in the middle and upper troposphere.

4. Conclusion

Dropsonde descents made during the DEEPWAVE campaign over New Zealand and the surrounding open sea have been used to verify the tropospheric RH simulated by regional configuration of the MetUM. Significant moist biases (up to 28%) in simulated RH were found in middle and upper troposphere. The moist bias was robust for different weather situations during DEEPWAVE. This moist bias was mainly caused by the moist bias in the simulated-specific humidity. The magnitude of the moist bias was determined by the relative errors of the specific humidity instead of its absolute errors. A moist bias with similar magnitude was also found in the Global UM, and very likely in ERA-Interim and NCEP-GFS analyses too.

Analysis showed that for both the Global UM and NZLAM the lower boundary conditions had a dry bias and evaporation from the lower boundary (mainly sea surface) is not a factor leading to the moist bias issue. Other possible factors that could cause the moist bias are the data assimilation configuration, precipitation and cloud processes, and the advection of water vapour. We briefly survey these mechanisms below.

Water vapour advection is determined by both the local winds and water vapour gradient, which are significantly affected by different weather systems. The continued presence of the moist bias during DEEPWAVE for the five weather situations suggests that water vapour advection may not be the main factor.

Given a reliable water vapour from the initial conditions, a moist bias may result from less precipitation and clouds. However, validation of NZLAM during the DEEPWAVE period shows both wet and dry biases in precipitation over New Zealand (Figure S2). This complicated distribution of precipitation bias is partly a result of the orographic effect on precipitation over New Zealand's mountainous regions. This issue and other errors in the parameterization and modelling of rainfall make it difficult to make a direct link between biases in upper troposphere RH and surface precipitation. In addition, the only slight increase in moist bias with forecast range, commensurate with typical forecast degradation with lead time, suggests that this is an initial condition problem. However, further research could be conducted to see if precipitation and cloud processes are a factor contributing to the moist bias.

The fact that the moist bias is observed from the very beginning of each model simulation and that all three data assimilating models investigated in this study show similar middle to upper tropospheric RH profiles, suggests that data assimilation may be making an important contribution to the bias. We think this is an area where more research is warranted.

Given a moist bias in the initial conditions, more clouds would be simulated. This would produce more precipitation with more latent heat release, and affect the surface energy budget by decreasing shortwave radiation and increasing downward longwave radiation reaching the surface. How significant these effects and the possible feedback will be left for future study.

Acknowledgements

We thank DEEPWAVE project that made this verification work possible. Constructive review comments by the three anonymous reviewers are greatly appreciated. This research was carried out under research collaboration SC0128 with the UK Met Office and funded by NIWA under its Hazards Research Programme (2014/15 and 2015/16 SCI).

Supporting information

The following supporting information is available:

Table S1. Weather situations and dominant incident wind directions (DIWD) to the NZLAM domain (Figure 1(a)) from NZLAM surface simulations around midnight, and the number of dropsondes associated with each weather situation during DEEPWAVE.

Figure S1. Cross-sections of RH (%) and potential temperature (K) across the South Island along A and B in Figure 1(a) using (a) observations of 8 dropsondes on 24 June 2014, (b) NZLAM simulations corresponding to the 8 dropsondes and (c) NZCSM simulations corresponding to the 8 dropsondes.

Figure S2. Mean error (sim − obs) in precipitation during Deepwave for NZLAM and NZCSM when compared against rain-gauge observations. Both models show significant biases, but apart from a consistent over-prediction in the south and east of the South Island the distribution of errors is quite different. The differences (and largest errors) are most obvious in areas of complex terrain where the higher resolution NZCSM tends to perform better.

References

Bao X, Zhang F. 2013. Evaluation of NCEP–CFSR, NCEP–NCAR, ERA-Interim, and ERA-40 reanalysis datasets against independent sounding observations over the Tibetan Plateau. *Journal of Climate* **26**: 206–214.

Bock O, Nuret M. 2009. Verification of NWP model analyses and radiosonde humidity data with GPS precipitable water vapor estimates during AMMA. *Weather Forecasting* **24**: 1085–1101.

Davies T, Cullen MJP, Malcolm AJ, Mawson MH, Staniforth A, White AA, Wood N. 2005. A new dynamical core for the Met Office's global and regional modelling of the atmosphere. *Quarterly Journal of the Royal Meteorological Society* **131**: 1759–1782.

Dee DP, Uppala SM, Simmons AJ, Berrisford P, Poli P, Kobayashi S, Andrae U, Balmaseda MA, Balsamo G, Bauer P, Bechtold P, Beljaars ACM, van de Berg L, Bidlot J, Bormann N, Delsol C, Dragani R, Fuentes M, Geer AJ, Haimberger L, Healy SB, Hersbach H, Hólm EV, Isaksen L, Kållberg P, Köhler M, Matricardi M, McNally AP, Monge-Sanz BM, Morcrette J-J, Park B-K, Peubey C, de Rosnay P, Tavolato C, Thépaut J-N, Vitart F. 2011. The ERA-Interim reanalysis: configuration and performance of the data assimilation system. *Quarterly Journal of the Royal Meteorological Society* **137**: 553–597.

Derbyshire SH, Beau I, Bechtold P, Grandpeix JY, Piriou JM, Redelsperger JL, Soares PMM. 2004. Sensitivity of moist convection to environmental humidity. *Quarterly Journal of the Royal Meteorological Society* **130**: 3055–3079.

Dunion JP, Velden CS. 2004. The impact of the Saharan air layer on Atlantic tropical cyclone activity. *Bulletin of the American Meteorological Society* **85**: 353–365.

Fritts DC, Smith RB, Taylor MJ, Doyle JD, Eckermann SD, Dörnbrack A, Rapp M, Williams BP, Pautet P-D, Bossert K, Criddle NR, Reynolds CA, Reinecke PA, Uddstrom M, Revell MJ, Turner R, Kaifler B, Wagner JS, Mixa T, Kruse CG, Nugent AD, Watson CD, Gisinger S, Smith SM, Lieberman RS, Laughman B, Moore JJ, Brown WO, Haggerty JA, Rockwell A, Stossmeister GJ, Williams SF, Hernandez G, Murphy DJ, Klekociuk AR, Reid IM, Ma J. 2015. The deep propagating gravity wave experiment (DEEPWAVE): an airborne and ground-based exploration of gravity wave propagation and effects from their sources throughout the lower and middle atmosphere. *Bulletin of the American Meteorological Society* **97**: 425–453.

Hartmann DL, Larson K. 2002. An important constraint on tropical cloud–climate feedbacks. *Geophysical Research Letters* **29**(20): 1951, doi: 10.1029/2002GL015835.

Held IM, Soden BJ. 2000. Water vapour feedback and global warming. *Annual Review of Energy and the Environment* **45**: 441–475.

Holloway CE, Neelin JD. 2009. Moisture vertical structure, column water vapour, and tropical deep convection. *Journal of the Atmospheric Sciences* **66**: 1665–1683.

Lorenc AC, Ballard SP, Bell RS, Ingleby NB, Andrews P, Barker DM, Bray JR, Clayton AM, Dalby LTD, Payne TJ, Saunders FW. 2000. The Met Office global three-dimensional variational data assimilation scheme. *Quarterly Journal of the Royal Meteorological Society* **126**: 2991–3012.

Newman KM, Schwartz CS, Liu Z, Shao H, Huang XY. 2015. Evaluating forecast impact of assimilating microwave humidity sounder (MHS) radiances with a regional ensemble Kalman filter data assimilation system. *Weather Forecast.* **30**: 964–983.

Noh YC, Sohn BJ, Kim Y, Joo S, Bell W. 2016. Evaluation of temperature and humidity profiles of unified model and ECMWF analyses using GRUAN radiosonde observations. *Atmosphere* **7**: 94.

Price J, Wood R. 2002. Comparison of probability density functions for total specific humidity and saturation deficit humidity, and consequences for cloud parametrization. *Quarterly Journal of the Royal Meteorological Society* **128**: 2059–2072.

Schneider T, O'Gorman PA, Levine X. 2010. Water vapour and the dynamics of climate changes. *Reviews of Geophysics* **48**: RG3001, doi:10.1029/2009RG000302.

Shine K, Sinha A. 1991. Sensitivity of the Earth's climate to height-dependent changes in the water vapour mixing ratio. *Nature* **354**: 382–384.

Webster S, Brown AR, Jones CP, Cameron DR. 2003. Improvements to the representation of orography in the Met Office Unified Model. *Quarterly Journal of the Royal Meteorological Society* **129**: 1989–2010.

Yang Y, Uddstrom M, Revell M, Andrews P, Turner R. 2012. Amplification of the impact of assimilating ATOVS radiances on simulated surface air temperatures over Canterbury by the Southern Alps, New Zealand. *Monthly Weather Review* **140**: 1367–1384.

A comparative study between bulk and bin microphysical schemes of a simulated squall line in East China

Lei Yin,[1,2] Fan Ping[1,3]* and Jiahua Mao[3]

[1] Laboratory of Cloud-Precipitation Physics and Severe Storms (LACS), Institute of Atmospheric Physics, Chinese Academy of Sciences, Beijing, China
[2] School of Earth Sciences, University of Chinese Academy of Sciences, Beijing, China
[3] School of Geography and Remote Sensing, Nanjing University of Information Science and Technology, Nanjing, China

*Correspondence to:

F. Ping, Laboratory of Cloud-Precipitation Physics and Severe Storms (LACS), Institute of Atmospheric Physics, Chinese Academy of Sciences, Beijing 100029, China.
E-mail: pingf@mail.iap.ac.cn

Abstract

A squall line occurred in East China during 12 July 2014 was simulated with the Weather Research and Forecasting (WRF) model using spectral bin and two-moment bulk microphysical parameterization scheme, respectively. Comparative study showed that significant differences existed in the dynamic, thermodynamic and microphysical structures of squall line between bulk and bin simulation results. The bulk scheme produced a well-organized but shorter radar structure while bin scheme simulated scattered but stronger radar echo which was more consistent with observation. Bulk scheme had a better performance in predicting the strong rainfall areas and amount. The strong rear-to-front (RTF) inflow and convective updrafts were identified in bulk scheme by comparison with weak RTF and updrafts in bin scheme. In addition, bulk simulated a deeper cold pool than bin. Much higher cloud droplet number concentration was simulated by bulk scheme, while higher raindrop mass and number concentration was generated by bin scheme. Detailed analysis and sensitivity tests are needed in future to further investigate the possible mechanisms that responsible for the distinctive results.

Keywords: squall line; microphysical parameterization scheme; SBM; bulk; comparative study

1. Introduction

Squall line is defined as 'a line of active thunderstorms, either continuous or with breaks, including contiguous precipitation areas resulting from the existence of the thunderstorms' (Glickman, 2000). Due to the distinctive geometrical and dynamical structure, squall lines are commonly simulated and tested with different explicit microphysics in the same dynamical framework (Lynn et al., 2005b; Seifert et al., 2006; Lynn and Khain, 2007; Khain et al., 2009; Li et al., 2009a; Morrison et al., 2009; Bryan and Morrison, 2012; Van Weverberg et al., 2012).

Two types of microphysical parameterization schemes were used in recent numerical models: bulk-microphysics scheme and spectral (bin) microphysics (SBM). In the traditional bulk scheme, the size distribution of hydrometeors is assumed as an empirical function and not changes during the simulation. It includes the one-moment (e.g. Kessler, 1969; Lin et al., 1983), two-moment (e.g. Thompson et al., 2008; Morrison et al., 2009) and three-moment (e.g. Milbrandt and Yau, 2005) bulk schemes. In the three-moment scheme, the predictive equation of radar reflectivity is added and the shape parameter μ ($N(D) = N_0 D^\mu e^{-\lambda D}$) becomes a fully prognostic variable. The simplification of bulk scheme makes it conceptually simple and computationally efficient and is widely used in numerical models, but on the other hand inevitably causes some limitations. For instance, the mean terminal velocity assumption of each type of hydrometeor which actually depends on the particle size may lead to errors in the spatial distribution of different particles (Lynn et al., 2005a). In addition, it does not solve the equation for diffusion growth of drops which is replaced by transformation of all supersaturated water vapor into cloud water mass and instead of solving the stochastic equation of collisions, semiempirical relationships for auto-conversion rates are used (Khain et al., 2009).

Another approach is the SBM. By comparison with bulk one, it uses dozens, even hundreds of mass bins to describe the size distributions of each type of hydrometeors as well as the cloud condensation nuclei (CCN). For example, the SBM implemented in Hebrew University Cloud Model (HUCM) solves prognostic equations for seven types of hydrometeors and CCN: water drops, three types of ice crystals (columnar, plate-like and dendrites), snowflakes (aggregates), graupel, hail/frozen drops. Each size distribution is represented by 33 mass bins. Lynn et al. (2005a) initially developed a fast version of SBM (SBM Fast) in which the number of size distributions decreased from eight to four (water drops, small ice particles, large ice particles and aerosol) and coupled it with a three-dimensional mesoscale model. One of their numerical experiments revealed the SBM fast had almost similar results to the full

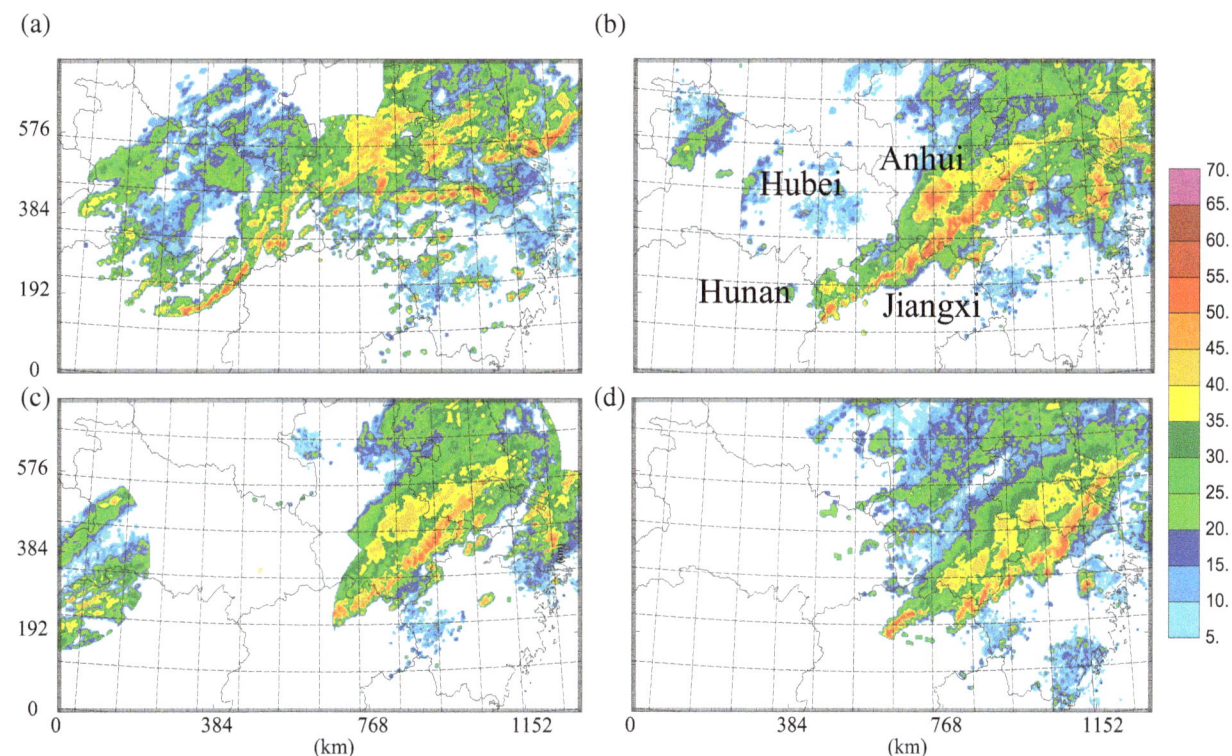

Figure 1. Radar mosaics at (a) 0500 UTC, (b) 0900 UTC, (c) 1000 UTC and (d) 1200 UTC on 12 July 2014.

version, which made it a good choice for our simulation experiments.

Several studies have been made to compare the bulk and SBM microphysics with the different dynamical frameworks. For instance, Lynn *et al.* (2005b) found the SBM Fast had a more realistic reproduction of radar reflectivity, surface rainfall and cloud structure than bulk schemes. By using a cloud-resolving model, Li *et al.* (2009a, 2009b) found the bulk scheme produced a multicell storm with rapid and strong evolution, while SBM produced a unicell storm with slow and weak evolution, which could be explained by different rain evaporation rate and fall velocities of precipitable ice particles between the two schemes. Note that similar results were also obtained by Khain *et al.* (2009). In terms of the Chinese cases, Fan *et al.* (2012) found a two-moment bulk scheme predicted much higher cloud droplet number and the opposite CCN effects on convection and heavy rain compared with SBM.

In summary, the differences between bulk and bin simulation results are case-dependent and no definite conclusions have been obtained. Sometimes SBM even had worse results than bulk scheme (Iguchi, 2014). On the other hand, it seems that most previous studies focused on idealized experiments or squall lines occurred in North America, where the weather background and atmospheric stratification were quite different from that in China (Meng *et al.*, 2013). How the bin microphysical scheme performs in a realistic severe convective cloud such as squall line in East China? Are there any differences from the previous results? In this article we will try to answer the two questions.

This article is organized as follows: Section 2 describes the brief description of the squall line of interest. Section 3 gives the design of numerical experiment. Section 4 compares the simulation results with observation and focuses on the differences between bin and bulk microphysical schemes. Section 5 briefly describes the results of another squall-line case. A summary and discussion is given in Section 6.

2. Case description

A severe squall line occurred in Anhui and Jiangsu province of East China on 12 July 2014 was studied in this work. Based on the radar mosaics composited by several S and C band Doppler radars in East China, the squall line was originated from scattered clouds at 0400 UTC 12 July 2014 and then merged into a thin bow echo near the border of Hubei, Hunan and Jiangxi Province at 0500 UTC (Figure 1(a)). As moving eastward, the northern part of squall line gradually enhanced and it matured at 0900 UTC (Figure 1(b)) with a 400-km long band of convective towers in the leading edge and an intense meso-β-scale convective cell behind it. After 1000 UTC (Figure 1(c)) a broad stratiform cloud started to expand to the rear of the dissipating squall line and when the squall line moved out to sea at around 1300 UTC, it finally decayed. The squall line lasted about 9 h and caused

(a)

(b)

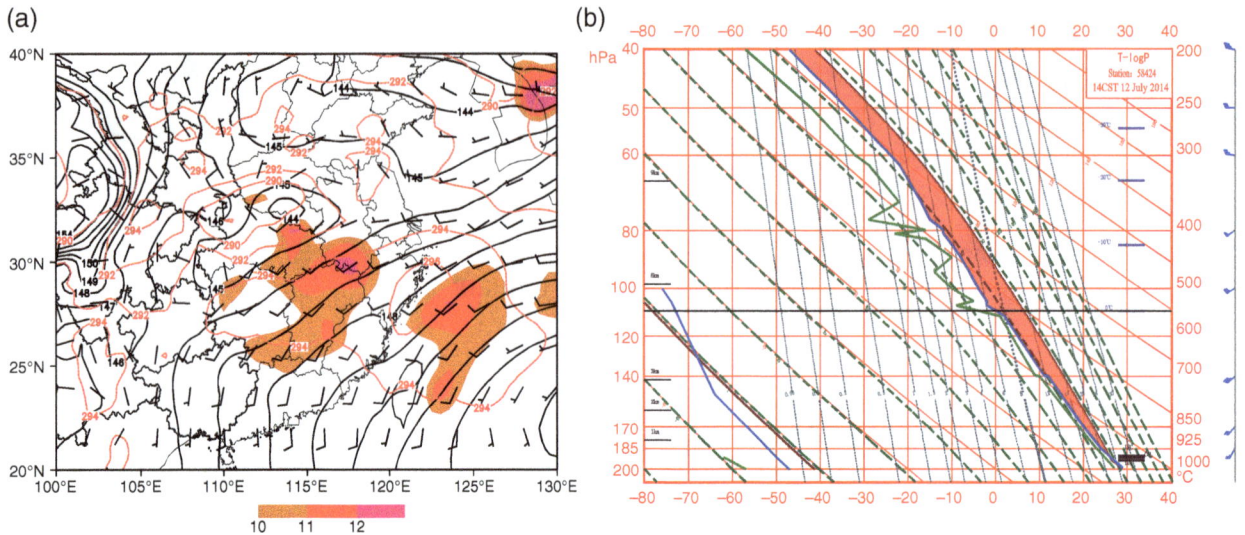

Figure 2. The (a) geopotential height (black contours, unit: 10 gpm), horizontal wind field (wind barb), temperature (red dashed contours, unit: K), and low-level jet (shaded, unit: m s^{-1}) of 850 hPa at 0000 UTC on 12 July. (b) Skew-T diagram in Anqing station at 0600 UTC 12 July 2014.

heavy rainstorms and strong winds in Anhui, Jiangsu, Hunan and Hubei provinces. For example, the 24-h total rainfall in Liuan city of Anhui Province was up to 168.9 mm.

Based on the Global Forecast System (GFS) reanalysis data from the National Centers for Environmental Prediction (NCEP) and radiosonde observations from China Meteorological Administration (CMA), the weather conditions of squall line are analyzed. From Figure 2(a), the squall line was formed to the southern side of a vortex at 850 hPa and the left front of a low-level jet (Figure 2(a)). Cold and dry air was carried from high latitude by the east-propagating vortex and met with the southwesterly warm and moist air, which triggered convections initially. From the skew-T plot at Anqing station in Anhui Province before squall line passed by (Figure 2(b)), the vertical wind shear within the 0–3 km layer was up to 16 m s^{-1} and the value of convective available potential energy (CAPE) was 2921.4 J kg^{-1}. The high CAPE and strong wind shear also provided favorable conditions to the longevity of convective activities.

3. Design of experiments

The Advanced Research WRF (ARW) v3.6.1 was used in our study to simulate the squall line of interest. The model was designed with three domains and two-way nesting. Each domain has 38 vertical levels with the model top at 50 hPa. Domain 1 has 300×240 grid points with a 13.5 km grid spacing, domain 2 has 361×301 grid points with the grid spacing of 4.5 km and domain 3 has 481×361 points with 1.5 km resolution. Two numerical experiments were conducted using Milbrandt 2-moment (bulk scheme) and HUJI SBM fast (bin scheme) microphysical schemes respectively with the same YSU planetary boundary scheme,

Noah Land surface Model, RRTM long radiation scheme, Dudhia shortwave radiation scheme and the improved Kain–Fritsch cumulus parameterization scheme (Tang, 2013). Note that cumulus convection was off in the finest domain. The time step is 60 s for domain 1, 20 s for domain 2 and 6 s for domain 3. It integrated for 15 h starting at 0000 UTC on 12 July 2014. The 3-h GFS data was chosen as the initial and boundary condition with the resolution of 0.5 degree.

The Milbrandt two-moment scheme used in our study is a default version in WRF3.6.1 and it predicts mass and number concentrations of cloud water, rain, ice, snow, graupel and hail. The size distribution of each hydrometeor type is represented by a gamma function with a fixed shape factor (Morrison and Milbrandt, 2011). The initial cloud droplet number concentration (N_c) is set to 500 cm^{-3} for polluted continental cases and is predicted during the simulation.

The fast version of SBM developed by Lynn et al. (2005a) includes four hydrometeor categories: water drops, ice/snow, graupel/hail and aerosol. The initial aerosol size distribution in SBM fast is determined by the power law function: $N_{CCN} = CS^k$, where N_{CCN} is the CCN number concentration (cm^{-3}), S is the supersaturation with respect to water (%), C and k are constants that depend on air mass type. Since the squall line occurred in East China, which was known as a polluted region, a type of continental aerosol concentration was used in our study ($C = 4000$ cm^{-3} and k is 0.308). The values of C and k were determined with reference to Fan et al. (2012). Based on their study, the total CCN number concentration is about 8600 cm^{-3}, which is close to observations of 10^4 cm^{-3} in Jinan (a city about 700 km away from Anhui province) during summer by Gao et al. (2007).

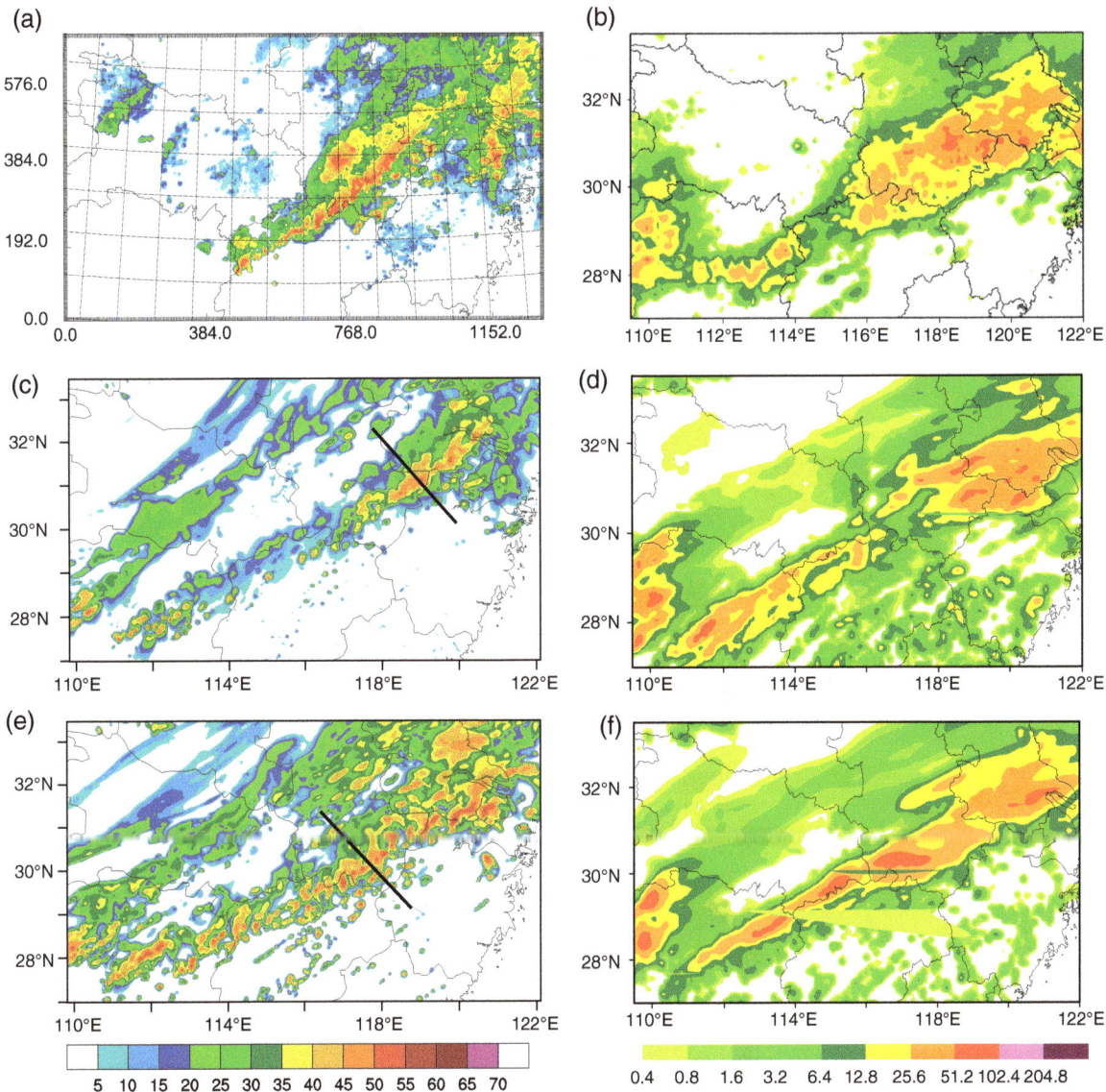

Figure 3. The radar reflectivity (unit: dbz) from (a) observation at 0900 UTC, (c) bulk scheme at 1000 UTC, (e) bin scheme at 1000 UTC and the 6-h total rainfall (unit: mm) from (b) observation, (d) bulk scheme, (f) bin scheme during the period of 0600 UTC to 1200 UTC 12 July 2014. The thick black line denotes the location of cross sections in Figure 4.

4. Comparison of bulk and bin simulation results

4.1. Radar reflectivity and surface rainfall

Figure 3 shows the observed and simulated radar reflectivity and surface rainfall. Since both bulk and bin scheme simulated a squall line occurred 1 h later than observation, we chose the different moment to represent the mature stage of squall line. In view of the radar reflectivity, the observed squall line (Figure 3(a)) matured at 0900 UTC 12 July, with a 400-km long band of convective clouds extended from northwestern Jiangxi to southern Anhui province. Besides, a meso-β-scale convective cell was observed around the border of Anqing and Liuan city in Anhui province. Figures 3(c) and (e) show simulated results by bulk and bin scheme. It seems that both of the two schemes captured the general features of the observed squall line,

including the orientation, moving direction and organizational mode. However, the bulk scheme (Figure 3(c)) produced weaker and shorter radar echoes and it underestimated the severe convections in southern part of the squall line. By comparison, the bin scheme (Figure 3(e)) simulated stronger and longer radar echoes which is more consistent with observation, although convections were scattered. Note that neither of the two schemes has simulated the strong convective cell in the trailing stratiform (TS) region from observation.

The right part of Figure 3 is the 6-h total rainfall from 0600 to 1200 UTC 12 July. The observed precipitation data is from the National Meteorological Information Center (Sheng *et al.*, 2013). In Figure 3(b), the observed rain band was southwest-northeast oriented and extended from southern Anhui to southern Jiangsu province. Strong precipitation mainly occurred around Wuhu and Yicheng city in Anhui

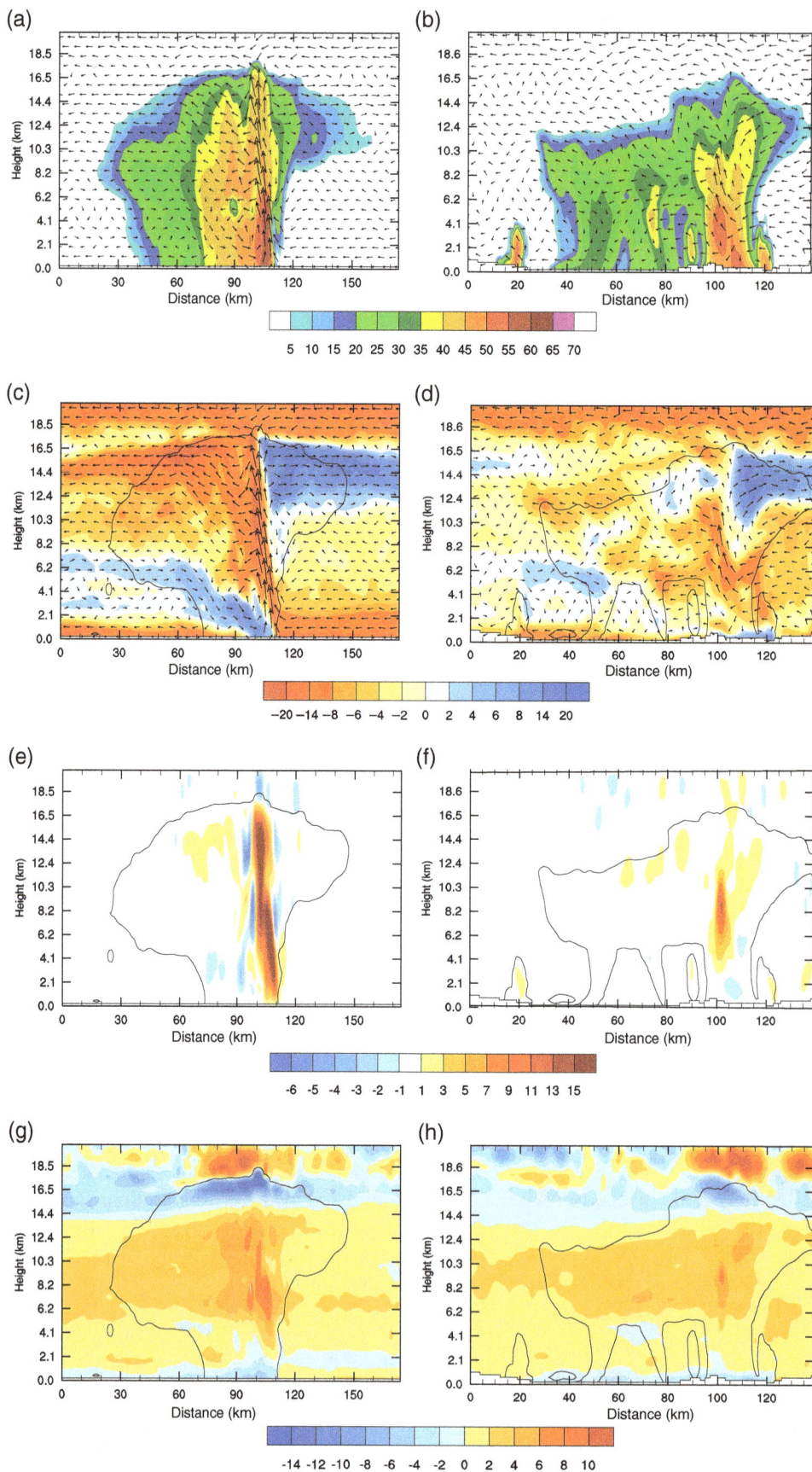

Figure 4. Vertical cross sections of (a, b) radar reflectivity (shading, unit: dbz) and storm-relative wind field (vector), (c, d) horizontal wind speed (shading, unit: m s^{-1}), (e, f) vertical velocity (unit: m s^{-1}), (g, h) potential temperature perturbation (unit: K). Figures (a, c, e, g) are for bulk scheme while figures (b, d, f, h) are for bin scheme. The black solid line is the water content of 0.2 g kg^{-1} which denotes the outline of the storm.

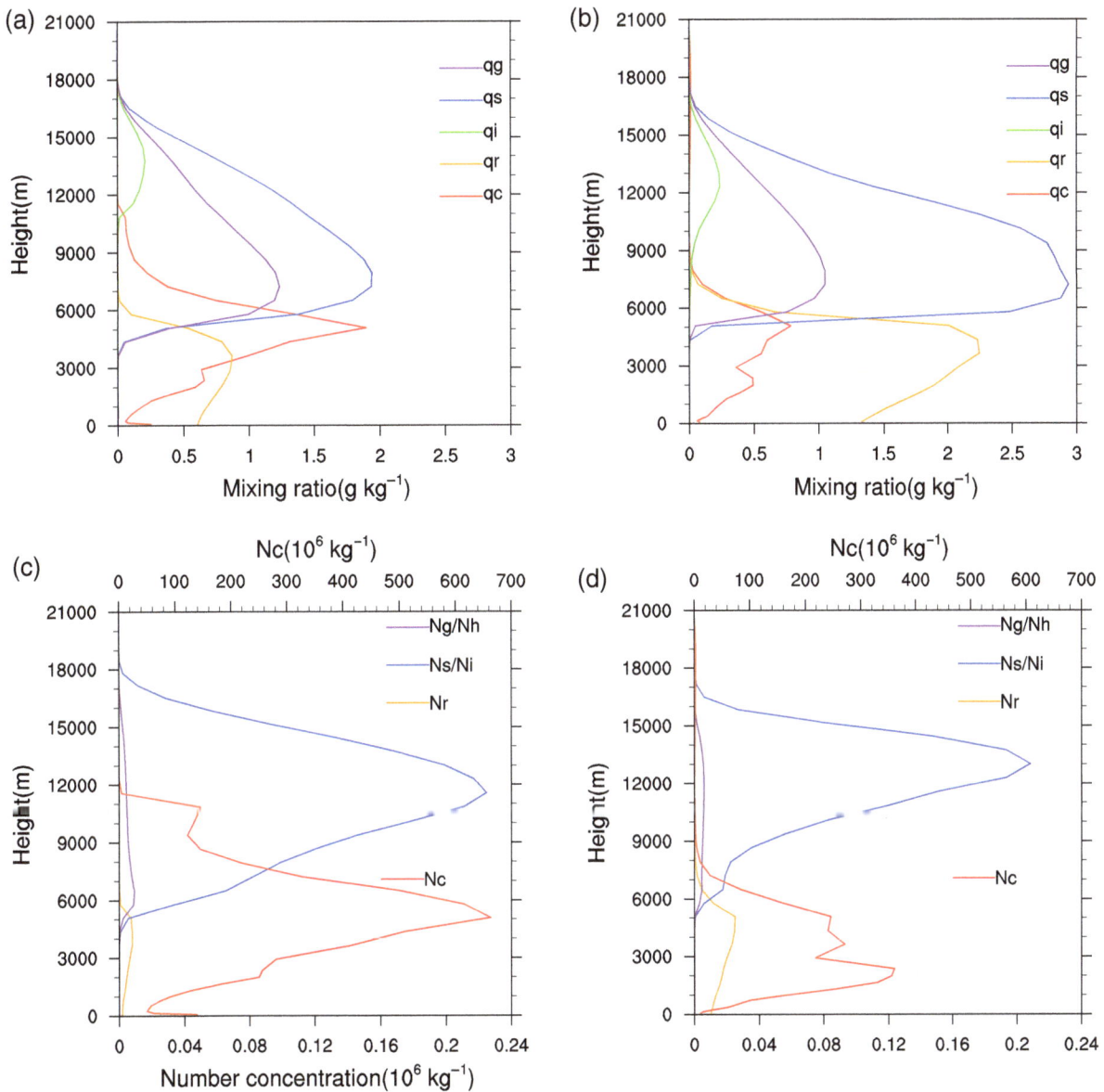

Figure 5. Vertical profiles of (a, b) mixing ratio (unit: $g\,kg^{-1}$) and (c, d) number concentration (unit: $10^6\,kg^{-1}$) for cloud droplet (red line), cloud ice (green line), snow (blue line), rain (orange line) and graupel (purple line) for (a, c) bulk scheme and (b, d) bin scheme.

province with 6-h rainfall amount in excess of 51.2 mm. From simulation results, the bulk scheme (Figure 3(d)) has basically reproduced the strong rainfall areas and amount in observation, whereas the bin scheme (Figure 3(f)) predicted three bogus rainfall centers in the southwestern part of rain band and it also underestimated the rainfall amount in Yicheng city.

In conclusion, significant differences existed in the simulated radar reflectivity and surface rainfall by bulk and bin microphysical schemes during the mature stage of squall line. The bulk scheme produced a shorter and weaker radar echo structure, while the bin scheme simulated a strong but scattered radar echo structure. In addition, the location and amount of rain band simulated by bulk scheme agreed better with observation than that by bin scheme.

4.2. Dynamic and thermodynamic structure

As the squall line we studied is a quasi-two-dimensional system, the vertical cross sections along the moving direction of storm were given in Figure 4. Here we just gave the simulated results because no vertical structure observations were available. In terms of the cross section of radar reflectivity, both of the bulk and bin scheme produced the radar structure of an intense convective core in the leading edge and several dissipating old cells followed by the wide-spread TS region. The structure of radar reflectivity is quite similar to the conceptual model proposed by Houze *et al.* (1989). The differences between bulk and bin simulation results are the strength of leading convective core and the wideness of stratiform cloud. The stratiform region in bin scheme was more broken than bulk scheme with some weak reflectivity zones embedded, which could be also seen from Figure 3. Whereas the bulk scheme simulated a more intense leading cell with higher echo tops than bin scheme. The radar structures of squall line simulated by bulk and bin scheme are also investigated by Lynn *et al.* (2005b) and Li *et al.* (2009a). Both of their results revealed bin scheme simulated a more realistic radar echo than bulk scheme.

Figures 4(c) and (d) show cross sections of storm-relative wind field which was obtained by subtracting the mean squall-line moving speed. It was seen that the airflows simulated by bulk scheme (Figure 4(c)) were featured by a storm-relative front-to-rear (FTR) inflow below the height of 2 km (red shading), a descending rear-to-front (RTF) inflow from 2 to 8 km (blue shading) and a FTR outflow on upper troposphere. The FTR in front of squall line ascended abruptly when approaching the surface gust front, then crossed the convective updrafts and eventually flowed out above the altitude of 8 km. The rear inflow descended when approaching the leading convective cell which carried the mid-level environmental air to the near ground. By comparison, the bin scheme (Figure 4(d)) produced much weaker low-level front inflow and mid-level rear inflow. Moreover, the rear inflow was discontinuous despite it still sank near the convective updraft. From the characteristics of vertical velocity, a very strong convective updraft was simulated by bulk scheme with maximum velocity over $15 \, \text{m s}^{-1}$ (Figure 4(e)). Around the strong upward motion area, several weak downdrafts were induced probably as a result of mass compensation. However, weaker updrafts and downdrafts were simulated in bin scheme (Figure 4(f)), which is consistent with the features of radar reflectivity. The overestimation of updrafts and production of much too strong convection by bulk scheme has been indicated in many previous studies (Tao *et al.*, 2007; Khain and Lynn, 2009; Li *et al.*, 2009a, 2009b; Fan *et al.*, 2012). The reasons for this feature have been discussed by Khain and Lynn (2009).

Apart from the storm dynamical structures, we also compared the simulated potential temperature perturbation, which was obtained by subtracting the initial temperature. From Figures 4(g) and (h), it was seen that both of bulk and bin scheme have simulated the three-layer structures of potential temperature perturbation, including the cold pool near the ground caused by rain evaporation, thick heating layer in the middle levels associated with non-adiabatic heating and cooling again above 14 km. The difference was that the bulk scheme (Figure 4(g)) had a deeper cold pool than bin (Figure 4(h)). The reason may be that stronger rear inflow was simulated in bulk scheme as shown in Figure 4(c) and more cold-dry air was carried from middle levels, which strengthened the cold pool in the near ground.

4.3. Microphysical structure

Figure 5 shows the vertical profiles of mixing ratio and number concentration for different type of hydrometeors. Both of the mixing ratio and number concentration are averaged over the area of domain 3 and then accumulated from 0000 to 1200 UTC. The most striking differences between bulk and bin results were the simulated cloud droplet (red line), rain water (orange line) and snow (blue line). The bulk scheme predicted much larger mixing ratio of cloud droplet (red line) with the maximum mass content of $2 \, \text{g kg}^{-1}$ compared with $0.75 \, \text{g kg}^{-1}$ in bin scheme. Similar characteristics were also seen from number concentration. The peak number concentration of cloud droplet in bulk scheme was about $660 \times 10^6 \, \text{kg}^{-1}$, while it was just about $360 \times 10^6 \, \text{kg}^{-1}$ in bin scheme. Correspondingly, rain number and mass concentration (orange line) in bulk scheme were drastically lower than those in bin scheme. This phenomenon may suggest that the conversion efficiency of cloud droplets to raindrops in Milbrandt scheme was much less than that in SBM. Similar results were also obtained by Fan *et al.* (2015). Another striking feature was that cloud droplet existed until ~12 km (close to homogeneous freezing level with a temperature of ~−38 °C) in bulk scheme, comparing to ~9 km in SBM. This phenomenon was also very obvious for the case on 31 March 2014. The reason for the different existing levels of cloud droplet might be that bulk scheme predicted much stronger updrafts than SBM and cloud droplets had no enough time to grow larger and convert to hydrometeors before reaching the homogeneous freezing level. Thus more cloud droplets were carried to the height of homogenous freezing and were instantly frozen into ice crystals (Xu *et al.*, 2011).

5. Another squall line case

Based on the analysis and discussions above, some preliminary results were obtained about the differences of bulk and bin scheme results. To generalize the conclusions of this study, another squall-line case which occurred over south China on 31 March 2014 was simulated and analyzed in this section. The corresponding plots are given in the Appendix. Here we just described the results briefly.

Figure A1 shows the observed and simulated radar reflectivity and rainfall just like Figure 3. It seems that the simulated squall line moved faster than observation and the radar features were very identical between the two schemes. From the surface rainfall, the observed rain band was in the northeastern and coastal areas of Guangdong province, while neither of two schemes has produced strong precipitation in the coastal area.

In terms of the storm dynamics and thermodynamics, the results of two squall lines had something in common, but still existed differences. For example, the bin scheme had wider but broken stratiform clouds than bulk one, while bulk scheme predicted stronger convective updrafts just as found in Figure 4. The differences of rear inflow and cold pool were not apparent between the two schemes.

As to the microphysical properties, a striking difference was that the second squall line had lower convective levels than first one, so the simulated hydrometeors were distributed blow 15 km. Moreover, the bulk scheme also predicted much higher cloud droplet number concentrations and lower raindrop concentrations than SBM. However, this time the differences of cloud droplet number concentrations between two schemes were not as large as that in the previous squall-line case.

6. Summary and discussion

A squall line that developed in East China during 12 July 2014 was simulated using WRF model with traditional bulk and spectral bin scheme respectively. In order to investigate the sensitivity of simulation results to different microphysical schemes, especially test the performance of newly incorporated SBM in WRF model, detailed comparative studies have been carried out.

By comparing the simulated radar reflectivity and surface rainfall with observation, we found that the bulk scheme produced a well-organized but shorter squall line, while the bin scheme produced stronger but scattered radar echoes, which was relatively more consistent with observation. Moreover, the bulk scheme had a better reproduction of the strong rainfall areas by comparison with the bin scheme.

In view of the dynamic and thermodynamic structures, both of bulk and bin scheme have simulated a radar structure with the leading convective tower and TS region. The difference was that bin produced wider convective cores and broken TS clouds than bulk one. The airflow in bulk scheme was characterized by the strong front inflow in low levels and intense rear inflow in middle levels compared with weak and discontinuous rear inflow in bulk scheme. Meanwhile, the bulk scheme simulated much stronger updrafts and deeper cold pool than SBM. In terms of the microphysical properties, much higher cloud droplet number concentrations were simulated by bulk scheme, while lower raindrop mass and number concentrations were generated than those in SBM.

As mentioned in introduction, a lot of comparative studies have been done between bulk and bin microphysical schemes using cloud-resolving models or mesoscale models. Most of the results proved bin scheme had a better performance than bulk scheme no matter in real-time or idealized simulations. Considering that seldom researches has been done in China on the sensitivity of squall lines to bulk and bin microphysics, our conclusions are very preliminary and case-dependent. More cases are needed in future to validate our results. Besides, detailed analysis and sensitivity tests are also required to investigate the mechanisms for the distinctive results between bulk and bin schemes.

Acknowledgements

This work is jointly supported by the National Basic Research Program of China (Grant No. 2013CB430105) and the National Natural Science Foundation of China (Grant No. 41675059, 41405059, 41375066, 40875031 and U1333130). The authors thank the anonymous reviewers for their suggestions that helped to improve the manuscript. Thanks also to Dr. Tang Xiba for his technical assistance.

Supporting information

The following supporting information is available:

Figure S1. Time series of domain-averaged cloud droplet number concentration (unit: 10^6 kg^{-1}).

References

Bryan GH, Morrison H. 2012. Sensitivity of a simulated squall line to horizontal resolution and parameterization of microphysics. *Monthly Weather Review* **140**: 202–225.

Fan J, Leung LR, Li Z, Morrison H, Chen H, Zhou Y, Qian Y, Wang Y. 2012. Aerosol impacts on clouds and precipitation in eastern China: results from bin and bulk microphysics. *Journal of Geophysical Research* **117**: D00K36.

Fan J, Liu YC, Xu KM, North K, Collis S, Dong X, Zhang GJ, Chen Q, Kollias P, Ghan SJ. 2015. Improving representation of convective transport for scale-aware parameterization: 1. Convection and cloud properties simulated with spectral bin and bulk microphysics. *Journal of Geophysical Research* **120**: 3485–3509.

Gao J, Wang J, Cheng SH, Xue LK, Yan HZ, Hou LJ, Jiang YQ, Wang WX. 2007. Number concentration and size distribution of submicron particles in Jinan urban area: characteristics in summer and winter. *Journal of Environmental Sciences* **19**(12): 1466–1473.

Glickman TS, Zenk W. 2000. *Glossary of Meteorology*, 2nd ed. American Meteorological Society: Boston: MA; 855 pp.

Houze RA Jr, Biggerstaff MI, Rutledge SA, Smull BF. 1989. Interpretation of Doppler weather radar displays of midlatitude mesoscale convective systems. *Bulletin of the American Meteorological Society* **70**: 608–619.

Iguchi T. 2014. Real case simulations using spectral bin cloud microphysics: Remarks on precedence research and feature activity. The 4th conference of ultra-high precision meso-scale weather prediction. Kobe, Japan.

Kessler E. 1969. On the distribution and continuity of water substance in atmospheric circulations. In *Meteorological Monographs*. American Meteorological Society: Boston, MA; 84 pp.

Khain AP, Lynn BH. 2009. Simulation of a supercell storm in clean and dirty atmosphere using weather research and forecast model with spectral bin microphysics. *Journal of Geophysical Research* **114**: D19209.

Khain AP, Leung LR, Lynn BH, Ghan S. 2009. Effects of aerosols on the dynamics and microphysics of squall lines simulated by spectral bin and bulk parameterization schemes. *Journal of Geophysical Research: Atmospheres* **114**: D22203.

Li X, Tao W-K, Khain AP, Simpson J, Johnson DE. 2009a. Sensitivity of a cloud-resolving model to bulk and explicit bin microphysical schemes. Part I: comparisons. *Journal of Atmospheric Sciences* **66**: 3–21.

Li X, Tao W-K, Khain AP, Simpson J, Johnson DE. 2009b. Sensitivity of a cloud-resolving model to bulk and explicit bin microphysical schemes. Part II: cloud microphysics and storm dynamics interactions. *Journal of Atmospheric Sciences* **66**: 22–40.

Lin YL, Farley HD, Orville HD. 1983. Bulk parameterization of the snow field in a cloud model. *Journal of Applied Meteorology* **22**(6): 1065–1092.

Lynn BH, Khain AP. 2007. Utilization of spectral bin microphysics and bulk parameterization schemes to simulate the cloud structure and precipitation in a mesoscale rain event. *Journal of Geophysical Research* **112**: D22205.

Lynn BH, Khain AP, Dudhia J, Rosenfeld D, Pokrovsky A, Seifert A. 2005a. Spectral (bin) microphysics coupled with a mesoscale model (MM5). Part I: model description and first results. *Monthly Weather Review* **133**(1): 44–58.

Lynn BH, Khain AP, Dudhia J, Rosenfeld D, Pokrovsky A, Seifert A. 2005b. Spectral (bin) microphysics coupled with a mesoscale model (MM5). Part II: simulation of a CaPE rain event with a squall line. *Monthly Weather Review* **133**(1): 59–71.

Meng Z, Yuan D, Zhang Y. 2013. General features of squall lines in East China. *Monthly Weather Review* **141**: 1629–1647.

Milbrandt JA, Yau MK. 2005. A multi-moment bulk microphysics parameterization: Part 2: a proposed three-moment closure and scheme description. *Journal of Atmospheric Science* **62**(9): 3065–3081.

Morrison H, Milbrandt J. 2011. Comparison of two-moment bulk microphysics schemes in idealized supercell thunderstorm simulations. *Monthly Weather Review* **139**(4): 1103–1130.

Morrison H, Thompson G, Tatarskii V. 2009. Impact of cloud microphysics on the development of trailing stratiform precipitation in a simulated squall line: comparison of one-and two-moment schemes. *Monthly Weather Review* **137**: 991–1007.

Seifert A, Khain A, Pokrovsky A, Beheng KD. 2006. A comparison of spectral bin and two-moment bulk mixed-phase cloud microphysics. *Atmospheric Research* **80**: 46–66.

Shen Y, Pan Y, Yu J-J, Zhao P, Zhou Z-J. 2013. Quality assessment of hourly merged precipitation product over China (in Chinese). *Transactions of Atmospheric Sciences* **36**(1): 37–46.

Tang XB. 2013. Improvement of cumulus convective parameterization schemes in regional model and its application on rainstorm and typhoon in China (in Chinese). PhD dissertation, The Institution of Atmospheric Physics, Chinese Academy of Science, Beijing; 49–50.

Tao WK, Li X, Khain A, Matsui T, Lang S, Simpson J. 2007. Role of atmospheric aerosol concentration on deep convective precipitation: cloud-resolving model simulations. *Journal of Geophysical Research* **112**: D24S18.

Thompson G, Field PR, Rasmussen RM, Hall WD. 2008. Explicit forecasts of winter precipitation using an improved bulk scheme. Part 2: implementation of a new snow parameterization. *Monthly Weather Review* **136**(12): 5095–5115.

Van Weverberg K, Vogelmann A, Morrison H, Milbrandt JA. 2012. Sensitivity of idealized squall-line simulations to the level of complexity used in two-moment bulk microphysics schemes. *Monthly Weather Review* **140**: 1883–1907.

Xu XH, Yu X, Dai J, Liu GH, Zhu YN, Yue ZG. 2011. Direct observation from sounding of the warming caused by homogeneous freezing in a severe storm (in Chinese). *Transactions of Atmospheric Sciences* **34**(4): 416–422.

Appendix

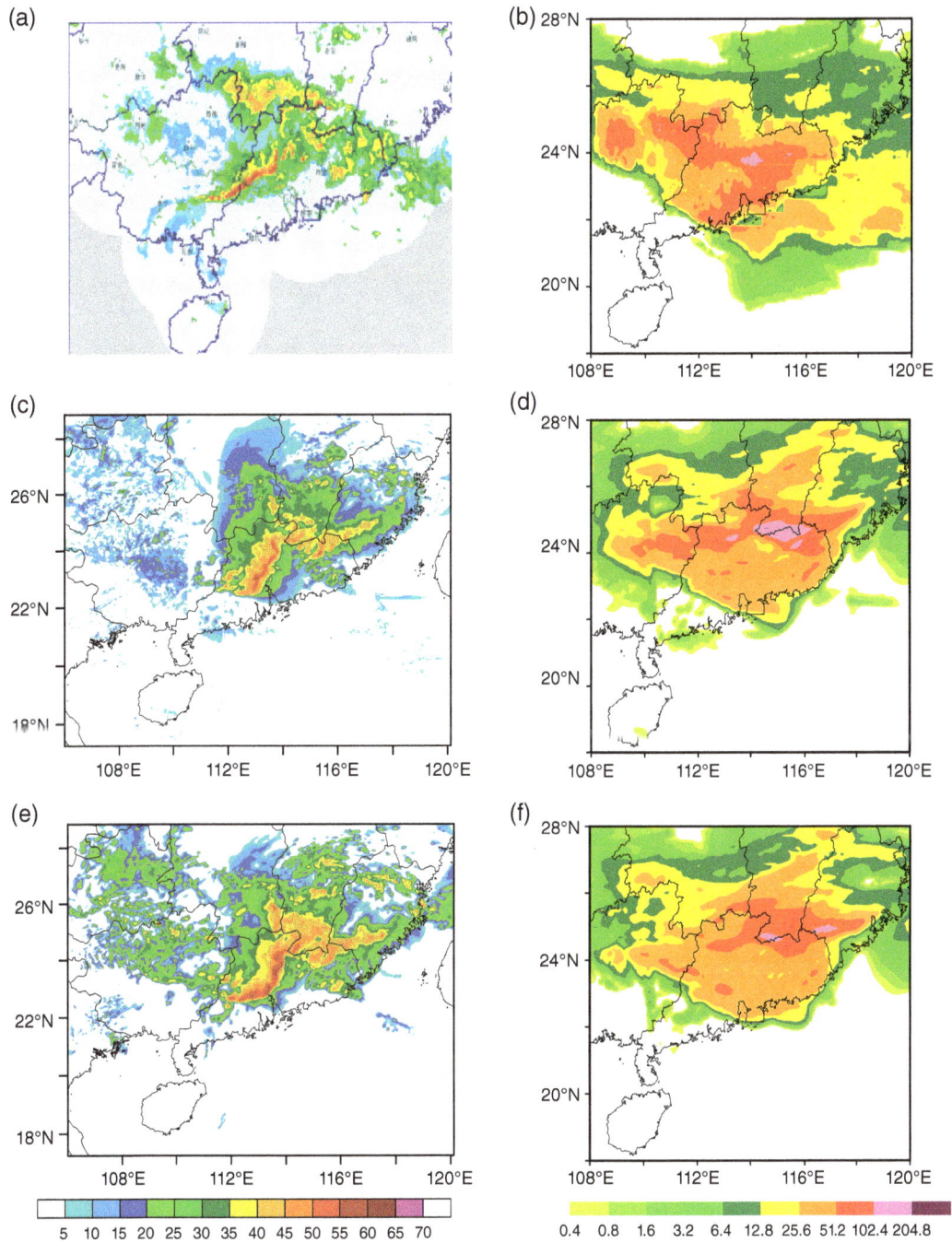

Figure A1 Radar reflectivity (unit: dbz) from (a) observation, (c) bulk scheme, (e) bin scheme at 0200 UTC on 31 March 2014 and 12-h total rainfall (unit: mm) from (b) observation, (d) bulk scheme, (f) bin scheme during the period from 1800 UTC on 30 March to 1200 UTC on 31 March 2014.

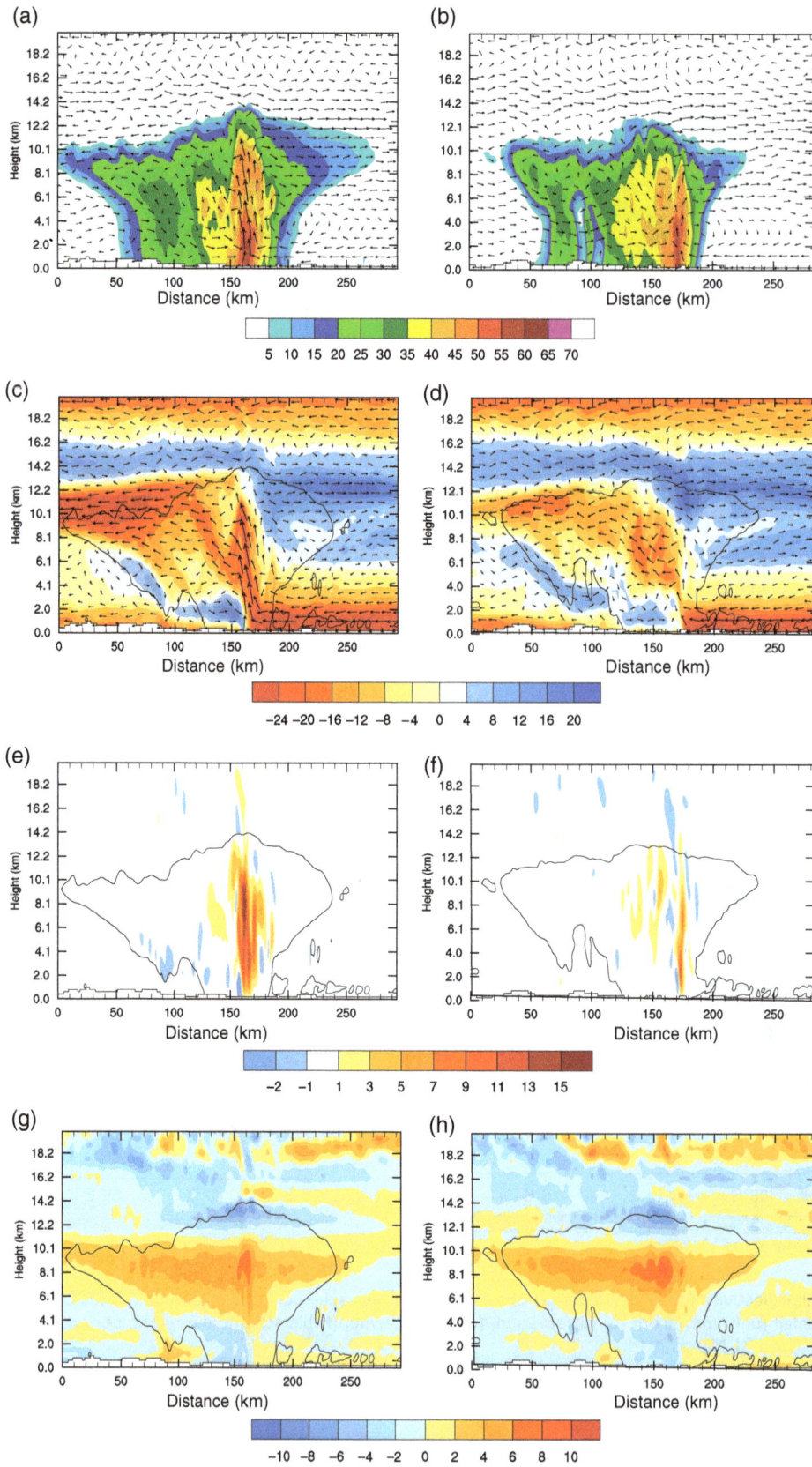

Figure A2 Same as Figure 4, but for the squall line occurred on 31 March 2014.

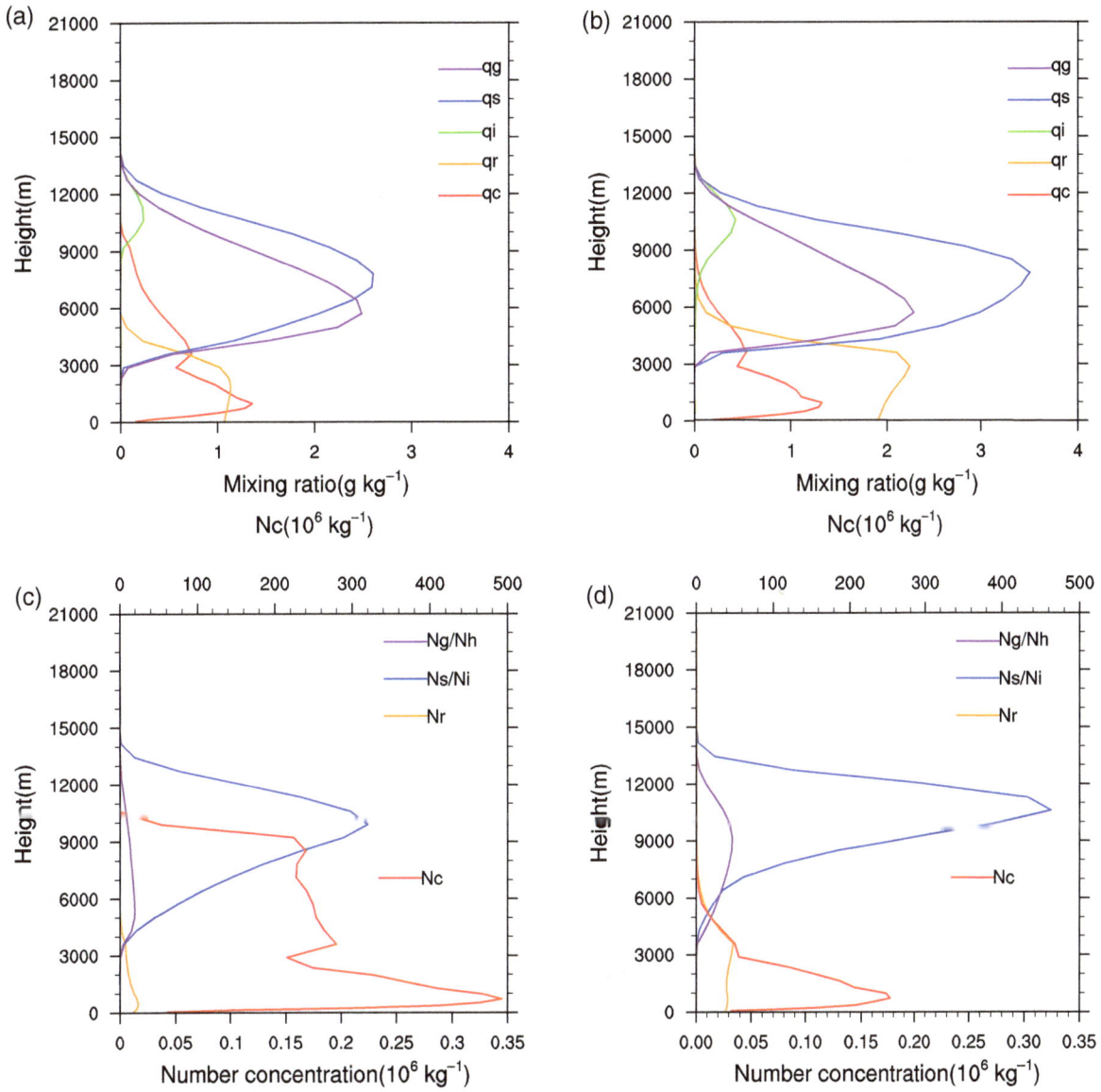

Figure A3 Same as Figure 5, but for the squall line occurred on 31 March 2014.

19

The role of carbonaceous aerosols on short-term variations of precipitation over North Africa

Jin-Ho Yoon,[1]* Philip J. Rasch,[2] Hailong Wang,[2] V. Vinoj[3] and Dilip Ganguly[4]

[1] School of Earth Sciences and Environmental Engineering, Gwangju Institute of Science and Technology, Gwangju, South Korea
[2] Pacific Northwest National Laboratory, Richland, WA, USA
[3] School of Earth, Ocean and Climate Sciences, Indian Institute of Technology Bhubaneswar, Odisha, India
[4] Center for Atmospheric Sciences, Indian Institute of Technology Delhi, New Delhi, India

*Correspondence to:
J.-H. Yoon, School of Earth
Sciences and Environmental
Engineering, Gwangju Institute of
Science and Technology, 123
Cheomdan Gwagi-Ro, Buk-Gu,
Gwangju, 61005, South Korea.
E-mail: yjinho@gist.ac.kr

Abstract

Subtropical North Africa has been subject to extensive droughts in the late 20th century, linked to changes in the sea surface temperature (SST). However, climate models forced by observed SSTs cannot reproduce the magnitude of the observed rainfall reduction. Here, we propose aerosol indirect effects (AIE) as an important positive feedback mechanism. Model results are presented using two sets of sensitivity experiments designed to distinguish the role of aerosol direct/semi-direct and indirect effects on regional precipitation. Changes in cloud properties due to the presence of carbonaceous aerosols are proposed as a key mechanism to explain the reduced rainfall over subtropical North Africa.

Keywords: Sahel rainfall; aerosol indirect effect; cloud lifetime effect

I. Introduction

Between 1950 and 1980, the Sahel, located between the tropical African rainforest and the Sahara, has experienced a large reduction in rainfall (e.g. Nicholson, 2001; Dai *et al.*, 2004). Various mechanisms have been proposed to explain this drought (e.g. Zeng *et al.*, 1999; Nicholson, 2001; Giannini *et al.*, 2003; Lu and Delworth, 2005; Hoerling *et al.*, 2006). The long-term change in sea surface temperature (SST) over the Atlantic and Indian Oceans is a commonly accepted hypothesis (Giannini *et al.*, 2003; Lu and Delworth, 2005; Hoerling *et al.*, 2006) for the rainfall change in Sahel. However, most atmospheric general circulation models forced by the observed SST changes cannot simulate the magnitude of this drought (Scaife *et al.*, 2009). Thus, alternative climate feedback processes such as atmosphere-land-vegetation feedbacks through soil moisture and vegetation albedo have been proposed as an amplifying mechanism (Charney *et al.*, 1975; Zeng *et al.*, 1999; Kucharski *et al.*, 2013).

A growing number of studies have also examined the potential impact of aerosols on Sahel rainfall (e.g. Held *et al.*, 2005; Ackerley *et al.*, 2011; Booth *et al.*, 2012; Wang *et al.*, 2012a; Chang, 2013; Hwang *et al.*, 2013; Dong *et al.*, 2014). These studies have particularly focused on how sulphate aerosols or dust in the Northern Hemisphere affected rainfall in North Africa by creating inter-hemispheric SST gradients. It has also been suggested that volcanic eruptions may be an important player in determining rainfall amount in North Africa (Haywood *et al.*, 2013). There are other ways that aerosols can be linked to droughts. For example,

prolonged droughts in the Amazon have been associated with an increase in biomass burning (e.g. Scholze *et al.*, 2006; Zeng *et al.*, 2008) that may increase emissions of carbonaceous aerosols, but there are very few studies that have examined the climatic feedback of carbonaceous aerosols on droughts in North Africa. In this study, we explore the potential impact of additional carbonaceous aerosols loading on regional precipitation over North Africa.

There are two broad categories of aerosol effects. Direct effects refer to the influence that particles have on atmospheric radiative transfer via scattering and absorption. Indirect and semi-direct effects refer to the influence particles have on cloud properties with subsequent effects on radiative heating. Climate modelling studies have historically focused on direct effects because of the difficulty in treating more complex indirect effects (e.g. Chung and Zhang, 2004; Chung and Seinfeld, 2005). However, the microphysical interaction of cloud, aerosols, and precipitation (the 'indirect effects') can be an important factor in the hydrological cycle and climate (e.g. Albrecht, 1989; Ramanathan *et al.*, 2005; Levy *et al.*, 2013). Our study uses a global climate model with a predictive aerosol lifecycle to investigate the hydrological impact of carbonaceous aerosols over North Africa.

Carbonaceous aerosols including black carbon (BC) and organic carbon (OC) can affect the regional hydrological cycle of tropical monsoon regions through direct, semi-direct, and indirect effects (Ramanathan *et al.*, 2005; Lau *et al.*, 2006; Schulz *et al.*, 2006; Jeong and Wang, 2010; Jiang *et al.*, 2013). This has been observationally confirmed by both field campaigns and satellite observations (Warner and Twomey, 1967; Rosenfeld, 1999; Koren *et al.*, 2004; Ramanathan *et al.*, 2005; Huang *et al.*, 2009a, 2009b, 2009c, 2009d; Martins *et al.*, 2009). Aerosols have the potential to either increase or reduce clouds and rainfall (Rosenfeld, 1999; Koren *et al.*, 2008), suggesting a complex relationship between aerosols, clouds, and precipitation (Jiang *et al.*, 2006; Stevens and Feingold, 2009).

Our work focuses on carbonaceous aerosols emitted from North Africa mainly by biomass burning process (Figure S1) and their impact on regional precipitation with an emphasis on the 'second indirect effect', or, cloud lifetime effect. We are interested in the fast response of the climate systems to perturbations in aerosol forcing rather than the slower response involving changes in SSTs (e.g. Hansen *et al.*, 2005; Ganguly *et al.*, 2012b). Thus, we will examine how this effect changes the liquid water content, the height, and the lifetime of clouds and how these changes in turn affect the regional precipitation over North Africa. Section 2 describes the model simulations and methods. Section 3 presents results and Section 4 provides a summary and concluding remarks.

2. Experimental design and methods

This study uses the Community Atmosphere Model version 5.1 (CAM5.1; Neale *et al.*, 2010) at 1.9° latitude × 2.5° longitude horizontal resolution with 30 vertical layers, the atmospheric component of the Community Earth System Model (CESM1.0.3). The model uses a modal aerosol treatment (Ghan *et al.*, 2012; Liu *et al.*, 2012) that allows interaction of aerosols and stratiform clouds (Park *et al.*, 2014). However, it is cautiously noted that aerosols interact with warm stratiform clouds, but not with convective or ice clouds. The 3-mode version of the modal aerosol module (MAM3) is used here; aerosols are divided into three size-dependent modes: Aitken, accumulation, and coarse mode. Aerosols interact with modelled meteorological conditions through both radiative and microphysical processes. The rapid radiative transfer method for general circulation models (Iacono *et al.*, 2008) provides the radiative transfer calculation, which interacts with internally mixed aerosols in each mode (Ghan and Zaveri, 2007). A two-moment formulation of microphysics scheme is used for stratiform clouds (Morrison and Gettelman, 2008), and aerosol activation is based on vertical velocity and aerosol properties (Abdul-Razzak and Ghan, 2000). CAM5.1 has been found to simulate the seasonal cycle of rainfall over the African continent better than earlier versions of this model (Neale *et al.*, 2010).

Our control scenario was run using the present-day (2000) aerosol emissions created for the Coupled Model Intercomparison Phase 5 (CMIP5) activity (Lamarque *et al.*, 2010), using the technique proposed in a study of Southern African aerosol forcing by Sakaeda *et al.* (2011). Carbonaceous aerosol emissions over

Table 1. A summary of sensitivity experiments.

	Standard runs Fully interactive: aerosol↔cloud↔radiation	No aerosol direct effects runs No aerosols seen by radiation: aerosol↔cloud
Control	Case 1	Case 5
No carbonaceous aerosol emissions	Case 2, no carb	Case 6, no carb-F
No OC emissions	Case 3, no OC	Case 7, no OC-F
No BC emissions	Case 4, no BC	Case 8, no BC-F

North Africa come primarily from biomass burning, and mainly from grass fires that have a distinct seasonal cycle (Figure S1). These fires reach a maximum/minimum during the dry/wet season, i.e. boreal winter/summer season. Therefore, our analysis focused primarily on biomass burning season, i.e. the boreal winter season (Figure S1). Present-day SSTs averaged over the period of 1982–2001 (Hurrell *et al.*, 2008) and greenhouse gases are used consistently in all of our experiments. Each experiment is run for 21 years with the last 20 years used in our analyses.

Aerosol direct and indirect effects have been estimated by a couple of ways: (1) comparing multiple climate models that have either both direct and indirect effects or direct effect only (e.g. Guo *et al.*, 2015) or (2) comparing two simulation configurations using different emissions scenarios (e.g. with and without carbonaceous emissions), with differences in these scenarios providing an estimate of the radiative forcing and climate response (e.g. Bollasina *et al.*, 2011). As the climate itself has changed with the aerosols and clouds, it is difficult to clearly isolate the role of direct and indirect effects from other climate feedback processes. In this study, we have used a different approach by preventing aerosol to interact with radiation and allowing aerosols to affect only the cloud microphysics in one set of experiments.

Two sets of experiments with CAM5.1 were performed (Table 1). In the first experiment set (the 'standard runs'), carbonaceous aerosols were allowed to interact with both the microphysical and radiative processes of clouds; aerosol direct, semi-direct, and indirect effects operate simultaneously. In the second set of experiments, the number and mass of aerosols were set as zero for the radiative transfer calculation, but the predicted aerosols were allowed to interact only with the cloud microphysics (the 'no aerosol direct effects runs'). A similar experiment with CAM4, an earlier version of CAM5, was done by Sakaeda *et al.* (2011).

In addition to control simulations for the standard and the removed direct effects experiments, we also explored three emission scenarios in which carbonaceous aerosols emissions over North Africa (defined by the red rectangles shown in the panels of Figure 1) were set to zero for all seasons: (1) a scenario with no carbonaceous aerosol emissions (both BC and OC

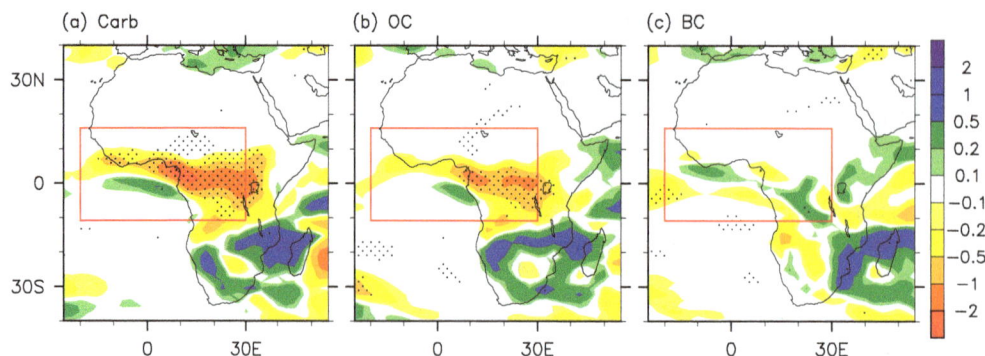

Figure 1. Total rainfall change (mm day^{-1}) during northern winter season (DJF) over Africa due to (a) OC and BC together (cases 1 and 2), (b) OC (cases 1–3), and (c) BC (cases 1–4) in the standard runs. Dotted areas represent statistically significant differences at the 95% confidence level compared to the control run. A red box indicates the domain used for area-averaged rainfall.

excluded, listed as 'no carb'), (2) a scenario without OC emissions (listed as 'no OC'), and (3) a scenario without BC emissions (listed as 'no BC'). These scenarios were run for both the 'standard' set and 'no direct aerosol effect' set of simulations. In this way, we could identify the relative importance of BC and OC on aerosol direct, semi-direct, and indirect effects.

3. Results

Figure 1 shows the change in simulated rainfall relative to the standard simulation (case 1 in Table 1) due to the exclusion of carbonaceous aerosols (cases 2–4) during the boreal winter season. Rainfall decreases about 1 mm day^{-1} over tropical Africa with the introduction of OC and BC (Figure 1(a)), implying that carbonaceous aerosols have the potential to reduce regional rainfall by about 25% of its climatology (Figure 2). Although an increase in rainfall is seen in some regions outside of the analysis domain (e.g. Madagascar), these increases are not statistically significant (Figure S4). Details of statistical significance are provided in the Supporting Information. Comparison between the simulations with only OC (case 3 in Table 1) and BC (case 4 in Table 1) suggests that OC plays a more important role in reducing rainfall in this area than BC (Figure 1(b)) while BC alone does not produce any systematic pattern of rainfall increase or decrease (Figure 1(c)).

The seasonal cycle of area-average rainfall change (%) over our analysis domain (20°W–30°E, 10°S–15°N), outlined by a solid red box in Figure 1, exhibits a large reduction (up to 25%) during boreal winter [December-January-February (DJF)] by carbonaceous aerosols (Figure 2). Note that only rainfall over land is considered in this analysis and that grey shading represents 25th and 75th percentile of precipitation anomaly from the 20-year mean in the control experiment (case 1). It is clear that adding both BC and OC over tropical and northern Africa can suppress rainfall by more than 20% during boreal winter season. Rainfall reduction due to OC is up to 15%, while that by BC is within natural variability except during February.

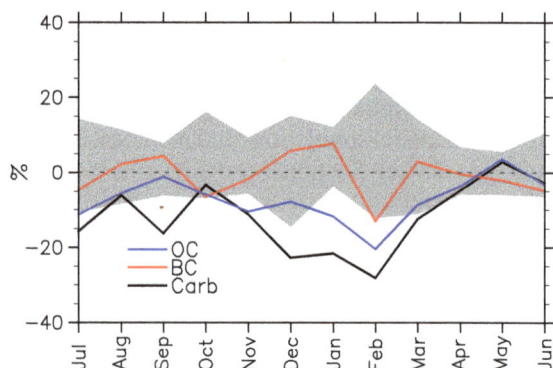

Figure 2. Rainfall change (%) compared to the control run (case 1) over tropical and Northern Africa indicated as a red box in Figure 1 due to OC and BC (cases 1 and 2; black), OC (cases 1–3; blue), and BC (cases 1–4; red). Grey shading area represents 25th and 75th percentile of precipitation anomaly from the 20-year mean in the control experiment (case 1) representing natural variability of model.

The effects of OC emissions dominate (Figure 2). This can be explained by the fact the total mass and number of OC are about four times greater than those of BC over this region (Figure S1) and also by the efficiency of OC in affecting regional water cycle. There are a number of effects associated with BC: heating in the lower troposphere increases stability near the surface and decreases it aloft, changing the environment and convection; the warmer air also tends to evaporate cloud drops more readily, suppressing convection (Ackerman et al., 2000; Koren et al., 2008). According to our simulations, these effects combined produce insignificant changes to rainfall in CAM5.1.

Preventing aerosols to interact with radiation in the model (cases 5–8 in Table 1) indicates that carbonaceous aerosols have a greater impact on precipitation through the aerosol indirect effects (AIE) (Figure 3). Rainfall reduction associated with case 6 is >2 mm day^{-1} larger (Figure 3(a)) than the control (case 5 in Table 1). This is consistent with the hypothesis that AIE can 'slow down' the regional hydrological cycle for reason discussed in the next paragraph. These results also show that BC decreased rainfall (Figure 3(c)), illustrating the competition between

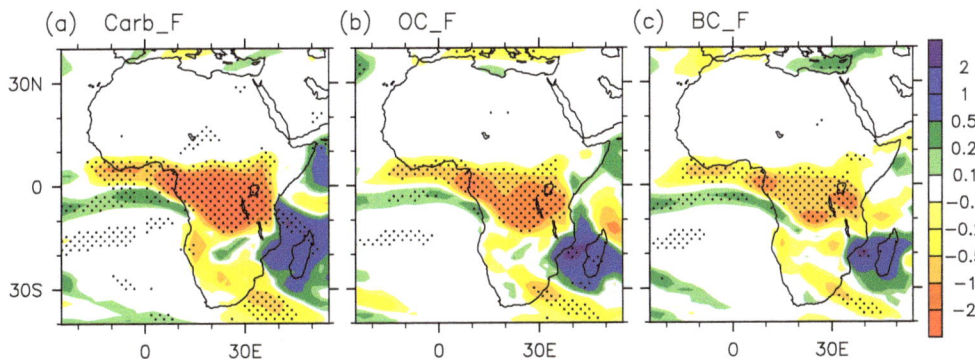

Figure 3. Same as Figure 1 except for no aerosol loading for radiation, i.e. cases 5 and 6 in (a), cases 5–7 in (b), and cases 5–8 in (c).

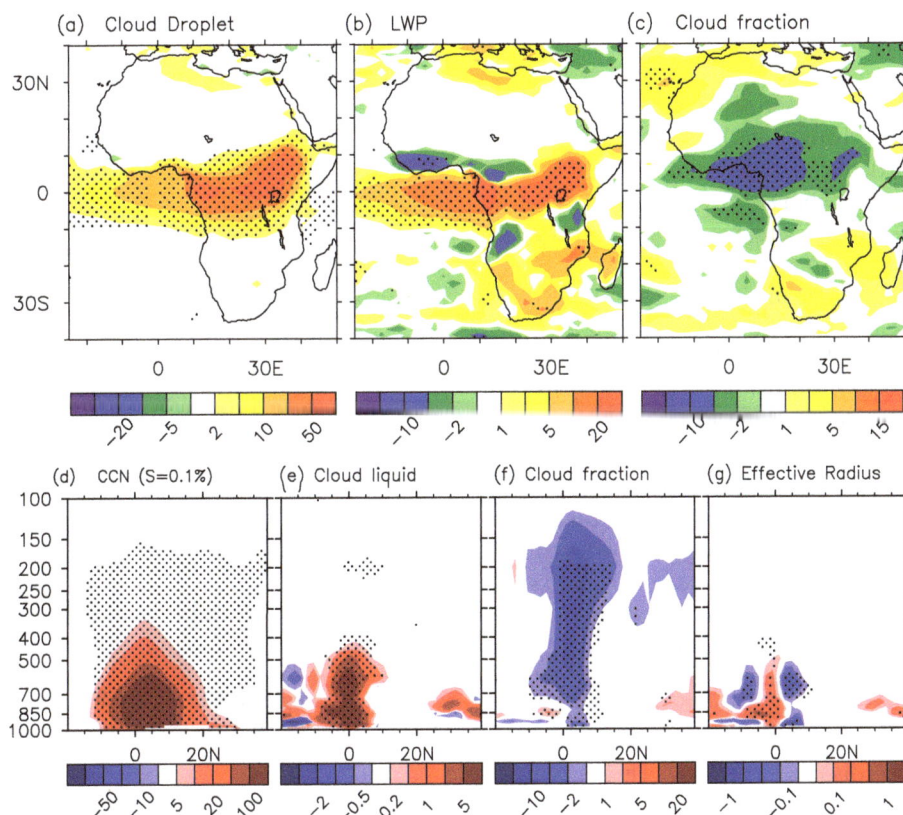

Figure 4. Spatial plots of change in cloud droplet number concentration (cm^{-3}) in (a), liquid water path (g m^{-2}) in (b), and total cloud fraction (%) in (c), and latitude-pressure (hPa) cross-section averaged over the longitudinal sections from 20°W to 30°E of CCN at 0.1 supersaturation (cm^{-3}) in (d), cloud liquid (g kg^{-1}) in (e), cloud fraction (%) in (f), and effective radius of liquid (μm) in (g) during northern winter season due to carbonaceous aerosols (cases 1 and 2). Cloud droplets, CCN, liquid water path, and cloud liquid increase while total cloud fraction and effective radius over land decrease.

direct/semi-direct and indirect effects of theses species in CAM5.1.

Several mechanisms can be considered to explain the reduction in regional rainfall by carbonaceous aerosols. Among such mechanisms is the cloud 'lifetime effect', i.e. more aerosols in the atmosphere lead to more but smaller cloud droplets and eventually reduce the efficiency of rain production (Albrecht, 1989; Ramaswamy et al., 2001; Lohmann and Feichter, 2005; Stevens and Feingold, 2009), allowing clouds to persist for a longer time. Experiments with CAM5.1 demonstrate that this mechanism can be very important with carbonaceous

aerosols, especially OC. Figure 4 illustrates how AIE of carbonaceous aerosols suppress regional hydrological cycle. Figure 4(a) and (d) shows an increase in the number concentrations of Cloud Condensation Nuclei (CCN) and cloud droplets. Figure 4(b) and (e) displays more liquid water inside clouds (liquid water path) while Figure 4(g) indicates smaller effective radius despite a reduced cloud fraction, especially over land (Figures 4(d) and (f) and S9). These results are largely consistent with the cloud lifetime effect depicted in Figure 1 of Stevens and Feingold (2009). However, cloud fraction simulated by CAM5.1 decreases in

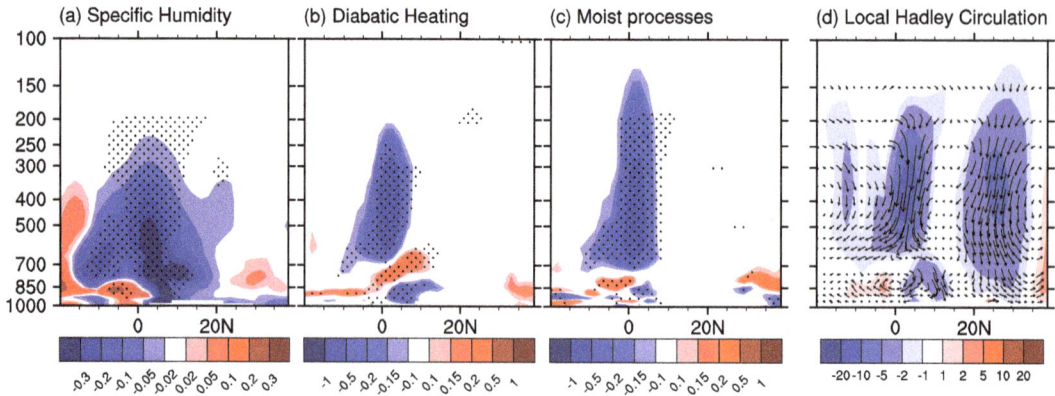

Figure 5. Same as Figure 4 except for specific humidity (g kg^{-1}) in (a), total diabatic heating (K day^{-1}) in (b), heating due to moist process (K day^{-1}) in (c), and local Hadley circulation depicted by $(v_D, -\omega)$ with colour shading in $-\omega$ (mb day^{-1}). v_D is divergent component of meridional wind (m s^{-1}) and ω is pressure velocity.

response to carbonaceous aerosols (Figure 4(c)), which is likely caused by changes in the circulation patterns (Figure 5(d)). Particularly, a clear reduction in upward motion, maintained by both a weaker convergence zone near the surface and with divergence aloft, appears to be a main dynamical response to carbonaceous aerosols over North Africa. This in turn produces a reduction in atmospheric humidity (Figure 5(a)), which can be by two reasons: one is reduction of Surface Air Temperature (SAT) due to aerosol loading and clouds, and the other is divergent at lower troposphere, and associated cloud cover.

The pathway described earlier suggests that cloud lifetime effect can have a significant impact on climate and that the atmospheric hydrological cycle can be changed by regional carbonaceous aerosols emissions via changes to the atmospheric circulation at both local and continental/global scales (Figures 5(d) and S3). The clear reduction in total diabatic heating (Figure 5(b)) is primarily driven by moist process (Figure 5(c)) despite a significant warming in shortwave radiation due to carbonaceous aerosols (Figure S2(a)). This appears to be closely linked to an anomalous downward motion, which is opposite to the local Hadley circulation (Figure 5(d)) associated with a Matsuno-Gill-type circulation anomaly at continental and global scales (Figure S3) (Matsuno, 1966; Gill, 1980).

A remaining question is why the other seasons also exhibit decreasing rainfall, similar to that in the boreal winter (Figure S3), despite a weaker aerosol forcing over North Africa. First, changes in cloud microphysical properties are in the same direction as those in winter but with a weaker intensity (not shown). Second, rainfall may be reduced due to soil moisture memory. Reduction of rainfall during the preceding boreal winter drives change in land-surface properties, especially the soil moisture, which can carry information to next seasons (Figures 6 and S11). In other words, the strong AIE-induced rainfall change occurs during boreal winter. During the rest of the season, both AIE and reduced soil moisture can contribute rainfall decrease. A similar mechanism was proposed to explain how El Niño

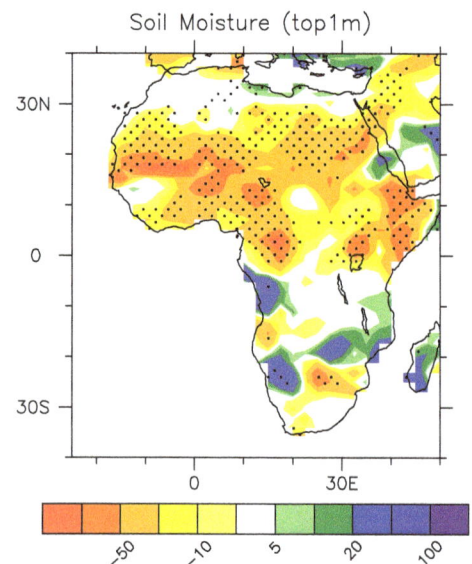

Figure 6. Change in annual mean soil moisture (mm) at top 1 m due to carbonaceous aerosols (cases 1 and 2).

and Southern Oscillation (ENSO) induced winter rainfall has impact on regional water cycle during boreal summer over the United States despite the demise of a particular ENSO (Seager *et al.*, 2005). Further study and more experiments are needed to fully understand the feedback of soil moisture changes to rainfall.

4. Conclusion

Carbonaceous aerosols emitted from biomass burning over North Africa slow down the regional hydrological cycle in CAM5.1 simulations. This is largely a consequence of the cloud lifetime effect of aerosols. More carbonaceous aerosols, and in particular, more OC, produce more cloud droplets with smaller sizes, thus increasing the condensed water path and the opacity of the clouds, which in turn reduces the energy reaching the surface (see Figure 4 and a summary schematic diagram in Figure S13). The response of the global hydrological cycle is constrained by the change in water vapour mixing ratio in the lower troposphere in most of coupled climate models, for example, those discussed by Held and Soden (2006). However, our result indicates that aerosols and associated cloud response can also play an important role in controlling regional and global hydrological cycle.

The effect of BC is more complicated than that of OC. BC acts almost identically to OC when only AIEs are active in the 'no aerosol direct effect' experiments (cases 5–8 in Table 1). However, direct and semi-direct effects of BC likely counterbalance these results over North Africa resulting in no significant change in regional rainfall.

State-of-the-art climate models participating in CMIP5 activity are now capable of simulating both direct/semi-direct and indirect effects of aerosols. Our analysis suggests that aerosol indirect effect (cloud lifetime effect) can be important in the regional atmospheric hydrological cycle, on top of the Earth's energy budget (Liu *et al.*, 2012). However, this scenario becomes more complicated when changes in the rainfall and local meteorology also affect land-surface properties and regional/global atmospheric circulation as a consequence (Figure 2). A more systematic or hierarchical approach is needed to analyse fully coupled climate simulations by including various feedback loops.

There are three important caveats to this work. First, accurately capturing regional aerosol optical and microphysical properties with the standard CAM5.1 (and other global models) is still a challenge (e.g. Wang *et al.*, 2012b, 2013). For example, the aerosol optical depth simulated by CAM5.1 is too low compared to observation over the Indian monsoon region (e.g. Ganguly *et al.*, 2012a). Also, aerosols directly interact only with the stratiform clouds in CAM5.1, but not with the convective clouds, as in many other global climate models. This can be an important issue because aerosols in convective clouds have been proposed to invigorate convection further (e.g. Rosenfeld *et al.*, 2008, 2014). A couple of attempts to simulate interaction between aerosols and convective clouds have been made using a regional (Lim *et al.*, 2014) and a global climate model (Song and Zhang, 2011; Song *et al.*, 2012). However, precipitation change due to aerosol forcing in the models that include interaction with both stratiform and convective clouds appears to be very complicated and more work is needed to tease out those effects. Second, the experiments in this study were done with fixed SSTs. Therefore, the longer timescale pathways through which aerosol can influence SSTs (e.g. Evan *et al.*, 2009) and associated precipitation changes remains to be investigated with a coupled climate model. Third, the intensity of the aerosol indirect effect simulated by global climate model still has large uncertainty (e.g. Wang *et al.*, 2012b).

Acknowledgements

We would like to acknowledge support from the US Department of Energy, Office of Science, Biological and Environmental Research (BER) through the Earth System Modelling programme. Comments from anonymous reviewers are helpful in improving the manuscript. We also thank Drs Carl Berkowitz and Zhao Chun at Pacific Northwest National Laboratory (PNNL) for providing helpful comments on an earlier version of the manuscript. PNNL is operated for the US Department of Energy by Battelle Memorial Institute under contract DE-AC06-76RLO1830. JHYoon was also partly supported by the funding from the Korean Polar Research Institute through the grant of PE16100.

Supporting information

The following supporting information is available:

Figure S1. Present-day emission of black carbon (a) and organic carbon (b) from all different sources over Northern Africa (20°W–50°E, EQ–20°N). The total mass of OC is about four to five times more than that of BC during its major burning season, i.e. northern winter. This is consistent with total number of particles emitted (not shown).

Figure S2. Same as Figure 5 except for heating due to shortwave radiation process in (a), and heating due to long-wave radiation process (K day^{-1}) in (b).

Figure S3. Change in lower tropospheric wind at 700 mb and rainfall due to carbonaceous aerosols over Northern Africa in (a) and long-wave stream function with divergent wind at 700 mb in (b) during northern winter season. Same label is used to shade rainfall in (a) as Figure 1. Contour interval of stream function is $1.0 \times 10^{-5} \text{ m}^{-1} \text{ s}^{-1}$ with positive value shaded. Strong divergence at the lower troposphere centred at the tropical Africa and weak trade winds along the equatorial Atlantic Ocean are simulated. Carbonaceous aerosols induce change in atmospheric moist processes and total diabatic heating through aerosol indirect effect (AIE), which results in atmospheric circulation response not only in regional but also global scales well explained by Gill-type solution (Matsuno, 1966; Gill, 1980).

Figure S4. Seasonal evolution of anomalous rainfall (mm day^{-1}) forced by carbonaceous aerosols (cases 1 and 2) including December-January-February (DJF), March-April-May (MAM), June-July-August (JJA), September-October-November (SON), and annual mean.

Figure S5. Seasonal evolution of precipitation simulated by the control experiment of CAM5.1.

Figure S6. Seasonal evolution of precipitation from CRU (Harris *et al.*, 2014).

Figure S7. Seasonal evolution of AOD simulated by the control experiment of CAM5.1.

Figure S8. Seasonal evolution of AOD from MISR (Wu *et al.*, 2011).

Figure S9. Change in effective radius at 850 hPa due to carbonaceous aerosols (cases 1 and 2).

Figure S10. Seasonal evolution of AOD–AOD_dust simulated by the control experiment of CAM5.1.

Figure S11. Seasonal evolution of soil moisture change (mm) at top 1m due to carbonaceous aerosols (cases 1 and 2).

Figure S12. Seasonal evolution of anomalous net surface radiation ($W\,m^{-2}$) forced by carbonaceous aerosols (cases 1 and 2) including DJF, MAM, JJA, SON, and annual mean.

Figure S13. A schematic diagram of carbonaceous aerosol's impact on regional atmospheric hydrological cycle over tropical Africa simulated by CAM5.1. More carbonaceous aerosols, especially organic carbon, produce more cloud droplets, more liquid in clouds and liquid water path, smaller drops, and slow down rainfall process. This is consistent with traditional AIE shown in Stevens and Feingold (2009). Difference is found in response of total cloud fraction. In our case, it is due to various reasons, such as change in atmospheric circulation as well as increased condensation to cloud drops.

References

Abdul-Razzak H, Ghan SJ. 2000. A parameterization of aerosol activation 2. Multiple aerosol types. *Journal of Geophysical Research – Atmospheres* **105**: 6837–6844.

Ackerley D, Booth BBB, Knight SHE, Highwood EJ, Frame DJ, Allen MR, Rowell DP. 2011. Sensitivity of twentieth-century Sahel rainfall to sulfate aerosol and CO(2) forcing. *Journal of Climate* **24**: 4999–5014.

Ackerman AS, Toon OB, Stevens DE, Heymsfield AJ, Ramanathan V, Welton EJ. 2000. Reduction of tropical cloudiness by soot. *Science* **288**: 1042–1047.

Albrecht BA. 1989. Aerosols, cloud microphysics, and fractional cloudiness. *Science* **245**: 1227–1230.

Bollasina MA, Ming Y, Ramaswamy V. 2011. Anthropogenic aerosols and the weakening of the South Asian summer monsoon. *Science* **334**: 502–505.

Booth BBB, Dunstone NJ, Halloran PR, Andrews T, Bellouin N. 2012. Aerosols implicated as a prime driver of twentieth-century North Atlantic climate variability (vol. 484, p. 228). *Nature* **485**: 534.

Chang EKM. 2013. CMIP5 projection of significant reduction in extratropical cyclone activity over North America. *Journal of Climate* **26**: 9903–9922.

Charney J, Stone PH, Quirk WJ. 1975. Drought in Sahara – biogeophysical feedback mechanism. *Science* **187**: 434–435.

Chung SH, Seinfeld JH. 2005. Climate response of direct radiative forcing of anthropogenic black carbon. *Journal of Geophysical Research – Atmospheres* **110**: D11102.

Chung CE, Zhang GJ. 2004. Impact of absorbing aerosol on precipitation: dynamic aspects in association with convective available potential energy and convective parameterization closure and dependence on aerosol heating profile. *Journal of Geophysical Research – Atmospheres* **109**: D22103.

Dai AG, Lamb PJ, Trenberth KE, Hulme M, Jones PD, Xie PP. 2004. The recent Sahel drought is real. *International Journal of Climatology* **24**: 1323–1331.

Dong B, Sutton RT, Highwood E, Wilcox L. 2014. The impacts of European and Asian anthropogenic sulfur dioxide emissions on Sahel rainfall. *Journal of Climate* **27**: 7000–7017.

Evan AT, Vimont DJ, Heidinger AK, Kossin JP, Bennartz R. 2009. The role of aerosols in the evolution of tropical North Atlantic Ocean temperature anomalies. *Science* **324**: 778–781.

Ganguly D, Rasch PJ, Wang H, Yoon J-H. 2012a. Climate response of the South Asian monsoon system to anthropogenic aerosols. *Journal of Geophysical Research – Atmospheres* **117**: D13209.

Ganguly D, Rasch PJ, Wang H, Yoon J. 2012b. Fast and slow responses of the South Asian monsoon system to anthropogenic aerosols. *Geophysical Research Letters* **39**: L18804.

Ghan SJ, Zaveri RA. 2007. Parameterization of optical properties for hydrated internally mixed aerosol. *Journal of Geophysical Research – Atmospheres* **112**: D10201.

Ghan SJ, Liu X, Easter RC, Zaveri R, Rasch PJ, Yoon J-H, Eaton B. 2012. Toward a minimal representation of aerosols in climate models: comparative decomposition of aerosol direct, semidirect, and indirect radiative forcing. *Journal of Climate* **25**: 6461–6476.

Giannini A, Saravanan R, Chang P. 2003. Oceanic forcing of Sahel rainfall on interannual to interdecadal time scales. *Science* **302**: 1027–1030.

Gill AE. 1980. Some simple solutions for heat-induced tropical circulation. *Quarterly Journal of the Royal Meteorological Society* **106**: 447–462.

Guo L, Turner AG, Highwood EJ. 2015. Impacts of 20th century aerosol emissions on the South Asian monsoon in the CMIP5 models. *Atmospheric Chemistry and Physics* **15**: 6367–6378.

Hansen J, Sato M, Ruedy R, Nazarenko L, Lacis A, Schmidt GA, Russell G, Aleinov I, Bauer M, Bauer S, Bell N, Cairns B, Canuto V, Chandler M, Cheng Y, Del Genio A, Faluvegi G, Fleming E, Friend A, Hall T, Jackman C, Kelley M, Kiang N, Koch D, Lean J, Lerner J, Lo K, Menon S, Miller R, Minnis P, Novakov T, Oinas V, Perlwitz J, Rind D, Romanou A, Shindell D, Stone P, Sun S, Tausnev N, Thresher D, Wielicki B, Wong T, Yao M, Zhang S. 2005. Efficacy of climate forcings. *Journal of Geophysical Research – Atmospheres* **110**: D18104.

Haywood JM, Jones A, Bellouin N, Stephenson D. 2013. Asymmetric forcing from stratospheric aerosols impacts Sahelian rainfall. *Nature Climate Change* **3**: 660–665.

Held IM, Soden BJ. 2006. Robust responses of the hydrological cycle to global warming. *Journal of Climate* **19**: 5686–5699.

Held IM, Delworth TL, Lu J, Findell KL, Knutson TR. 2005. Simulation of Sahel drought in the 20th and 21st centuries. *Proceedings of the National Academy of Sciences of the United States of America* **102**: 17891–17896.

Hoerling M, Hurrell J, Eischeid J, Phillips A. 2006. Detection and attribution of twentieth-century northern and southern African rainfall change. *Journal of Climate* **19**: 3989–4008.

Huang J, Adams A, Wang C, Zhang C. 2009a. Black carbon and West African monsoon precipitation: observations and simulations. *Annales Geophysicae* **27**: 4171–4181.

Huang J, Zhang C, Prospero JM. 2009b. Large-scale effect of aerosols on precipitation in the West African monsoon region. *Quarterly Journal of the Royal Meteorological Society* **135**: 581–594.

Huang JF, Zhang CD, Prospero JM. 2009c. Aerosol-induced large-scale variability in precipitation over the tropical Atlantic. *Journal of Climate* **22**: 4970–4988.

Huang JF, Zhang CD, Prospero JM. 2009d. African aerosol and large-scale precipitation variability over West Africa. *Environmental Research Letters* **4**: 15006, doi: 10.1088/1748-9326/4/1/015006.

Hurrell JW, Hack JJ, Shea D, Caron JM, Rosinski J. 2008. A new sea surface temperature and sea ice boundary dataset for the Community Atmosphere Model. *Journal of Climate* **21**: 5145–5153.

Hwang Y-T, Frierson DMW, Kang SM. 2013. Anthropogenic sulfate aerosol and the southward shift of tropical precipitation in the late 20th century. *Geophysical Research Letters* **40**: 2845–2850.

Iacono MJ, Delamere JS, Mlawer EJ, Shephard MW, Clough SA, Collins WD. 2008. Radiative forcing by long-lived greenhouse gases: calculations with the AER radiative transfer models. *Journal of Geophysical Research – Atmospheres* **113**: D13103.

Jeong GR, Wang C. 2010. Climate effects of seasonally varying biomass burning emitted carbonaceous aerosols (BBCA). *Atmospheric Chemistry and Physics* **10**: 8373–8389.

Jiang HL, Xue HW, Teller A, Feingold G, Levin Z. 2006. Aerosol effects on the lifetime of shallow cumulus. *Geophysical Research Letters* **33**: L14806.

Jiang Y, Liu X, Yang X-Q, Wang M. 2013. A numerical study of the effect of different aerosol types on East Asian summer clouds and precipitation. *Atmospheric Environment* **70**: 51–63.

Koren I, Kaufman YJ, Remer LA, Martins JV. 2004. Measurement of the effect of Amazon smoke on inhibition of cloud formation. *Science* **303**: 1342–1345.

Koren I, Martins JV, Remer LA, Afargan H. 2008. Smoke invigoration versus inhibition of clouds over the Amazon. *Science* **321**: 946–949.

Kucharski F, Zeng N, Kalnay E. 2013. A further assessment of vegetation feedback on decadal Sahel rainfall variability. *Climate Dynamics* **40**: 1453–1466.

Lamarque JF, Bond TC, Eyring V, Granier C, Heil A, Klimont Z, Lee D, Liousse C, Mieville A, Owen B, Schultz MG, Shindell D, Smith SJ, Stehfest E, Van Aardenne J, Cooper OR, Kainuma

M, Mahowald N, Mcconnell JR, Naik V, Riahi K, van Vuuren DP. 2010. Historical (1850–2000) gridded anthropogenic and biomass burning emissions of reactive gases and aerosols: methodology and application. *Atmospheric Chemistry and Physics* **10**: 7017–7039.

Lau KM, Kim MK, Kim KM. 2006. Asian summer monsoon anomalies induced by aerosol direct forcing: the role of the Tibetan Plateau. *Climate Dynamics* **26**: 855–864.

Levy H II, Horowitz LW, Schwarzkopf MD, Ming Y, Golaz J-C, Naik V, Ramaswamy V. 2013. The roles of aerosol direct and indirect effects in past and future climate change. *Journal of Geophysical Research – Atmospheres* **118**: 4521–4532.

Lim K-SS, Fan J, Leung R, Ma P-L, Singh B, Zhao C, Zhang Y, Zhang G, Song X. 2014. Investigation of aerosol indirect effects using a cumulus microphysics parameterization in a regional climate model. *Journal of Geophysical Research – Atmospheres* **119**: 906–926.

Liu X, Easter RC, Ghan SJ, Zaveri R, Rasch P, Shi X, Lamarque JF, Gettelman A, Morrison H, Vitt F, Conley A, Park S, Neale R, Hannay C, Ekman AML, Hess P, Mahowald N, Collins W, Iacono MJ, Bretherton CS, Flanner MG, Mitchell D. 2012. Toward a minimal representation of aerosol direct and indirect effects: model description and evaluation. *Geoscientific Model Development* **5**: 31.

Lohmann U, Feichter J. 2005. Global indirect aerosol effects: a review. *Atmospheric Chemistry and Physics* **5**: 715–737.

Lu J, Delworth TL. 2005. Oceanic forcing of the late 20th century Sahel drought. *Geophysical Research Letters* **32**: L22706.

Martins JA, Silva Dias MAF, Goncalves FLT. 2009. Impact of biomass burning aerosols on precipitation in the Amazon: a modeling case study. *Journal of Geophysical Research – Atmospheres* **114**: D02207.

Matsuno T. 1966. Quasi-geostrophic motions in the equatorial area. *Journal of the Meteorological Society of Japan* **44**: 19.

Morrison H, Gettelman A. 2008. A new two-moment bulk stratiform cloud microphysics scheme in the community atmosphere model, version 3 (CAM3). Part I: description and numerical tests. *Journal of Climate* **21**: 3642–3659.

Neale RB, Chen C-C, Gettelman A, Lauritzen PH, Park S, Williamson DL, Conley AJ, Garcia R, Kinnison D, Lamarque J-F, Marsh D, Mills M, Smith AK, Tilmes S, Vitt F, Morrison H, Cameron-Smith P, Collins WD, Iacono MJ, Easter RC, Ghan S, Liu X, Rasch PJ, Taylor MA. 2010. Description of the NCAR Community Atmosphere Model (CAM 5.0). *NCAR Technical Note*. National Center for Atmospheric Research, Boulder, CO.

Nicholson SE. 2001. Climatic and environmental change in Africa during the last two centuries. *Climate Research* **17**: 123–144.

Park S, Bretherton CS, Rasch PJ. 2014. Integrating cloud processes in the community atmosphere model, version 5. *Journal of Climate* **27**: 6821–6856.

Ramanathan V, Chung C, Kim D, Bettge T, Buja L, Kiehl JT, Washington WM, Fu Q, Sikka DR, Wild M. 2005. Atmospheric brown clouds: impacts on South Asian climate and hydrological cycle. *Proceedings of the National Academy of Sciences of the United States of America* **102**: 5326–5333.

Ramaswamy V, Boucher O, Haigh J, Hauglustaine D, Haywood J, Myhre G, Nakajima T, Shi GY, Solomon S. 2001. Radiative forcing of climate change. In Climate Change 2001: The Scientific Basis, Houghton JT, Ding Y, Griggs DJ, Noguer M, Van Der Linden PJ, Dai X, Maskell K, Johnson CA (eds). Contribution of Working Group I to the Third Assessment Report of the Intergovernmental Panel on Climate Change, Cambridge and New York, NY.

Rosenfeld D. 1999. TRMM observed first direct evidence of smoke from forest fires inhibiting rainfall. *Geophysical Research Letters* **26**: 3105–3108.

Rosenfeld D, Lohmann U, Raga GB, O'Dowd CD, Kulmala M, Fuzzi S, Reissell A, Andreae MO. 2008. Flood or drought: how do aerosols affect precipitation? *Science* **321**: 1309–1313.

Rosenfeld D, Sherwood S, Wood R, Donner L. 2014. Climate effects of aerosol–cloud interactions. *Science* **343**: 379–380.

Sakaeda N, Wood R, Rasch PJ. 2011. Direct and semidirect aerosol effects of southern African biomass burning aerosol. *Journal of Geophysical Research – Atmospheres* **116**: D12205.

Scaife AA, Kucharski F, Folland CK, Kinter J, Bronnimann S, Fereday D, Fischer AM, Grainger S, Jin EK, Kang IS, Knight JR, Kusunoki S, Lau NC, Nath MJ, Nakaegawa T, Pegion P, Schubert S, Sporyshev P, Syktus J, Yoon JH, Zeng N, Zhou T. 2009. The CLIVAR C20C project: selected twentieth century climate events. *Climate Dynamics* **33**: 603–614.

Scholze M, Knorr W, Arnell NW, Prentice IC. 2006. A climate-change risk analysis for world ecosystems. *Proceedings of the National Academy of Sciences of the United States of America* **103**: 13116–13120.

Schulz M, Textor C, Kinne S, Balkanski Y, Bauer S, Berntsen T, Berglen T, Boucher O, Dentener F, Guibert S, Isaksen ISA, Iversen T, Koch D, Kirkevag A, Liu X, Montanaro V, Myhre G, Penner JE, Pitari G, Reddy S, Seland O, Stier P, Takemura T. 2006. Radiative forcing by aerosols as derived from the AeroCom present-day and pre-industrial simulations. *Atmospheric Chemistry and Physics* **6**: 5225–5246.

Seager R, Kushnir Y, Herweijer C, Naik N, Velez J. 2005. Modeling of tropical forcing of persistent droughts and pluvials over western North America: 1856–2000. *Journal of Climate* **18**: 4065–4088.

Song X, Zhang GJ. 2011. Microphysics parameterization for convective clouds in a global climate model: description and single-column model tests. *Journal of Geophysical Research – Atmospheres* **116**: D02201.

Song X, Zhang GJ, Li JLF. 2012. Evaluation of microphysics parameterization for convective clouds in the NCAR community atmosphere model CAM5. *Journal of Climate* **25**: 8568–8590.

Stevens B, Feingold G. 2009. Untangling aerosol effects on clouds and precipitation in a buffered system. *Nature* **461**: 607–613.

Wang C, Dong S, Evan AT, Foltz GR, Lee S-K. 2012a. Multidecadal covariability of North Atlantic sea surface temperature, African dust, Sahel rainfall, and Atlantic hurricanes. *Journal of Climate* **25**: 5404–5415.

Wang M, Ghan S, Liu X, L'Ecuyer TS, Zhang K, Morrison H, Ovchinnikov M, Easter R, Marchand R, Chand D, Qian Y, Penner JE. 2012b. Constraining cloud lifetime effects of aerosols using A-Train satellite observations. *Geophysical Research Letters* **39**: L15709.

Wang H, Easter RC, Rasch PJ, Wang M, Liu X, Ghan SJ, Qian Y, Yoon JH, Ma PL, Vinoj V. 2013. Sensitivity of remote aerosol distributions to representation of cloud–aerosol interactions in a global climate model. *Geoscientific Model Development* **6**: 765–782.

Warner J, Twomey S. 1967. Production of cloud nuclei by cane fires and effect on cloud droplet concentration. *Journal of the Atmospheric Sciences* **24**: 704.

Zeng N, Neelin JD, Lau KM, Tucker CJ. 1999. Enhancement of interdecadal climate variability in the Sahel by vegetation interaction. *Science* **286**: 1537–1540.

Zeng N, Yoon JH, Marengo JA, Subramaniam A, Nobre CA, Mariotti A, Neelin JD. 2008. Causes and impacts of the 2005 Amazon drought. *Environmental Research Letters* **3**: 014002, doi: 10.1088/1748-9326/3/1/014002.

20

Interaction between moisture transport and Kelvin waves events over Equatorial Africa through ERA-interim

Sinclaire Zebaze,[1]*[iD] André Lenouo,[2] Clément Tchawoua,[1] Amadou T. Gaye[3] and François M. Kamga[1,4]

[1]Department of Physics, Faculty of Science, University of Yaoundé I, Yaoundé, Cameroon
[2]Department of Physics, Faculty of Science, University of Douala, Douala Cameroon
[3]Laboratoire de Physique de l'Atmosphère Siméon Fongang, ESP-Université Cheikh Anta Diop, Dakar, Sénégal
[4]Faculté des Sciences et de Technologie, Université des Montagnes, Banganté, Cameroon

*Correspondence to
S. Zebaze, Department of
Physics, Faculty of Science,
University of Yaoundé I, P.O. Box
812 Yaoundé, Cameroon.
E-mail: zebaze.s@gmail.com

Abstract

This study examines the moisture transport variability and its interaction between rainfall and Kelvin wave's events over Equatorial Africa using 1979–2010 ERA-interim reanalysis data, precipitation from Global Precipitation Climatology Project and outgoing long-wave radiation. Kelvin waves events influenced in Congo basin varies within each rainy season, as the intertropical convergence zone moves through the region. The moisture flux is calculated for the entire tropospheric column (1000–300 hPa) over Central Africa (5–10°N; 5–30°E). Analysis of mean monthly fluxes shows a progressive penetration of the flux into Gulf of Guinea (5°S–5°N; 0–15°E). Mean seasonal values of moisture components across boundaries indicate that the zonal component is the largest contributor to mean moisture over Central Africa, while the meridional component contributes the most over the Gulf of Guinea. Lag correlation between precipitation and moisture is largely dominated over land with a coefficient greater than 0.5, while moisture increases with enhanced phase of Kelvin waves.

Keywords: ERA-interim; moisture flux; Kelvin waves; Central Africa

1. Introduction

Climatology of Central Africa is particularly result from localized convection which is increasing by the passage of intertropical convergence zone (ITCZ). The moisture flux and convergence over Central Africa was first described by Matsuyama *et al.* (1994). These authors said that during March–April rainy season, vertically integrated vapor is transported westward. From June to July, the moisture flux is divergent over the region. Pokam *et al.* (2011) analyzed annual cycle of the vertically integrated moisture convergence for the full, lower and upper tropospheric layers. They showed that the full column vertically integrated moisture convergence exhibits a bimodal distribution with maximum during the rainy seasons.

Lélé *et al.* (2015) evaluated the ocean–land transport of moisture for rainfall in West Africa and showed a pronounced south-west incursion of the flux over West Africa during April–June, while during July–September, the southerly transport weakens, but westerly transport is enhanced. Most studies evaluated the variability of moisture transport and it relationship with precipitation, large-scale or small-scale but the interaction between moisture and equatorials waves like Kelvin waves has not yet been studied and clearly understood in this region. Based on these previous studies, it will be more useful to evaluate the impact of Kelvin waves on moisture variability. The main question is: what would be the impact of Kelvin wave's

events upon the moisture transport over Central Africa? The aim of this study is to explore the interaction between moisture flux, rainfall and Kelvin waves events at seasonal and annual time scales and evaluate their modification in the climate system over Central Africa. Datasets and methods used in this study are briefly described in Section 2. Section 3 investigates the characteristic of moisture flux field and Kelvin wave's events over Central Africa, modulation of moisture flux field by Kelvin wave's events and precipitation and annual cycle of moisture flux convergence associated with Kelvin wave's activity. Finally, Section 4 summarizes the main results and indicates some prospects for future work.

2. Data and methods

2.1. Data

The study area extends from Eastern Atlantic (5°N–Eq.; 30–10°W), Gulf of Guinea (5°S–5°N; 0–15°E) and Central Africa (5°S–10°N; 5–30°E). The topography of Central Africa is always varied, including highlands, mountains and plateaus. Several observational and reanalysis datasets such as outgoing long-wave radiation (OLR) (Grueber and Krueger, 1974; Liebmann and Smith, 1996; Straub and Kiladis, 2002), Global Precipitation Climatology Project and European Centre for Medium-Range Weather Forecasts (ECMWF) Interim reanalysis datasets (Simmons *et al.*,

Table 1. List of observational datasets used in this study.

Dataset	Origin/platform	Horizontal resolution	Temporal resolution	Vertical levels	Selected variables
NOAA/OLR	Satellite	$2.5° \times 2.5°$	Daily mean from 1979 to 2010	Top of atmosphere	OLR
(ECMWF) ERA-I	Radiosonde, satellite, model forecast	$0.75° \times 0.75°$	4 times daily (6-hourly) from 1979 to 2010	1000–300 hPa	Winds, specific humidity
GPCP	GPCP polar satellite precipitation data center – emission/observation	$2.5° \times 2.5°$	Monthly mean from 1979 to 2010	Surface	Precipitation

2007) are used in this study. Table 1 summarizes the information of observational datasets.

2.2. Methods

In this study, Kelvin wave filtering is used for the period of 2.5–20 days, and eastward wavenumber 1–14 (Wheeler and Kiladis, 1999; Wheeler *et al.*, 2000; Kiladis *et al.*, 2009). The filtering is carried out with OLR-National Oceanic and Atmospheric Administration (NOAA) data, field is not separated into symmetric and a dissymmetric components (Straub and Kiladis, 2002). Based on the methodology used by Zebaze *et al.* (2015), a time series was developed based on a selected regional zone over Central Africa region (5°S–10°N; 5–30°E), selecting all days where the minimum Kelvin-filtered negative OLR anomalies were less than −1.5 standard deviations in magnitude during the 1979–2010 March–June (MAMJ) seasons. The year 1992 appeared to be the weaken Kelvin year event with only 3 events occurred, while 1999 appears as the most active Kelvin year with 12 events appear. These 2 years are used as indicators for possible interaction between Kelvin wave's events and moisture signatures. The Equations (1) and (2) below represent zonal and meridional flux components:

$$Q_u = \frac{1}{g} \int_{P_0}^{P_s} uq \, dP \qquad (1)$$

$$Q_v = \frac{1}{g} \int_{P_0}^{P_s} vq \, dP \qquad (2)$$

where, u and v are zonal and meridional components of the wind, respectively (their units are $m\,s^{-1}$). The units of Q_u and Q_v are $kg\,m^{-1}\,s^{-1}$. g is the gravitational acceleration ($9.81\,m.s^{-2}$), q is specific humidity ($g\,kg^{-1}$), P is the pressure, P_s the surface pressure and P_0 pressure at the top of the atmospheric layer (Rao *et al.*, 1999). The horizontal moisture divergence was obtained from Q_v and Q_u, which provided the vertically integrated meridional and zonal moisture fluxes. In the results, a negative value of divergence is convergence, while positive value is divergence. Moisture flux divergence is calculated using following equation:

$$MCF = \frac{1}{g} \left[dy \int_{P_0}^{P_s} \left(q \frac{\partial u}{\partial x} + u \frac{\partial q}{\partial x} \right) dp \right.$$

$$\left. + dx \int_{P_0}^{P_s} \left(q \frac{\partial v}{\partial y} + u \frac{\partial q}{\partial y} \right) dp \right] \qquad (3)$$

where, Moisture Flux Convergence (MCF) is moisture flux in the entire tropospheric column (1000–300 hPa). The units of MFC are $s^{-1}\,g\,kg^{-1}$.

Consider N pairs of observations on two time series X_t and Y_t (precipitation and moisture transport). Following Chatfield (2004), the sample cross-covariance function (ccvf) is given by:

$$c_{XY}(k) = \frac{1}{N} \sum_{t=1}^{N-k} \left(X_t - \overline{X} \right) \left(Y_{t+k} - \overline{Y} \right)$$

$$[k = 0, 1, \dots (N-1)] \qquad (4a)$$

$$c_{XY}(k) = \frac{1}{N} \sum_{t=1-k}^{N} \left(X_t - \overline{X} \right) \left(Y_{t+k} - \overline{Y} \right)$$

$$[k = -1, -2, \dots - (N-1)] \qquad (4b)$$

where, N is the series length, X and Y are the sample means, and k is the lag. The sample lag-correlation function is the ccvf scaled by the variances of the two series:

$$r_{XY}(k) = \frac{c_{XY}(k)}{\sqrt{c_{XX}(0) \, c_{YY}(0)}} \qquad (5)$$

where, $c_{XX}(0)$ and $c_{YY}(0)$ are the sample variances of X_t and Y_t.

3. Results and discussions

3.1. Characteristic of moisture flux field and Kelvin waves events over Central Africa

The propagation regime of the Central Africa integrated horizontal moisture flux during 1979–2010 is illustrated in Figure 1. Between March and April (Figures 1(a) and (b)), convection is located over the Gulf of Guinea, between the southern part of the Congo basin and southern part of 10°N. In these northern spring months, moisture field propagated over the East (located over southern part of the region), the large part of the flow is retraced through the Gulf of Guinea. Results showed that vectors flux are small compared to mean value out of the Gulf of Guinea, the moisture transport over this area is very important. Increase in moisture flux over Congo basin region is relatively accompanied the progression of the ITCZ from its southern position near the Guinea coast in December and progresses to around 10°N in April, associated

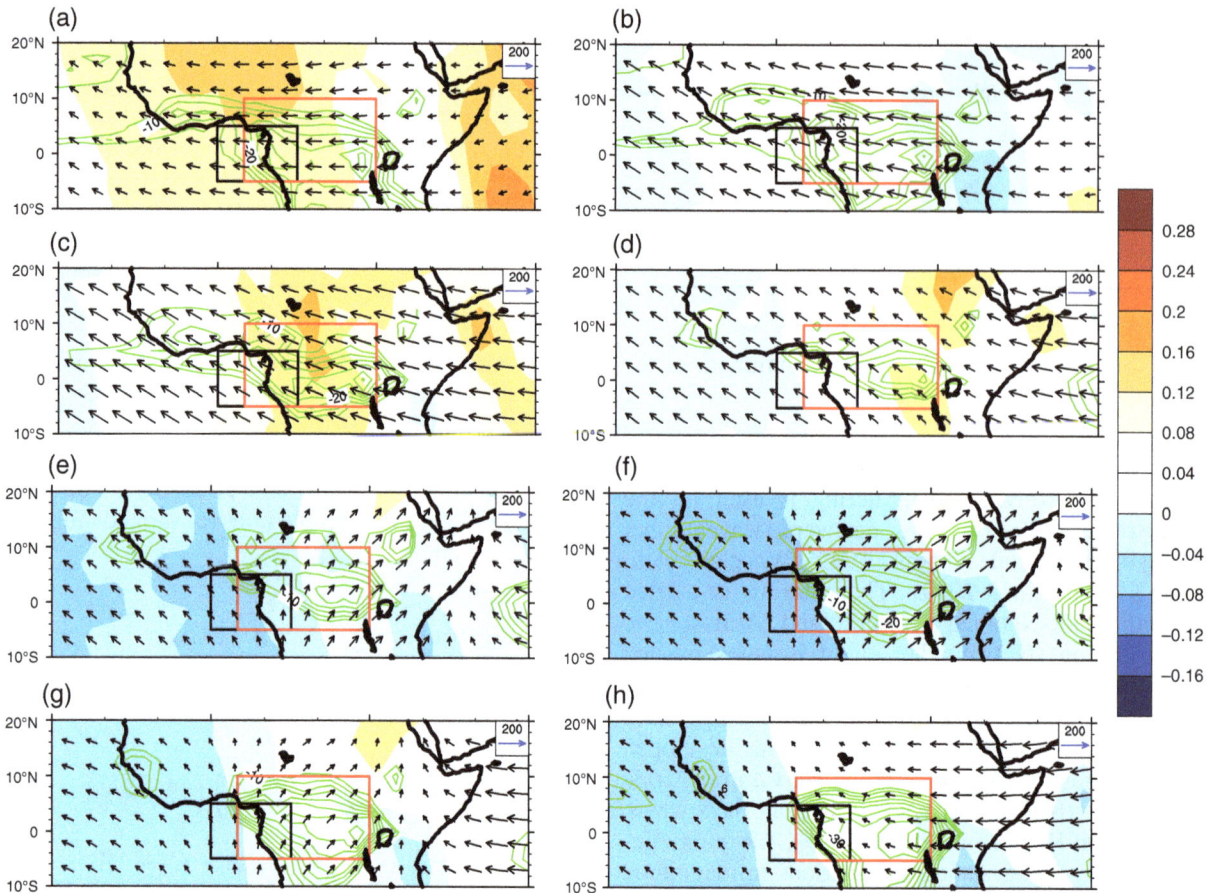

Figure I. Monthly mean moisture transport (vectors; $kg\,m^{-1}\,s^{-1}$), moisture flux divergence/convergence (shading; $10^{-6}\,s^{-1}\,g\,kg^{-1}$) vertically integrated in layers 1000–300 hPa and NOAA OLR anomalies (green contours; $<-5\,Wm^{-2}$, only the enhanced phase is shown for clarity) for (a) March, (b) April, (c) May, (d) June, (e) July, (f) August, (g) September and (h) October averaged from 1979 to 2010. Boxes indicate Central Africa (5°S–10°N; 5–30°E, red). Gulf of Guinea (5°S–5°N; 0–15°E, black).

convection is clearly observed over this part of the region. Congo basin is more associated with convergence during March, May and June (Figures 1(a)–(c)), more convective activity is also denoted in the region. The easterly moisture transport over Central Africa is clearly observed and this propagation is associated with a dipole. The zone of predominance of moisture is no longer compared to the convection which persists over the region. In April (Figure 1(b)), there are clearly an easterly anomaly resulting for the propagation characteristics while westerlies Kelvin waves are observed at that period. Finally, the presence of Kelvin waves can favors formation of moisture which propagated in opposite side. These propagation characteristics were shown by Kamsu *et al.* (2014). This plot also displays the climatology of moisture flow fields, propagation characteristics and associated convection during July–September (Figures 1(e)–(g)). Moisture field weaken with progression of month and change in the intensity of flow is clearly observed when convection decreases and reaches southern part of the region. The change in direction of moisture is also observed during these months while their intensity is not associated with convective activity. These results are comparable to the previous studies which showed that the Gulf of

Guinea region is an important source of moisture for the tropical Africa (Giannini *et al.*, 2003; Lélé *et al.*, 2015).

The mean geographical distribution of the Kelvin-filtered OLR-NOAA variance [averaged over 1979–2010 for MAMJ] shows a peak activity around 2.5°N band over Central Africa. Previous studies suggested that the strongest Kelvin wave's signatures are observed over 2.5°N (Janicot *et al.*, 2008). The highest variance is also progressed toward the eastern Pacific, Central America and the Atlantic ITCZ. The Kelvin-filtered OLR variance is not symmetric with respect to the equator over this region, consistent with previous analyses of Kelvin wave variance (Straub and Kiladis, 2002). Kelvin wave activity is clearly present over tropical Africa with a peak variance associated with convective activity. The strong convection persists over the region and this convection signature covers almost all the Congo forest. Based on composite methodology, 1980 and 1992 were weak Kelvin year's event with only 3 events occurred although 1984 and 1999 appear as the most active Kelvin years with 12 events; 1992 and 1999 are used as indicators for possible interaction between Kelvin wave's events and moisture signatures.

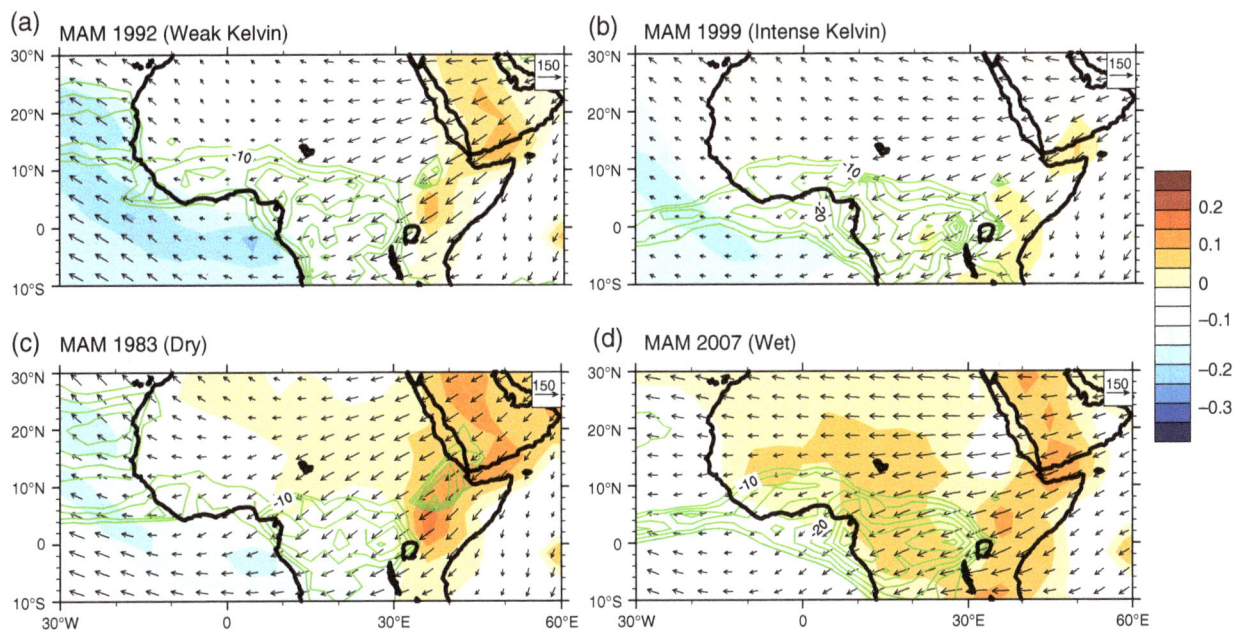

Figure 2. March–April–May (MAM). moisture transport (vectors; $kg\,m^{-1}\,s^{-1}$), moisture flux divergence/convergence (shading; $10^{-6}\,s^{-1}\,g\,kg^{-1}$) vertically integrated in layers 1000–300 hPa and NOAA OLR anomalies (green contours; $<-5\,Wm^{-2}$, only the enhanced phase is shown for clarity) during: (a) 1992 weak Kelvin year, (b) 1999 intense Kelvin year, (c) 1983 dry year and (d) 2007 wet year.

3.2. Interaction between moisture flux field, Kelvin waves events and precipitation

During March–May season, moisture flux propagated westward during wet year (Figure 1(d)) and intense Kelvin year (Figure 2(c)). This wettest year is associated with divergence and westward propagating flux. Over Central Africa, these perturbations are centered on the equator and meridional fluxes are low. Enhanced convection occupied more sectors compared to previous season. The vertically integrated horizontal flux is in phase with the easterly wind for low surface pressure showed in previous studies (Nguyen and Duvel, 2008) which immediately associates with the minimum OLR over the Congo basin. This is physically consistent because a maximum in convective activity is associated with stronger convergence. In particular, there is a clear weakening and a latitudinal expansion of the convective anomaly over the region.

High fluxes dominate continental region during June–August, zone of enhanced convection are seen over continental area. The moisture flux propagated northward when it is located over the north part of enhanced convection, the ITCZ displacement and the associated Central Africa rainfall suppression were more pronounced. During the 2007 wet year, the moisture flux transport is located over 10°N. During the years 1992 and 1983, there is a weak convective anomaly over Congo basin. This would be associated with a negative latent heating anomaly (a cooling anomaly) or equivalently a reduction in the total heating and this heating was associated with divergence. The large zone of moisture flux convergence is observed

over the ocean. A large area convective activity was observed over Central Africa during intense Kelvin year (1999) and the associated convection extends from Atlantic Ocean to the eastern sector of equatorial Africa. The associated moisture flux propagation characteristic is opposite to the one observed during March–May. Convection occupied large zone during spring while in monsoon, convection is not large. The quasi-dipole is also observed during weak Kelvin year while, divergence is seen during driest and wettest years over the region.

3.3. Annual cycle of moisture flux convergence associated with Kelvin wave's activity

Figure 3 presents the annual cycle of the atmospheric zonal (Figure 3(b)) and meridional (Figure 3(c)) components of moisture flux in the entire tropospheric column (1000–300 hPa) and associated convergence (Figure 3(d)) averaged over 5–25°E trough a Hovmöller diagram between 1979 and 2010. The Kelvin wave variability is also presented (Figure 3(a)). The increase in moisture flux convergence is seen over continental region (between 5°S and 5°N), while the moisture flux divergence zone is seen to the south of 5°S and north of 5°N (Figure 3(d)). The northward progression of moisture flux convergence across the Gulf of Guinea coast started in March when ITCZ is centered over equator and progressively migrated trough the north (after 5°N). Two dipoles are observed in zonal flux (Figure 3(b)). The first (Eastward) propagated from December to April and this propagation is clearly associated with dry season. Along the Gulf of Guinea there is some evidence of a convection belt which

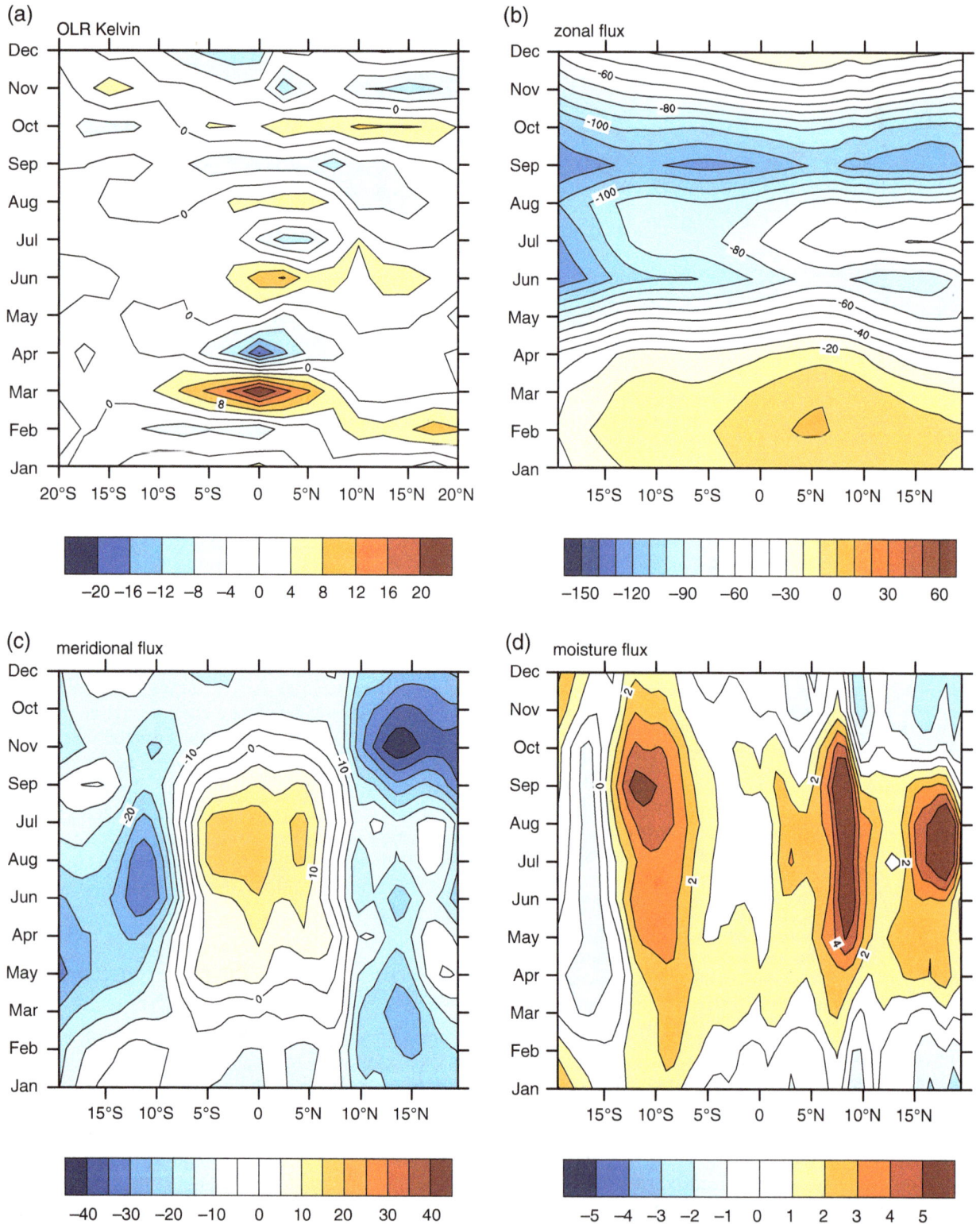

Figure 3. Annual cycle (1979–2010) of NOAA-interpolated OLR filtered for the Kelvin band anomalies (a, amplified 100 times, W m^{-2}), zonal flux (b, kg m^{-1} s^{-1}), meridional flux (c, kg m^{-1} s^{-1}) and moisture flux divergence/convergence (d, amplified 100 times, 10^{-6} s^{-1} kg^{-1}), integrated in layers 1000–300 hPa.

increased convergence over the ocean and divergence over land during the spring season. At that time Kelvin waves and associated convection are located over equator forming a convective belt over the region. During June–August season, the field gradually progressed over the continent. The moisture transport clearly denotes south-westerly and north-easterly propagation characteristics. Many phenomenons are responsible for the presence of the quasi-dipole observed in moisture flux convergence regime. The belt of convergence over land is accompanied by a large zone of moisture flux divergence over the ocean.

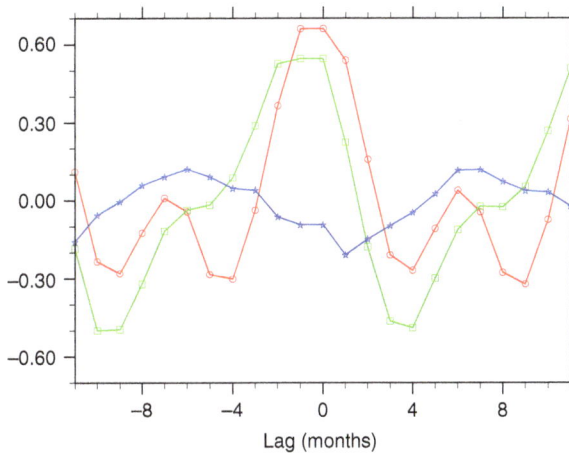

Figure 4. Temporal lag correlation (in month) between monthly mean precipitation and monthly mean moisture flux. Positive/negative lag corresponds to precipitation leading/preceding moisture flux. Mean values over 3 boxes: Eastern Atlantic: (5°N–Eq.; 30–10°W, blue, star); Gulf of Guinea (5°S–5°N; 0–15°E, square, green) and Central Africa (5°S–10°N; 5–30°E, red, circle).

The position of the flux reveals that moisture transport over Central Africa is regulated by an enhanced evaporation coming from tropical Atlantic Ocean at the beginning of Kelvin wave's active phase which positively interacts with convection over Atlantic. Previous studies (Lélé et al., 2015) showed that Atlantic Ocean is the mayor source of water vapor and the propagated characteristic is mostly progresses from equator to north and south of the continent and this created a bimodal regime over the continental area. But this cannot explain the fact that propagation characteristics suddenly changed with season, this may be related to the presence of Kelvin waves over the region which influenced propagation of dynamic structures which progressed in opposite direction. To explore the precipitation and moisture transport relationship, mean temporal lag-correlation coefficients are calculated with a time lag varying between ±9 months. Positive/negative lag corresponds to moisture changing after/before precipitation.

Figure 4 shows this result. The maximum correlation between precipitation and moisture flux over the Central Africa occurs with a lag between 0 and 2 months and no lag over the ocean. The maximum correlation over the Gulf of Guinea occurs with a positive lag of about 1–2 months (moisture is larger 2 months after precipitation). Correlation between precipitation and moisture is largely dominated over land with a coefficient greater than 0.5. Over the ocean, correlation between precipitation and moisture is lower than correlation over land; this correlation is not significant (lower than 0.3). This result confirms past study which showed that along the coastline and Gulf of Guinea, over land, the precipitation is well correlated to moisture convergence (Bielli and Roca, 2010). Results indicate that over the study area different geographical precipitation regimes are related to different relationships with the moisture flux and the local sources, both in terms of temporal and

spatial scales. The scale dependence and the main role of each source indicate that the analysis should be help to clarify the functioning of the moisture control in the climate system.

4. Summary and conclusion

First objective of this study was to understand the moisture transport variability accordingly to wet and dry condition, its interaction with Kelvin wave's events. The pre-monsoon months (March–May) are characterized by westerly moisture field over the Gulf of Guinea and are associated with intense Kelvin wave's activity. But relatively weak flow incursion is observed over Gulf of Guinea associated with the northward evolution of the ITCZ. Results also showed that precipitation is strongly correlated with moisture transport over Central Africa in monthly time scale while, Kelvin waves which is almost persists in synoptic time scale has no correlation with moisture. In spite of fact that this correlation between Kelvin wave and moisture is too low, the wave creates better conditions for moisture incursion over the region. Lag correlation reveals that the maximum moisture transport occurs 1–2 months after the maximum precipitation occurs over the region. The zonal component of moisture is the largest contributor to mean moisture over Central Africa, while the meridional component contributes the most over the Gulf of Guinea. Wet condition and better precipitation occur over the region. It becomes necessary to quantify the contribution of Kelvin waves to moisture variability in synoptic and interannual time scale. Final point will identify the time scale regime which most contributed for the moisture initiation over Central Africa.

Acknowledgements

The authors are grateful to the European Centre for Medium-Range Weather Forecasts for the use of the ECMWF-Interim dataset. This study was supported by the French component of AMMA. Based on French initiative, AMMA was built by an international scientific group and is currently funded by a large number of agencies, especially from France, UK, US and Africa. It has been beneficiary of a major financial contribution from the European Community's Sixth Framework Research Program. Detailed information on scientific coordination and funding is available on the AMMA International website http://www.amma-international.org. The authors are also grateful to the two reviewers for their thoughtful and constructive reviews, which helped improving the manuscript.

References

Bielli S, Roca R. 2010. Scale decomposition of atmospheric water budget over West Africa during the monsoon 2006 from NCEP/GFS analyses. *Climate Dynamics* **35**(1): 143–157.

Chatfield C. 2004. *The Analysis of Time series, An Introduction*, 6th ed. Chapman Hall/CRC: New York, NY; 333.

Giannini A, Saravanan R, Chang P. 2003. Oceanic forcing of Sahel rainfall on interannual to interdecadal time scales. *Science* **302**: 1027–1030. https://doi.org/10.1126/science.1089357.

Grueber A, Krueger AF. 1974. The status of the NOAA outgoing longwave radiation data set. *Bulletin of the American Meteorological Society* **65**: 958–962.

Janicot S, Mounier F, Diedhiou A. 2008. Les ondes atmosphériques d'échelle synoptique dans la mousson d'Afrique de l'Ouest et centrale: ondes d'est et onde de Kelvin. *Sécheresse* **19**: 13–22.

Kamsu TPH, Janicot S, Monkam D, Lenouo A. 2014. Convection activity over the Guinean coast and Central Africa during northern spring from synoptic to intra-seasonal timescales. *Climate Dynamics* **43**: 3377–3401. https://doi.org/10.1007/s00382-014-2111-y.

Kiladis GN, Wheeler MC, Haertel PT, Straub KH, Roundy PE. 2009. Convectively coupled equatorial waves. *Reviews of Geophysics* **47**: RG2003. https://doi.org/10.1029/2008RG000266.

Lélé IM, Lance ML, Lamb PJ. 2015. Analysis of low-level atmospheric moisture transport associated with the West African Monsoon. *Journal of Climate* **28**: 4414–4430. https://doi.org/10.1175/JCLI-D-14-00746.1.

Liebmann B, Smith CA. 1996. Description of a complete (interpolated) outgoing longwave radiation dataset. *Bulletin of the American Meteorological Society* **77**: 1275–1277.

Matsuyama H, Oki T, Shinoda M, Masuda K. 1994. The seasonal change of the water budget in the Congo River Basin. *Journal of the Meteorological Society of Japan* **32**: 281–299.

Nguyen TTH, Duvel JP. 2008. Synoptic wave perturbations and convective systems over equatorial Africa. *Journal of Climate* **21**: 6372–6388.

Pokam MW, Djiotang TLA, Mkankam KF. 2011. Atmospheric water vapor transport and recycling in Equatorial Central Africa through NCEP/NCAR reanalysis data. *Climate Dynamics* **38**: 1715–1729.

Rao VB, Chapa SR, Franchito SH. 1999. Decadal variation of atmosphere–ocean interaction in the tropical Atlantic and its relationship to the Northeast-Brazil rainfall. *Journal of the Meteorological Society of Japan* **77**: 63–75.

Simmons AS, Uppala DD, Kobayashi S. 2007. ERA Interim: New ECMWF reanalysis products from 1989 onwards. ECMWF Newsletter No. 110, ECMWF, Reading, United Kingdom, 25–35.

Straub KH, Kiladis GN. 2002. Observations of convectively coupled Kelvin waves in the eastern Pacific ITCZ. *Journal of the Atmospheric Sciences* **59**: 30–53.

Wheeler M, Kiladis G. 1999. Convectively coupled equatorial waves: analysis of clouds and temperature in the wavenumber–frequency domain. *Journal of the Atmospheric Sciences* **56**: 374–399.

Wheeler M, Kiladis GN, Webster PJ. 2000. Large scale dynamical fields associated with convectively coupled equatorial waves. *Journal of the Atmospheric Sciences* **57**: 613–640.

Zebaze S, Lenouo A, Tchawoua C, Janicot S. 2015. Synoptic Kelvin type perturbation waves over Congo basin over the period 1979–2010. *Journal of Atmospheric and Solar-Terrestrial Physics* **130**: 43–56.

21

Advection errors in an orthogonal terrain-following coordinate: idealized 2-D experiments using steep terrains

Xun Zou,[3] Yiyuan Li,[1]* Jinxi Li[1] and Bin Wang[1,2]

[1] LASG, Institute of Atmospheric Physics, Chinese Academy of Sciences, Beijing, China
[2] CESS, Tsinghua University, Beijing, China
[3] School of Atmospheric Sciences, Nanjing University, China

*Correspondence to:
Y. Li, LASG, Institute of
Atmospheric Physics, Chinese
Academy of Sciences, 40
Huayanli, Chaoyang District,
Beijing 100029, China.
E-mail: liyiyuan@mail.iap.ac.cn

Abstract

This study illustrated the importance of smoothed vertical layers and orthogonal grids in an orthogonal terrain-following coordinate (an OTF-coordinate) in reducing advection errors over steep terrain. Three coordinates, namely, classic terrain-following coordinate, hybrid terrain-following coordinate (HTF-coordinate) and OTF-coordinate, were employed in Schär-type advection experiments for various terrains. The results demonstrated that the OTF-coordinate could diminish the grids with high skewness in the HTF-coordinate over steep terrain; the orthogonal grids share the same importance in reducing advection errors with smoothed vertical layers. Therefore, the advection errors in the OTF-coordinate are considerably reduced than those in the corresponding HTF-coordinate over steep terrain.

Keywords: terrain-following coordinate; advection errors; smoothed vertical layers; orthogonal computational grids; skewness; steep terrain

1. Introduction

The classic terrain-following coordinate (the CTF-coordinate; Phillips, 1957; Gal-Chen and Somerville, 1975) can turn a complex earth surface into a coordinate surface, so as to simplify the lower boundary of a numerical model. At present, the CTF-coordinate has been widely applied in atmospheric and oceanic models (Bleck, 2002; Wang *et al.*, 2004; Davies *et al.*, 2005; Madec, 2008; Skamarock *et al.*, 2008, 2012; Wallcraft *et al.*, 2009; Skamarock *et al.*, 2012; Schättler *et al.*, 2013). The steep vertical layers and the non-orthogonal vertical computational grids of the CTF-coordinate over steep terrain, however, induce significant advection errors in a model (Thompson *et al.*, 1985; Sharman *et al.*, 1988; Pielke, 2002; Sankaranrayanan and Spaulding, 2003; Steppeler *et al.*, 2003; Ji *et al.*, 2005; Li *et al.*, 2012; Mesinger *et al.*, 2012). Many methods have been proposed to reduce the advection errors in the CTF-coordinate. The most popular and successful one is the hybrid terrain-following coordinate (the HTF-coordinate; Arakawa and Lamb, 1977; Simmons and Burridge, 1981; Simmons and Strüfing, 1983; Schär *et al.*, 2002; Klemp, 2011; Li *et al.*, 2011), which has the smoothed vertical layers over steep terrain. An innovative method called cut-cell method featured with the orthogonal Cartesian grids above terrain and irregular grids near the terrain has been introduced to overcome the problem of non-orthogonal grids in the vertical (Adcroft *et al.*, 1997; Yamazaki and Satomura, 2010; Lock *et al.*, 2012; Adcroft, 2013; Steppeler *et al.*, 2013; Good *et al.*, 2014).

Recently, an orthogonal terrain-following coordinate (an OTF-coordinate; Li *et al.*, 2014) has been designed to reduce the advection errors in the CTF-coordinate. In details, the OTF-coordinate can reduce the advection errors via smoothing the vertical layers, such as in a HTF-coordinate. More importantly, it can create orthogonal grids in the vertical, thereby further reducing the advection errors compared to the corresponding HTF-coordinate. The experiments with three kinds of terrains implemented by Li *et al.* (2015) indicated that the more complex the terrain is, the greater reduction of advection errors by the OTF-coordinate is. All the experiments implemented by Li *et al.* (2014, 2015) however, adopted terrains with gentle slopes only. Though the results consistently attested that both the smoothed vertical layers and orthogonal grids could reduce the advection errors, the former contributes more than the later.

This study aims to distinguish the relative importance between the smoothed vertical layers and orthogonal grids created by the OTF-coordinate over steep terrain. First, a new kind of terrain is designed with much steeper slope than the terrain employed by Li *et al.* (2014). Then, a set of 2-D linear advection experiments is conducted in the CTF-coordinate, HTF-coordinate and OTF-coordinate using the newly designed terrains. Finally, the relative importance of smoothed vertical layers and orthogonal grids in reducing the advection errors is shown by comparing the advection errors in these three coordinates. Moreover, the cause for different effects of orthogonal grids over different kinds of terrains is analyzed by comparing the number of grids with high skewness in the HTF-coordinate and OTF-coordinate over various terrains.

2. Design of a steep terrain

In order to increase terrain slope, a new kind of terrain is designed based on the five-crest wavelike terrain proposed by Schär *et al.* (2002). The expression of the new terrain is given by

$$h(x) = \sin^2\left(\frac{n\pi x}{2a}\right) h^*(x), \quad (1)$$

where

$$h^*(x) = \begin{cases} h_0 \sin^2\left(\frac{\pi x}{2a}\right) & |x| \leq a \\ 0 & |x| \geq a \end{cases} \quad (2)$$

h_0 is the maximum height of the terrain, a is the half-width of the terrain, and n is the number of terrain peaks. Note that the maximum $h(x)$ equals h_0, only if n is twice of an odd number.

The three kinds of terrains proposed by Schär *et al.* (2002) are shown in Figure 1(a)–(c), the one used in Li *et al.* (2014) is shown in Figure 1(c), and three kinds of new terrains designed in this paper are illustrated in Figure 1(e)–(g). Specifically, the new terrains have the same width but double the number of the peaks, respectively, corresponding to the ones proposed by Schär *et al.* (2002), and then lead to the increase of the slope as well as the number of the maximum slope that is mostly relevant to the high skewness of the computational grids (Figure 1(d) and (h)).

3. Parameters in the 2-D linear advection experiments

For consistency, we use the same parameters as Li *et al.* (2014), except for the terrain, the time step and the time integration scheme. Specifically, the definition $\sigma = H_t \frac{H_t - z}{H_t - h}$ proposed by Gal-Chen and Somerville (1975) is adopted in the CTF-coordinate, where H_t is the top of model and h represents the terrain. The rotation parameter $b = \left(\frac{H_t - z}{H_t - h}\right)^n$ $(n = 20)$ is chosen for the OTF-coordinate. The definition $\sigma = z - \left(\frac{H_t - z}{H_t - h}\right)^n$ $(n = 24)$ is used in the HTF-coordinate to create similar vertical layers as those in the OTF-coordinate. Note that, the vertical layers are similar, but not identical, to those in the HTF-coordinate and OTF-coordinate, and the slopes of the vertical layers in these coordinates are shown in Figure S1, Supporting Information.

Three kinds of terrains shown in Figure 1(e)–(g) are used in the following experiments, and named as two-, six- and ten-crest terrain, respectively. The ten-crest terrain is shown in Figure 2 as an example. The domain of the experiments, fixed with 0–300 km in the horizontal and 0–25 km in the vertical, is also shown in Figure 2.

In addition, the definitions of tracer q and horizontal velocity u follow the advection experiments designed by Schär *et al.* (2002). The expression of q is given by

$$q(x,z) = q_0 \cdot \begin{cases} \cos^2\left(\frac{\pi}{2} \cdot r\right), & r \leq 1 \\ 0 \end{cases}, \quad (3)$$

where $q_0 = 1$, $r = \sqrt{\left(\frac{x - x_0}{A_x}\right)^2 + \left(\frac{z - z_0}{A_z}\right)^2}$, $x_0 = 100$ km, $A_x = 25$ km, $A_z = 3$ km, and $z_0 = 6$ km. The expression of u is given by

$$u(z) = u_0 \cdot \begin{cases} 1 & , \quad z_2 \leq z \\ \sin^2\left(\frac{\pi}{2} \cdot \frac{z - z_1}{z_2 - z_1}\right) & , \quad z_1 \leq z \leq z_2 \\ 0 & , \quad z \leq z_1 \end{cases}, \quad (4)$$

where $u_0 = 10$ m s^{-1}, $z_1 = 2$ km, and $z_2 = 3$ km.

The 2-D linear advection equation used in all the experiments is given by

$$\frac{\partial}{\partial t}\left(J^{-1}q\right) + \frac{\partial}{\partial X}\left(J^{-1}Uq\right) + \frac{\partial}{\partial Z}\left(J^{-1}Wq\right) = 0, \quad (5)$$

where q represents the tracer, U and W are the velocities, J^{-1} is the inverse of the Jacobian of the coordinate transformation, and X and Z are the horizontal and vertical coordinates in each coordinate system. The discretization of Equation (5) using the fourth-order Runge–Kutta scheme in time and centered scheme in space is described as follows:

$$F(q^n) = -\frac{J^{-1}_{i+1,k}U_{i+1,k}q^n_{i+1,k} - J^{-1}_{i-1,k}U_{i-1,k}q^n_{i-1,k}}{2\Delta X}$$
$$- \frac{J^{-1}_{i,k+1}W_{i,k+1}q^n_{i,k+1} - J^{-1}_{i,k-1}W_{i,k-1}q^n_{i,k-1}}{2\Delta Z} \quad (6)$$

$$\frac{J^{-1}_{i,k}\left(q^{n+\frac{1}{2}*}_{i,k} - q^n_{i,k}\right)}{\Delta t/2} = F(q^n) \quad (7)$$

$$\frac{J^{-1}_{i,k}\left(q^{n+\frac{1}{2}**}_{i,k} - q^n_{i,k}\right)}{\Delta t/2} = F\left(q^{n+\frac{1}{2}*}\right) \quad (8)$$

$$\frac{J^{-1}_{i,k}\left(q^{n+1*}_{i,k} - q^n_{i,k}\right)}{\Delta t} = F\left(q^{n+\frac{1}{2}**}\right) \quad (9)$$

$$\frac{J^{-1}_{i,k}\left(q^{n+1}_{i,k} - q^n_{i,k}\right)}{\Delta t} = \frac{1}{6}\left[F(q^n) + 2F\left(q^{n+\frac{1}{2}*}\right)\right.$$
$$\left. + 2F\left(q^{n+\frac{1}{2}**}\right) + F\left(q^{n+1*}\right)\right] \quad (10)$$

In all the experiments, $\Delta X = 0.5$ km and $\Delta Z = 1.0$ km. And each experiment is integrated for 6400 time steps with $\Delta t = 1.5625$ s. In addition, the periodic boundary condition is applied in the horizontal, while the rigid-lid boundary condition is applied in the vertical.

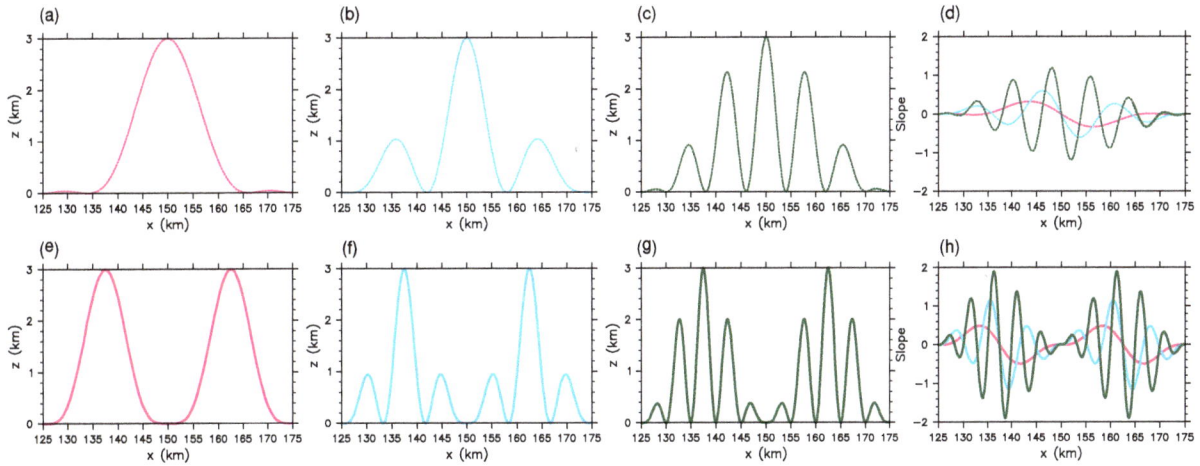

Figure 1. Six kinds of terrains used in this study. (a)–(c) are the terrains proposed by Schär et al. (2002) with $\lambda = 32$, 16 and 8 km. (e)–(g) are the terrains designed in this paper with $n = 2$, 6 and 10 in Equation (1). In both types of terrains, $h_0 = 3$ km and $a = 50$ km. (d) and (h) are the slopes of one-, three- and five-crest terrains and two-, six- and ten-crest terrains, respectively.

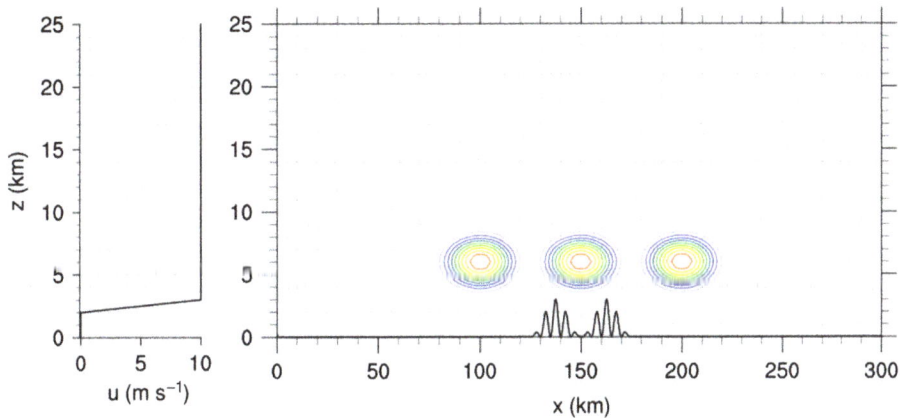

Figure 2. The wind field, the analytical solution of tracer and terrain used in the 2-D linear advection experiments. The left panel is the vertical profile of the given u field. The colored contours in the right panel represent the tracer q with contour interval of 0.1, and the thick black curve represents the newly designed ten-crest wavelike terrain.

4. Comparison of advection errors in different coordinates over various terrains

First, the vertical computational girds of the CTF-coordinate, HTF-coordinate and OTF-coordinate over two-, six- and ten-crest terrains as well as their skewnesses are shown in Figures 3 and 4. For each coordinate, there are 301 grids in the horizontal and 51 grids in the vertical, and every other grid line is shown in Figure 3. The skewness is calculated by maximum $\left|\frac{\theta - 90^0}{90^0}\right|$, where θ is the angle between two neighboring faces of a grid. Note that the smaller the skewness is, the more orthogonal the grid is, and the higher the grid quality is.

Compared to the CTF-coordinate (black grid lines in Figure 3), the HTF-coordinate (red grid lines in Figure 3) enjoys the smoothed vertical layers (also shown in Figure S1); and compared to the HTF-coordinate, the OTF-coordinate employs the orthogonal grids as shown in Figures 3 and 4. Consequently, the effect of smoothed vertical layers

on reducing the advection errors can be obtained through comparing the advection errors between the HTF-coordinate and CTF-coordinate; while the comparison of the advection errors between the OTF-coordinate and HTF-coordinate can explain the effect of orthogonal grids.

Second, the root mean square errors (RMSEs) of the advection term in the CTF-coordinate, HTF-coordinate and OTF-coordinate in the whole integration over two-, six- and ten-crest terrains are all shown in Figure 5 (the relevant absolute errors at the end of integration are shown in Figure S2). Over the two-crest terrain, the RMSE in the HTF-coordinate is smaller than that in the CTF-coordinate (Figure 5(a)); while the RMSE in the OTF-coordinate is comparable with that in the HTF-coordinate (Figure 5(b)). It reveals that the smoothed vertical layers contribute more to reduce the advection errors than the orthogonal grids over the two-crest terrain, which is consistent with Li et al. (2014, 2015) as shown in Figure S3.

However, over the six- and ten-crest terrains, the RMSEs in the HTF-coordinate are much smaller than

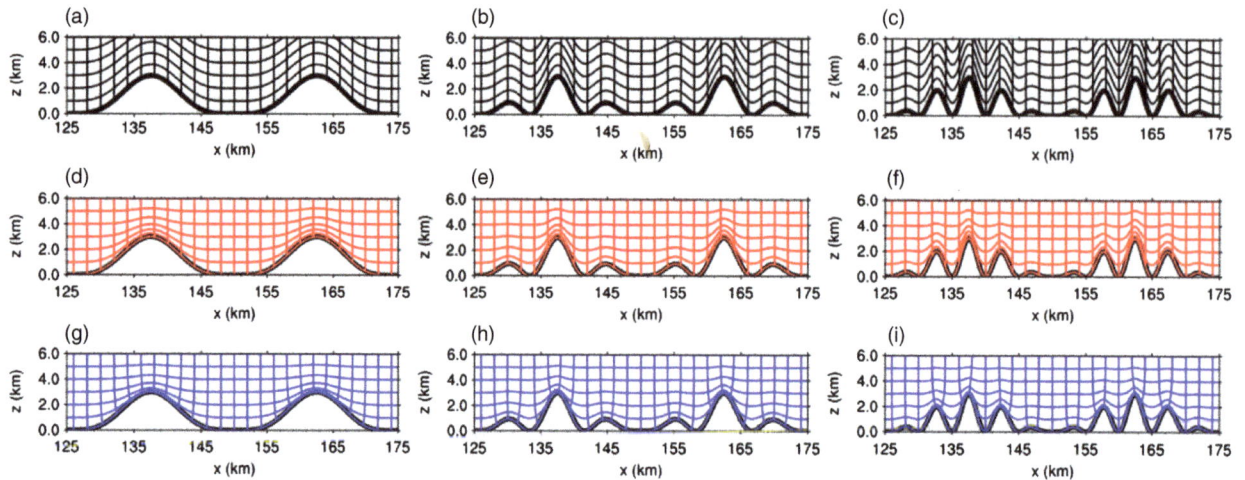

Figure 3. Computational grids of the CTF-coordinate, HTF-coordinate and OTF-coordinate in the vertical over three kinds of terrains. Black curves in (a)–(c) are the grid lines in the CTF-coordinate; red curves in (d)–(f) are the grid lines in the HTF-coordinate; and blue curves in (g)–(i) are the grid lines in the OTF-coordinate.

those in the CTF-coordinate (Figure 5(c) and (e)); while the RMSEs in the OTF-coordinate are also much smaller than those in the HTF-coordinate (Figure 5(d) and (f)). Therefore, the effect of orthogonal grids is as important as the effect of smoothed vertical layers over the six- and ten-crest terrains.

Third, the RMSEs in the CTF-coordinate, HTF-coordinate and OTF-coordinate at the end of the integration obtained in this study using the newly designed steep terrains as well as the results obtained by Li *et al.* (2014, 2015) using relative gentle terrains are summarized in Table 1. According to the order of magnitude of RMSE in each coordinate, these six experiments can be classified into three types (in black, light blue and orange in Table 1).

Type one includes the experiments using one- and two-crest terrains (in black in Table 1). The RMSEs in all three coordinates are of the same order of magnitude, namely, both effects of smoothed vertical layers and orthogonal grids on reducing the advection errors are relatively inconspicuous over gentle terrain.

Type two includes the experiments using three- and five-crest terrains (in light blue in Table 1). The RMSEs in the HTF-coordinate and OTF-coordinate are of the same order of magnitude, but both of them are at least one order of magnitude smaller than those in the CTF-coordinate. It shows that the effect of smoothed vertical layers on reducing the advection errors is evident over steep terrain; namely, the effect of smoothed vertical layers over steep terrain is greater than the effect of orthogonal grids.

Type three includes the experiments using six- and ten-crest terrains (in orange in Table 1). The RMSEs in the HTF-coordinate are one order of magnitude smaller than those in the CTF-coordinate; however, the RMSEs in the OTF-coordinate are also one order of magnitude smaller than those in the HTF-coordinate. Explicitly, the effect of orthogonal grids is as important as

the effect of smoothed vertical layers on reducing advection errors over very steep terrain.

Finally, in order to explain different effects of orthogonal grids over various terrains as categorized in Table 1, the numbers of grids with certain skewness in the HTF-coordinate and OTF-coordinate are counted (Table 2). The numbers of a certain value of skewness in the OTF-coordinate are constantly smaller than those in the HTF-coordinate; namely, the effect of orthogonal grids on reducing the advection errors is achieved through diminishing the skewness of the computational grids.

For the one- and two-crest terrain experiments (in black in Table 2), none of the grids has skewness greater than 0.3. However, Thompson *et al.* (1985) proposed that the truncation error due to the non-orthogonal grids could be significantly reduced when the grid angle is smaller than 45° (skewness greater than 0.5). Therefore, the orthogonal grids have small effect on reducing the advection errors over one- and two-crest terrains. For the three-crest terrain experiments (in light blue in Table 2), there are only eight grids with skewness greater than 0.3 in the HTF-coordinate. Again, the effect of orthogonal grids is inconspicuous.

For the five- and six-crest terrain experiments (in magenta and orange in Table 2, respectively), the numbers of grids with skewness greater than 0.5 are five and eight in the HTF-coordinate, respectively. Therefore, the effect of orthogonal grids should be conspicuous. There results reveal the considerable effect of orthogonal grids over six-crest terrain, but not consistent with the relatively low effect over five-crest terrain, indicating that the effect of orthogonal grids may be sensitive to the number of the grids with high skewness. For the ten-crest terrain experiments (in orange in Table 2), there are 12 grids with skewness greater than 0.6, and 32 grids with skewness greater than 0.5 in the HTF-coordinate. Accordingly, the effect of orthogonal grids is significant.

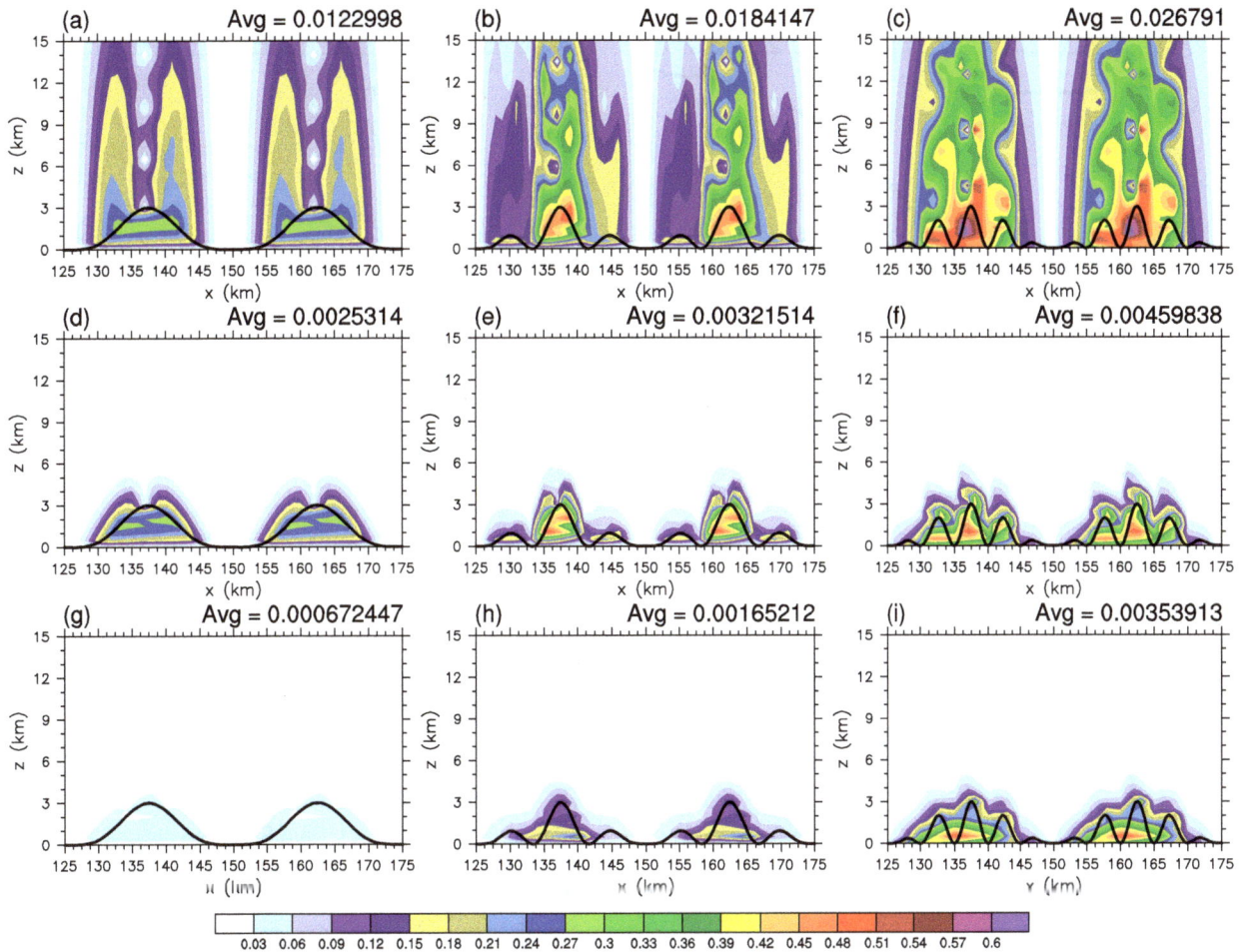

Figure 4. Skewness of vertical grids in the CTF-coordinate, HTF-coordinate and OTF-coordinate for three kinds of terrains. Shading represents skewness. The thick black curve represents the two-, six- and ten-crest terrains in left, middle and right panels, respectively. The shading below the thick black curve is due to the interpolation by plotting software.

In addition, the RMSEs in Table 1 are consistent with the skewness shown in Figures 4 and S4, namely, having the maximum in the CTF-coordinate and the minimum in the OTF-coordinate. Therefore, just as the orthogonal grids, the smoothed vertical layers can also diminish the skewness of computational grids, and therefore reduce the advection errors.

In conclusion, both the smoothed vertical layers and orthogonal grids can diminish the skewness of the computational grids, therefore reducing the advection errors (Figures 4 and S4, Table 1). However, the effect of orthogonal grids is sensitive to the number of grids with high skewness (Tables 1 and 2). Specifically, the effect of orthogonal grids on reducing the advection errors is considerable only when it can diminish the number of grids with high skewness. As a result, over gentle terrain, the grids in the HTF-coordinate are all with small skewness, and then the effect of orthogonal grids on reducing the advection errors is very small. Over very steep terrain, numerous grids in the HTF-coordinate have high skewness; therefore, the effect of orthogonal

grids is considerable. Accordingly, the effect of orthogonal grids is as important as the smoothed vertical layers on reducing the advection errors over very steep terrain.

5. Conclusion and discussion

In order to explore the advection errors in the OTF-coordinate associated with smoothed vertical layers as well as orthogonal grids over steep terrain, this study used the CTF-coordinate, HTF-coordinate and OTF-coordinate to implement a set of 2-D linear advection experiments with terrains of various slopes. First, a new kind of terrain was designed based on the wavelike terrain proposed by Schär *et al.* (2002) to increase steepness (Figure 1(e)–(g)). Then, the CTF-coordinate, HTF-coordinate and OTF-coordinate were implemented in the 2-D linear advection experiments with three kinds of newly designed terrains (two-, six- and ten-crest terrains).

Comparing the advection errors between the CTF-coordinate and HTF-coordinate, as well as those between the HTF-coordinate and OTF-coordinate, we

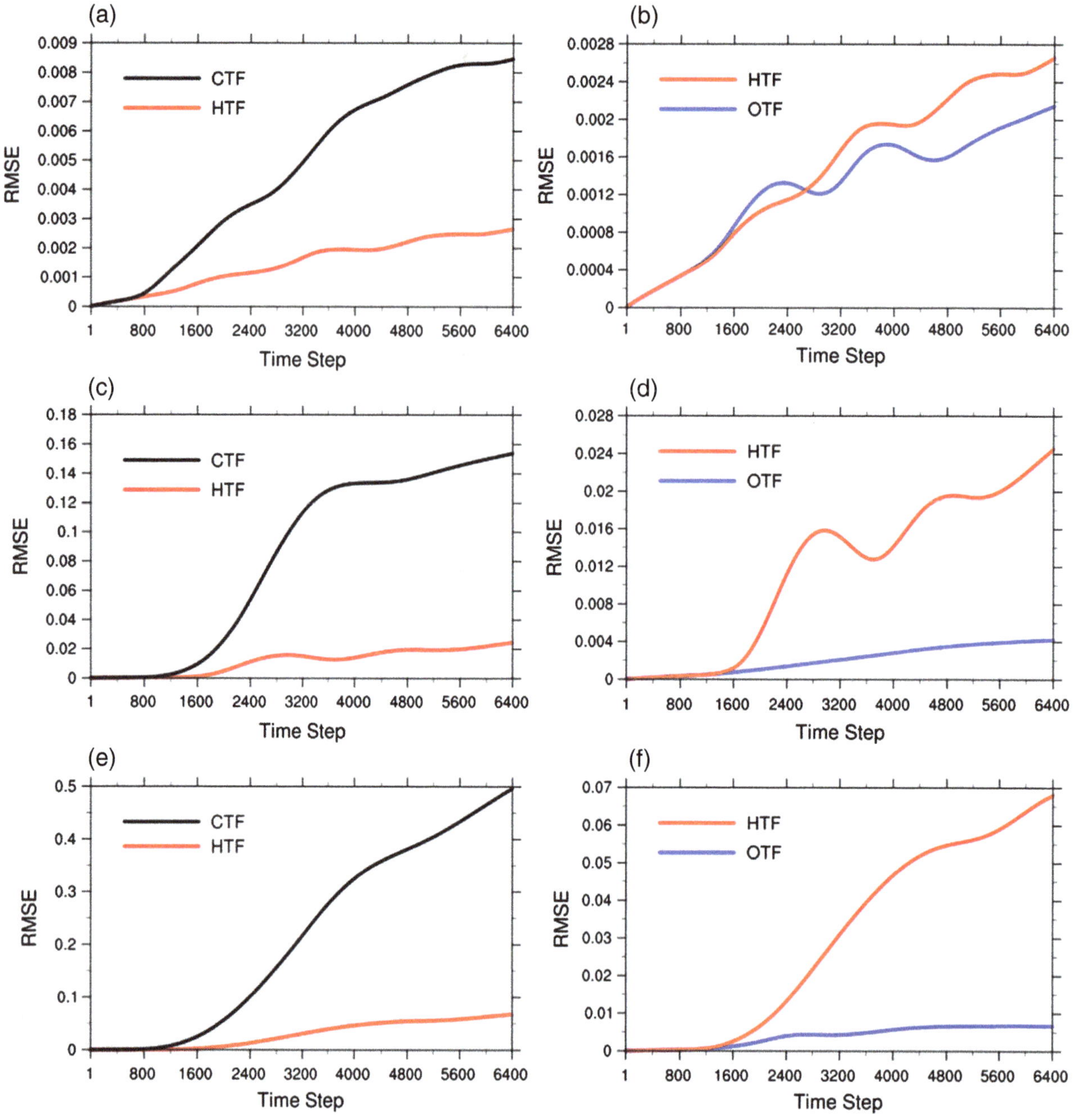

Figure 5. RMSEs of advection term in three coordinates for three kinds of terrains. Black, red and blue curves represent the RMSEs in the CTF-coordinate, HTF-coordinate and OTF-coordinate, respectively.

verified that both smoothed vertical layers and orthogonal grids can consistently reduce the advection errors over various terrains (Figure 5). However, the relative importance of these two approaches in reducing the advection errors depends on terrain slope (Figure 5 and Table 1). The experimental results showed that: (1) over gentle terrain, both effects of smoothed vertical layers and orthogonal grids are relatively inconspicuous; (2) over steep terrain, the effect of smoothed vertical layers becomes significant; therefore, it becomes a more important factor than the orthogonal grids; (3) over very steep terrain, the effect of orthogonal grids is as important as that of smoothed vertical layers. Therefore, the advection errors in the OTF-coordinate can be considerably reduced than those in the HTF-coordinate

over very steep terrain (Figure 5(d) and (f); in orange in Table 1).

The effect of orthogonal grids is sensitive to the number of grids with high skewness (Tables 1 and 2) as proposed by Thompson *et al.* (1985). The effect is considerable only when the orthogonal grids can diminish the number of grids with high skewness, while the grids with high skewness are sensitive to the steepness of terrain. Therefore, the effect of orthogonal grids is considerable over steep terrain. The effect of smoothed vertical layers on reducing the advection errors is also achieved through diminishing the skewness of grids (Figures 4 and S4, Table 1).

In addition, the vertical layers in the OTF-coordinate are smoother than those in the HTF-coordinate in this

Table 1. RMSEs of advection term in three coordinates for six kinds of terrains at the end of the integration.

Kind of terrain	CTF-coordinate	RMSE	
		HTF-coordinate (representing the effect of smoothed vertical layers)	OTF-coordinate (representing the effect of orthogonal grids)
One-crest[a]	$3.36*10^{-3}$	$2.18*10^{-3}$	$2.14*10^{-3}$
Three-crest[a]	$1.01*10^{-2}$	$2.32*10^{-3}$	$2.15*10^{-3}$
Five-crest[a]	$1.59*10^{-1}$	$6.78*10^{-3}$	$2.61*10^{-3}$
Two-crest	$8.47*10^{-3}$	$2.66*10^{-3}$	$2.15*10^{-3}$
Six-crest	$1.54*10^{-1}$	$2.46*10^{-2}$	$4.21*10^{-3}$
Ten-crest	$4.97*10^{-1}$	$6.80*10^{-2}$	$6.72*10^{-3}$

[a]Results in the CTF-coordinate, HTF-coordinate and OTF-coordinate in one-, three- and five-crest terrain experiments are calculated in this paper using the same parameters as in Li et al. (2015) except for the time integration scheme, which is to use the fourth-order Runge–Kutta scheme instead of leapfrog scheme.

Table 2. Numbers of grids with certain skewness in the HTF-coordinate and OTF-coordinate for six kinds of terrains[a].

Kind of terrain	Number of grids with certain skewness											
	HTF-coordinate						OTF-coordinate					
	>0.1	>0.2	>0.3	>0.4	>0.5	>0.6	>0.1	>0.2	>0.3	>0.4	>0.5	>0.6
One-crest	84	2	0	0	0	0	0	0	0	0	0	0
Three-crest	93	38	8	0	0	0	0	0	0	0	0	0
Five-crest	173	96	46	25	5	0	87	10	4	0	0	0
Two-crest	162	58	0	0	0	0	0	0	0	0	0	0
Six-crest	154	90	52	30	8	0	80	4	4	0	0	0
Ten-crest	194	126	86	58	32	12	178	88	48	26	10	0

[a]Experiments indicated by light blue and orange are consistent with the category in Table 1, and those by magenta have differences with the category in Table 1.

study (Figure S1), which partly enhanced the effect of orthogonal grids. The experiments using identical vertical layers in the HTF-coordinate and OTF-coordinate are being implemented in an ongoing study. Further analyses are needed to explain the relationship between grid quality in the vertical and advection errors over various terrains, through using more indexes such as smoothness and aspect ratio. Moreover, the effect of orthogonal terrain-following grids created by the OTF-coordinate needs to be compared with that of the vertical grids created by the cut-cell method in terms of reducing the advection errors.

Acknowledgements

This work is jointly supported by the National Basic Research Program of China (973 Program, Grant No. 2015CB954102) and the National Natural Science Foundation of China (41305095). Zou wishes to thank Mr. Tianyuan Zhu and Miss Chen Chang for improving the presentation of the paper. Y. Li wishes to acknowledge helpful discussion about the advection errors in the OTF-coordinate with Prof. Jianhua Lu.

Supporting information

The following supporting information is available:

Figure S1. Slopes of vertical layers in three coordinates for two-, six- and ten-crest terrains. Figure S2. Absolute errors of advection term in three coordinates for two-, six- and ten-crest terrains. Figure S3. RMSEs of advection term in three coordinates for one-, three- and five-crest terrains. Figure S4. Skewness of vertical grids in the CTF-coordinate, HTF-coordinate and OTF-coordinate for three kinds of terrains.

References

Adcroft A. 2013. Representation of topography by porous barriers and objective interpolation of topographic data. *Ocean Modelling* **67**: 13–27.

Adcroft A, Hill C, Marshall J. 1997. Representation of topography by shaved cells in a height coordinate ocean model. *Monthly Weather Review* **125**: 2293–2315.

Arakawa A, Lamb VR. 1977. Computational design of the basic dynamical processes of the UCLA general circulation model. In *Methods in Computational Physics: Advances in Research and Applications*, Chang J (ed), Vol. **17**. Academic Press,; 173–265.

Bleck R. 2002. An oceanic general circulation model framed in hybrid isophycnic-Cartesian coordinates. *Ocean Modelling* **37**: 55–88.

Davies T, Cullen MJP, Malcolm AJ, Mawson MH, Staniforth A, White AA, Wood N. 2005. A new dynamical core for the Met Office's global and regional modelling of the atmosphere. *Quarterly Journal of the Royal Meteorological Society* **131**: 1759–1782.

Gal-Chen T, Somerville RCJ. 1975. On the use of a coordinate transformation for the solution of the Navier–Stokes Equations. *Journal of Computational Physics* **17**: 209–228.

Good B, Gadian A, Lock S-J, Ross A. 2014. Performance of the cut-cell method of representing orography in idealized simulations. *Atmospheric Science Letters* **15**: 44–49.

Ji LR, Chen JB, Zhang DM, *et al.* 2005. Review of some numerical aspects of the dynamic framework of NWP model (in Chinese). *Chinese Journal of Atmospheric Sciences* **29**: 120–130.

Klemp JB. 2011. A terrain-following coordinate with smoothed coordinate surfaces. *Monthly Weather Review* **139**: 2163–2169.

Li C, Chen DH, Li XL. 2012. A design of height-based terrain-following coordinate in the atmospheric numerical model: theoretical analysis and idealized tests (in Chinese). *Acta Meteorologica Sinica* **70**: 1247–1259.

Li YY, Wang B, Wang DH. 2011. Characteristics of terrain-following sigma coordinate. *Atmospheric and Oceanic Science Letters* **4**: 157–161.

Li YY, Wang B, Wang DH, Li JX, Dong L. 2014. An orthogonal terrain-following coordinate and its preliminary tests using 2-D idealized advection experiments. *Geoscientific Model Development* **7**: 1767–1778.

Li YY, Li JX, Wang B, Zou X. 2015. Advection errors in an orthogonal terrain-following coordinate: idealized experiments (in Chinese). *Chinese Science Bulletin* **60**: 3144–3152, doi: 10.1360/N972015-00075.

Lock S-J, Bitzer H, Coals A, Gadian A, Mobbs S. 2012. Demonstration of a cut-cell representation of 3-D orography for studies of atmospheric flows over steep hills. *Monthly Weather Review* **140**: 411–424.

Madec G. 2008. *Nemo Ocean Engine*, Vol. **27**. Note du Pôle de modélisation Institut Pierre-Simon Laplace(IPSL): Guyancourt. ISSN No. 1288-1619.

Mesinger F, Chou S-C, Gomes JL, Jovic D, Bastos P, Bustamante JF, Lazic L, Lyra AA, Morelli S, Ristic I, Veljovic K. 2012. An upgraded version of the eta model. *Meteorology and Atmospheric Physics* **116**: 63–79.

Phillips NA. 1957. A coordinate system having some special advantages for numerical forecasting. *Journal of Meteorology* **14**: 184–185.

Pielke RA. 2002. *Mesoscale Meteorological Modeling*. Academic press: Cambridge, MA.

Sankaranrayanan S, Spaulding ML. 2003. A study of the effects of grid non-orthogonality on the solution of shallow water equations in boundary-fitted coordinate systems. *Journal of Computational Physics* **184**: 299–320.

Schär C, Leuenberger D, Fuhrer O, et al. 2002. A new terrain-following vertical coordinate formulation for atmospheric prediction models. *Monthly Weather Review* **130**: 2459–2480.

Schättler U, Doms G, Schraff C. 2013. A description of the nonhydrostatic regional COSMO-model part VII: user's guide. http://www.cosmo-model.org/content/model/documentation/core/cosmoUser Guide.pdf. (accessed 30 July 2015).

Sharman RD, Keller TL, Wurtele MG. 1988. Incompressible and anelastic flow simulations on numerically generated grids. *Monthly Weather Review* **116**: 1124–1136.

Simmons AJ, Burridge DM. 1981. An energy and angular-momentum conserving vertical finite-difference scheme and hybrid vertical coordinates. *Monthly Weather Review* **109**: 758–766.

Simmons AJ, Strüfing R. 1983. Numerical forecasts of stratospheric warming events using a model with a hybrid vertical coordinate. *Quarterly Journal of the Royal Meteorological Society* **109**: 81–111.

Skamarock WC, Klemp JB, Dudhia J, Gill DO, Barker DM, Duda MG, Huang X-Y, Wang W, Powers JG. 2008. A description of the advanced research WRF Version 3. NCAR Technical Note NCAR/TN-475+STR, 113 pp.

Skamarock WC, Klemp JB, Duda MG, Fowler LD, Park S-H, Ringler T. 2012. A multi scale nonhydrostatic atmospheric model using centroidal Voronoi tesselations and C-grid staggering. *Monthly Weather Review* **140**: 3090–3105.

Steppeler J, Hess R, Schättler U, Bonaventura L. 2003. Review of numerical methods for nonhydrostatic weather prediction models. *Meteorology and Atmospheric Physics* **82**: 287–301.

Steppeler J, Park SH, Dobler A. 2013. Forecasts covering one month using a cut-cell model. *Geoscientific Model Development* **6**: 875–882.

Thompson JF, Warsi ZUA, Mastin CW. 1985. *Numerical Grid Generation foundations and applications*. North Holland: Amsterdam.

Wallcraft AJ, Metzger EJ, Carroll SN. 2009. Software design description for the Hybrid coordinate ocean model (HYCOM) version 2.2. NRL/MR/7320-09-9166.https://hycom.org/attachments/063_metzger1-2009.pdf. (accessed 30 July 2015).

Wang B, Wan H, Ji Z, Zhang X, Yu R, Yu Y, Liu H. 2004. Design of a new dynamical core for global atmospheric models based on some efficient numerical methods. *Science in China Serial A* **47**(Supp): 4–21.

Yamazaki H, Satomura T. 2010. Nonhydrostatic atmospheric modeling using a combined Cartesian grid. *Monthly Weather Review* **138**: 3932–3945.

22

Assessment of the variability of pollutants concentration over the metropolitan area of São Paulo, Brazil, using the wavelet transform

M. Zeri,[1]* V. S. B. Carvalho,[2] G. Cunha-Zeri,[1] J. F. Oliveira-Júnior,[3] G. B. Lyra[3] and E. D. Freitas[4]

[1] Brazilian Center for Monitoring and Early Warnings of Natural Disasters (CEMADEN), São José dos Campos, Brazil
[2] Instituto de Recursos Naturais, Universidade Federal de Itajubá, Brazil
[3] Departamento de Ciências Ambientais, Instituto de Florestas, Universidade Federal Rural do Rio de Janeiro, Seropédica, Brazil
[4] Instituto de Astronomia, Geofísica e Ciências Atmosféricas, Universidade de São Paulo, Brazil

*Correspondence to:
M. Zeri, Brazilian Center for
Monitoring and Early Warnings
of Natural Disasters
(CEMADEN), Estrada Doutor
Altino Bondesan, 500 – Eugênio
de Melo, 12247-016, São José
dos Campos, São Paulo, Brazil.
E-mail:
marcelo.zeri@cemaden.gov.br

Abstract

The objective of this work was to investigate the mean and variability of a dataset of pollutant concentrations from measurements taken over the metropolitan area of São Paulo city, Brazil. Wavelet analysis was applied to the time series of pollutant concentrations, revealing the strongest harmonics influencing the signals. A mode of variability of 4–8 days was significant until the middle of the last decade and is likely associated with the approach and passage of meteorological systems. A dataset representing the number of frontal systems moving across the state helped to explain the interannual variability during wintertime. Years with fewer frontal systems had higher levels of pollutants in several locations. Weather events such as inversions and the passage of frontal systems influence the concentration of pollutants. Public policies on air quality should focus not only on reducing the long-term exposure of city-dwellers to the negative effects of pollutants but also account for the possible short-term effects of the weather on air quality.

Keywords: air pollution; wavelet analysis, meteorological systems; data analysis; industrial activity

1. Introduction

Air pollution is a common problem in cities around the world, particularly in metropolitan areas (Sharma *et al.*, 1983; deLeon *et al.*, 1996; Schwartz, 1996; Samet *et al.*, 2000; de Miranda *et al.*, 2002; Godoy *et al.*, 2009). The effects of air pollution – such as particulate matter with diameter $\leq 10\,\mu m$ (PM_{10}), sulfur dioxide (SO_2), carbon monoxide (CO), or ozone (O_3) – increase hospitalizations due to respiratory problems, lung cancer trends, acid rain, and the black dust covering building's façades (Fajersztajn *et al.*, 2013). The effects of air pollution over the metropolitan area of São Paulo (MASP), Brazil, are already linked to impacts on human health (Gonçalves *et al.*, 2005; Cançado *et al.*, 2006), as well as to feedbacks with atmospheric and climatic conditions including composition of aerosols (Castanho and Artaxo, 2001; Bourotte *et al.*, 2005), the urban heat island (UHI) effect, local circulation patterns (Silva Dias and Machado, 1997; Freitas *et al.*, 2007), and mesoscale circulations induced by topography and sea/land breezes (Oliveira *et al.*, 2003). In this work, a dataset of pollutants concentration was analyzed using statistical inference (means and variances) as well as wavelet analysis, to detect the most important modes of variability influencing the levels of pollution. In general, the concentrations have a daily cycle associated with traffic of vehicles or industrial activity.

There is also an annual trend, with a maximum in winter; the cold air makes the atmospheric surface layer shallow and not well mixed, increasing the concentrations. In addition, atmospheric inversions are sometimes observed in winter, when the vertical profile of air temperature makes it impossible to an air parcel to rise due to buoyancy. Here, we used daily averages, thus the daily cycle was filtered out. The wavelets helped to identify harmonics of several days, which contributed to modulate the concentrations beyond the daily cycle.

Observed variability in air pollutant concentrations was analyzed in the context of technological changes facing Brazil regarding air quality policies and the use of ethanol as a substitute to gasoline (Goldemberg, 2007). Brazilian federal agencies, such as the Federal Environmental Council – CONAMA, and state agencies continue to study environmental quality, establish pollutant concentration standards, and operate pollution monitoring networks. These agencies also enforce the use of new technologies to help reduce pollutant emissions, such as catalytic converters which chemically catalyze reactions to oxidize or reduce toxic pollutants into less toxic products and are required by CONAMA to be installed in all new cars from 1997 onwards. Additional examples that public policies decisions have on pollutants concentrations can be found in Andrade *et al.* (2015) and Carvalho *et al.* (2015). The oil crisis in the 1970s also had a significant impact on technology in

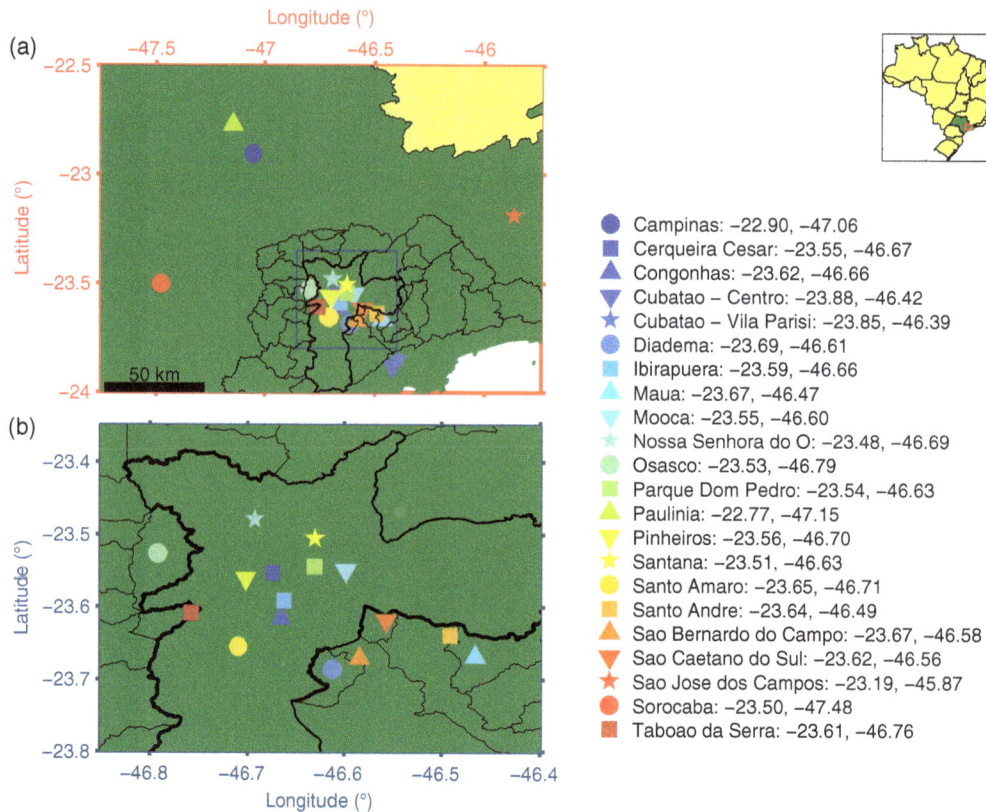

Figure 1. Map of stations and coordinates (latitude and longitude, °) within the metropolitan region of São Paulo (a). The São Paulo city limits are highlighted in bold in the zoomed in detail (b).

Brazil. The rising prices of gasoline compelled the federal government to implement an ethanol fuel industry for road transportation guaranteeing a supply of ethanol from sugarcane coupled with the auto industry building vehicles to run on the new fuel (Goldemberg, 2007). Currently, ethanol-only or flex fuel cars represent nearly 60% of the total vehicle fleet and contribute to 80% of new licensed vehicle in Brazil.

In this work, we report on the variability of pollutants in the MASP using time series of pollutants concentrations measured over 22 stations. The objective here was not a comprehensive study of annual or daily cycles or causes and effects of public policies, which was described in other studies (Salvo and Geiger, 2014; Andrade et al., 2015; Carvalho et al., 2015), but to complement the analysis in those studies by applying wavelet decomposition to the time series of pollutants concentrations.

2. Site and data

The MASP is located in São Paulo state (Figure 1), southeastern Brazil. The territorial area of MASP is of 7944 km², while the urbanized area covers 2139 km², including many cities which are adjacent to the border of São Paulo city (thick black line in bottom panel). For this study, cities outside the limits of MASP were also included, such as São José dos Campos, Cubatão, Sorocaba, Paulínia, and Campinas. The MASP is surrounded

by topographical features with altitudes up to 1100 m above sea level, such as the hills of *Serra do Mar*, to the south, and *Serra da Cantareira*, to the north. The climate is tropical wet with rainy summers and dry winters (Bourotte et al., 2005).

During summer, the region is influenced by South Atlantic convergence zone (SACZ) and mesoscale convective systems (MCS), which typically cause thunderstorms before sunset and nighttime fog (Silva Dias and Machado, 1997; Castanho and Artaxo, 2001; Freitas et al., 2007). During winter, the reduction in rainfall and the occurrence of thermal inversions in the atmospheric boundary layer (ABL) contribute to higher concentrations of SO_2, CO, and PM_{10} (Angevine et al., 1998; Martins et al., 2004; Barbaro et al., 2014). The dispersion of pollutants in this region is influenced by sea breeze and by valley-mountain circulations, due to the proximity to both the coast and the mountain range that runs parallel to the Atlantic Ocean known as *Serra do Mar* (Silva Dias and Machado, 1997; Oliveira et al., 2003; Carvalho et al., 2012). In addition, circulations associated with the UHI effect contribute to the general dispersion of pollutants (Freitas et al., 2007). Similar to other metropolitan areas of the world, emissions by industrial activity and vehicle traffic are the main source of anthropogenic pollutants in the MASP (Castanho and Artaxo, 2001).

Pollutant concentration datasets were obtained from the São Paulo Environmental Agency (CETESB) from stations primarily located within the urban perimeter.

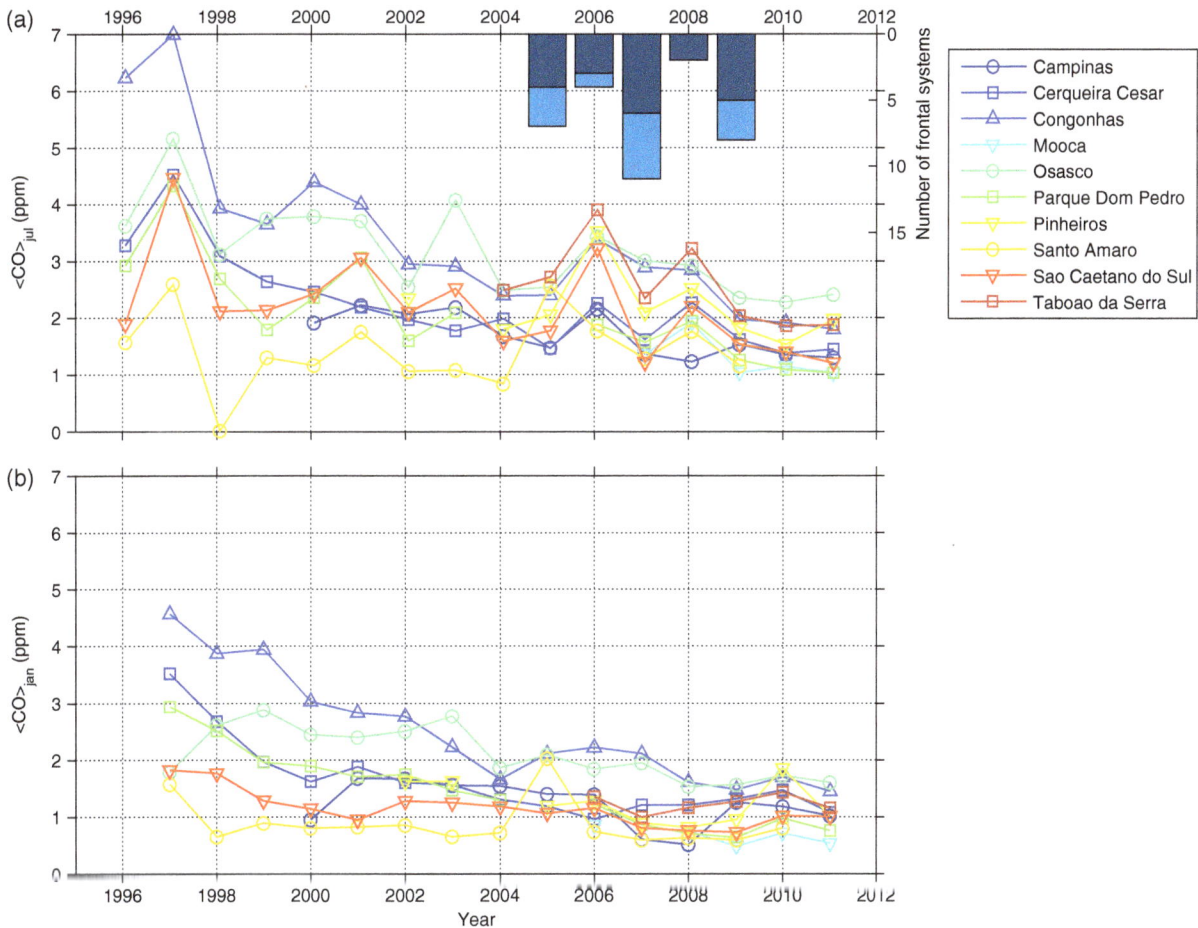

Figure 2. Evolution of monthly averages of CO (ppm) from 1996 to 2011. (a) July and (b) January. Bars denote the number of frontal systems reaching the coast (deep blue) and the countryside (light blue).

Most of the air quality stations used in this study are influenced by vehicle or/and industrial emissions with exception of the Ibirapuera station (Figures S2(e) and S5, Supporting Information), which is located in a park. Stations located outside the MASP, such as Campinas, Sorocaba, and São José dos Campos, had historically lower concentrations for some pollutants. Information regarding the number of licensed vehicles and the size of the ethanol fleet were obtained from the Brazilian Sugarcane Industry Association (UNICA).

The dataset of pollutants concentration from each station included hourly records on PM_{10}, O_3, CO, and SO_2, averaged in this work over 8- or 24-h windows, as recommend by World Health Organization (WHO) standards. In recent years, many studies have addressed both the variability, annual cycle, daily and weekday patterns (Carvalho *et al.*, 2015), and relationships of ethanol prices and public policies on this variability (Salvo and Geiger, 2014; Andrade *et al.*, 2015). In this work, we chose to work only with CO and SO_2, because they presented the highest change from 1996 to 2012 and also had continuous time series of concentrations which were suitable for wavelet analysis. Results on O_3 and PM_{10} are shown in Figures S1–S12 of Supporting Information.

To help explain some peaks in mean monthly concentrations from year to year, data on the number of frontal systems reaching the coast and the countryside were obtained from Climanálise (Climanálise, 2005). These data were plotted together with the monthly concentrations in Figures 2 and 3.

3. Methodology

Data analysis consisted of statistical inferences on time series of concentrations, such as averages, and also correlations with time, to give support to observed trends in pollutant concentrations. In general, wintertime is characterized by less mixing of pollutants, due to colder temperatures and the proximity of the South Atlantic Subtropical Anticyclone. To better identify interannual trends in concentrations, averages were calculated for both July (winter) and January (summer). This procedure enhances the long-term trends because annual maxima (and minima) are compared together. Because of the nature of the annual cycle, with a strong peak during winter but low concentration during summer, yearly averages tend to weaken the maxima, masking the effects of frontal systems on air pollution. It should be noted that not all pollutants were measured at all stations.

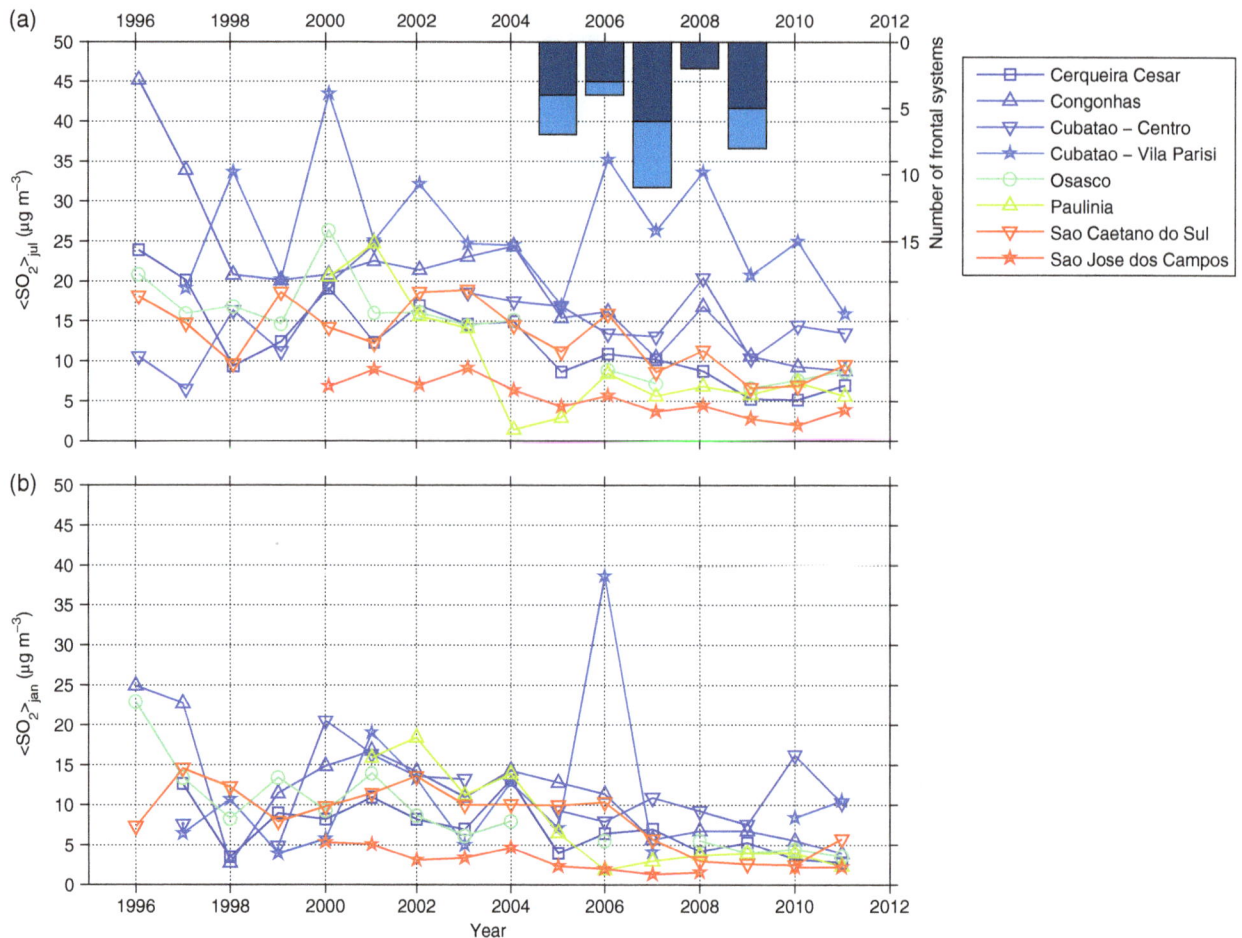

Figure 3. Evolution of monthly averages of SO_2 ($\mu g\,m^{-3}$) from 1996 to 2011. (a) July and (b) January. Bars denote the number of frontal systems reaching the coast (deep blue) and the countryside (light blue).

The full dataset was presented in Figures S1–S4 together with standards for each pollutant following the WHO or CONAMA. The standards used were 9 ppm for CO (maximum 8-h moving average), 150 $\mu g\,m^{-3}$ for PM_{10} (24-h mean), 100 $\mu g\,m^{-3}$ for O_3 (maximum 8-h moving average), and 20 $\mu g\,m^{-3}$ for SO_2, (24-h mean). The standards for PM_{10} and CO were established by CONAMA.

The time series of pollutant concentrations were analyzed using wavelet analysis, a technique that enables the most important frequencies influencing the variability of a signal to be inferred (Daubechies, 1992; Torrence and Compo, 1998). Although Fourier analysis can also be used to identify the most important harmonics in time series, wavelets make it possible to locate in time the influence of harmonics which are not stationary. In recent years, wavelet analysis has been used in many studies of geophysical data, such as river levels, turbulence over plant canopies, and pollutant concentrations (Collineau and Brunet, 1993; Sá *et al.*, 1998; Terradellas *et al.*, 2005; Zeri *et al.*, 2011). The wavelet decomposition works similar to a spectrum, separating the harmonics in a signal while assigning a 'wavelet power' to them, which is proportional to the overall variance. The most important harmonics will be the ones with high wavelet power. Mathematically,

the wavelet power is calculated from the convolution of a function (the wavelet mother) with portions of the signal. The wavelet mother chosen for this study was the Morlet (wavenumber 6), because it was shown to be appropriate to identify the variability of climatological data (Torrence and Compo, 1998). Wavelet analysis requires continuous time series. For this reason, gaps in the data were filled using linear interpolation. Because the wavelet power is calculated locally, the influence of interpolated gaps is easily identified in the scalograms. The wavelet power shown in the scalograms of Figures 5 and 6 (and Figures S1–S12) was calculated as the squared modulus of the wavelet coefficients, having units of signal variance.

4. Results, discussion, and conclusions

The concentration of some pollutants over the MASP is well below the standards (WHO) while others have been continuously surpassing the safe limits. The concentration of CO reached more than the limit of 9 ppm only in the beginning of the period analyzed here (1996–1997), except for two stations that are influenced heavily by heavy vehicle traffic (Congonhas and Cerqueira César). From 2004 to 2011, only 2 or 3 events of CO higher than

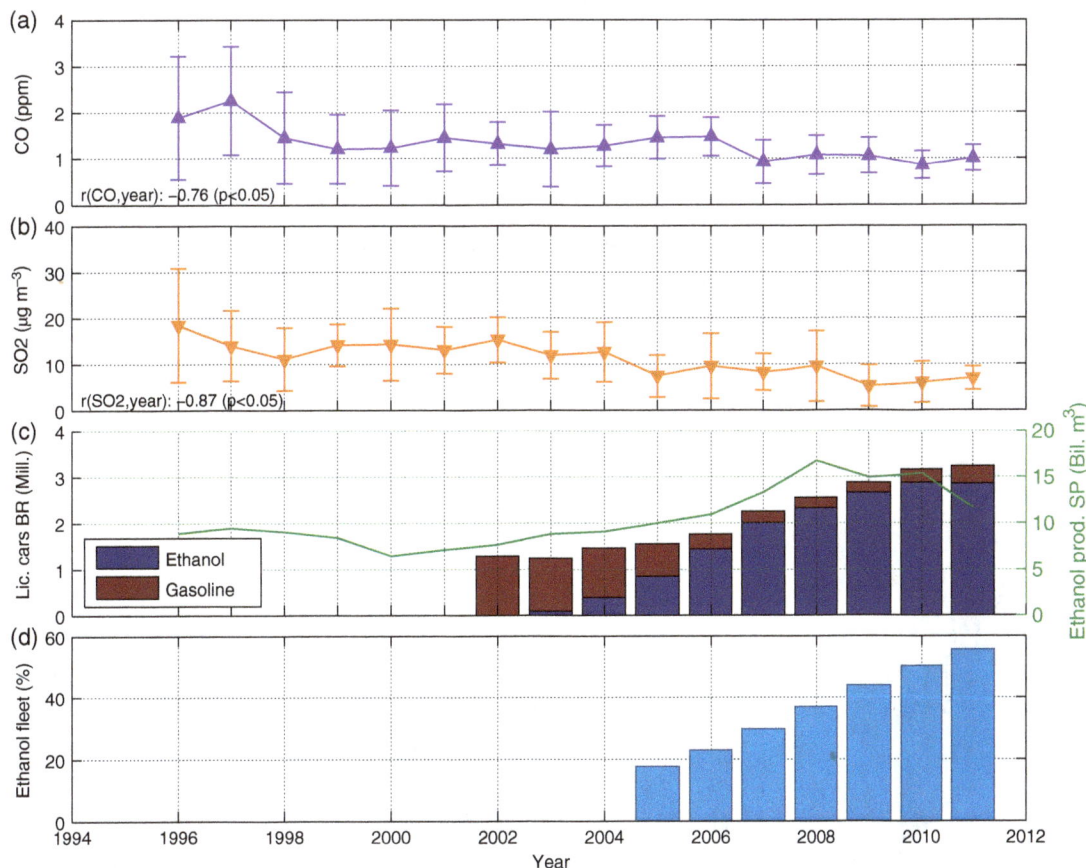

Figure 4. Evolution of average July concentration for CO (ppm) (a), and SO_2 ($\mu g\,m^{-3}$) (b). (c) Licensed cars in Brazil for gasoline and ethanol (left axis); evolution of ethanol production in São Paulo state (right axis). (d) Proportion of the fleet associated with ethanol cars.

9 ppm were observed (Figure S1). For PM_{10} (Figure S2), even wintertime peaks have been below the safe limit of 150 $\mu g\,m^{-3}$. The only station that deviated from these results is Cubatão (Figure S2(d)), where average concentrations are rarely below 70 $\mu g\,m^{-3}$. Cubatão has intense industrial activity associated with its chemical and petrochemical complex, and as a result it has the highest levels of pollution in the dataset. This is also reflected in the concentrations of SO_2 (Figure S3), with both stations in Cubatão frequently showing SO_2 above the 20 $\mu g\,m^{-3}$ limit. Finally, O_3 has been increasing in several stations and frequently reaching over the limit of 100 $\mu g\,m^{-3}$, creating concerns associated with this damaging pollutant (Figure S4).

The monthly averaged concentrations of CO and SO_2 (Figures 2 and 3) presented strong interannual variability, with peaks following minima and vice versa. The wintertime concentrations of SO_2 (Figure 3(a)) peaked in several cities in 2006 and later in 2008 with different amplitudes. These peaks and valleys during wintertime were also observed for other pollutants, including PM_{10} (Figure S5) and O_3 (Figure S6). The number of frontal systems reaching the coast and the countryside (bars) helps to explain this variability. In general, a frontal system brings rainfall for several days, washing out pollutants from the air. Indeed, July of 2006 and 2008 had the lowest number of frontal systems and the highest

values of mean monthly pollution concentration. The monthly concentrations of PM_{10} decreased for January (Figure S5(b)) and are approximately constant when averaged for July (wintertime), responding strongly only to months with lower rainfall (lower number of frontal systems, Figure S5(a)).

Overall, a decrease in CO and SO_2 concentrations from 1996 to 2011 is evident for most of the stations. However, the decreasing trend is stronger for July (Figures 2(a) and 3(a)). For some locations (Osasco, Cerqueira César), CO decreased by about 50% from 1996 to 2011 while the station near the airport of Congonhas, a site strongly influenced by vehicles emissions, decreased from 4.7 to 1.5 ppm in the period, a change of almost 70%. Similar changes were observed for SO_2 (Figure 3), with largest reductions observed in winter compared to summer. For PM_{10} and O_3 (Figures S5 and S6), a decreasing trend was observed only for summer in PM_{10} (from 1996 to 2002). This trend could be associated with public policies enforced to reduce vehicular emissions (Carvalho *et al.*, 2015). On the other hand, the summertime concentration of O_3 has been increasing (Figure S6) since 2007, which could be associated with the increasing fleet of vehicles using ethanol, producing more precursors to O_3 formation, particularly aldehydes (Salvo and Geiger, 2014).

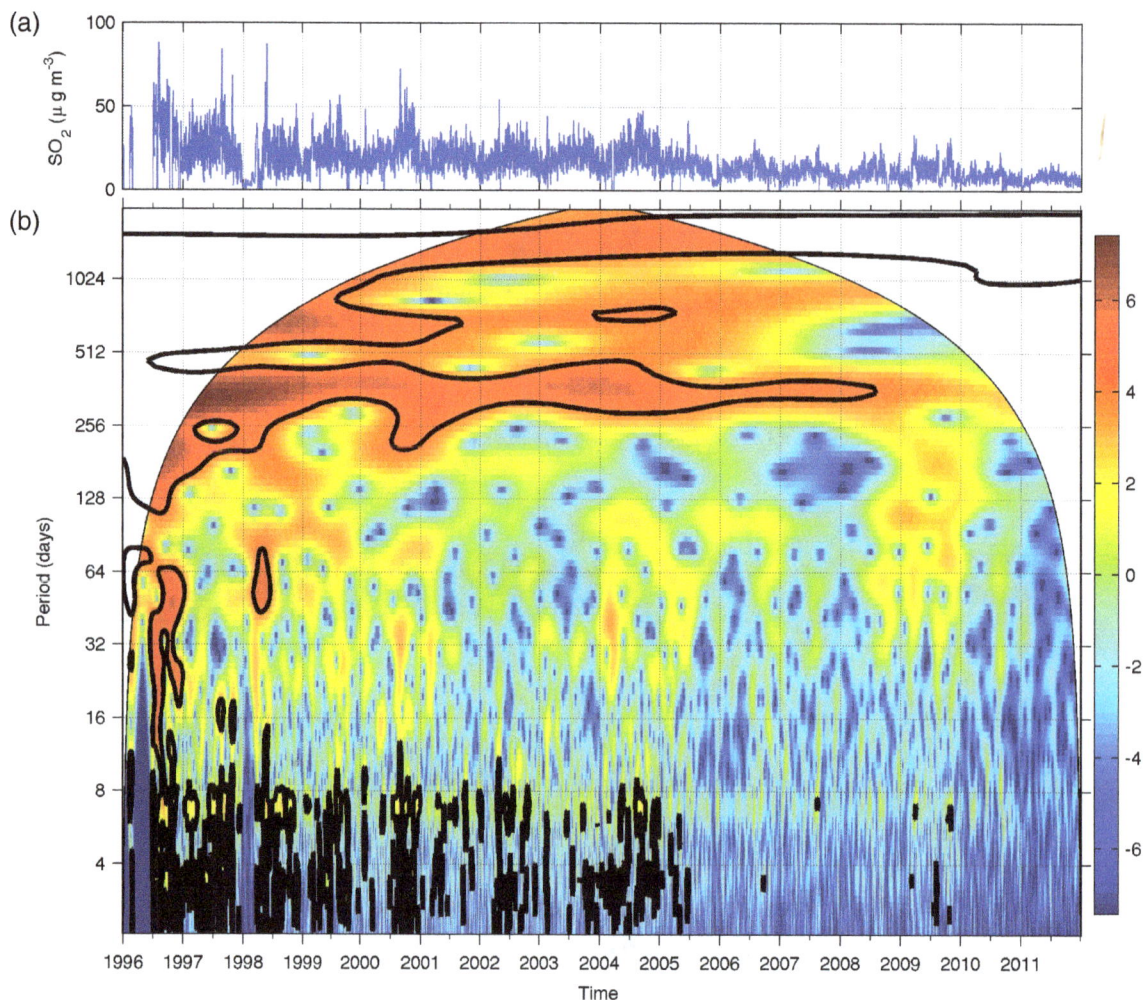

Figure 5. (a) Time series of the concentration of SO_2 ($\mu g\,m^{-3}$) for the station Congonhas and (b) wavelet decomposition for the time series of SO_2. The color scale is proportional to the series variance while bold contours enclose regions where the wavelet power is statistically significant, when compared to a random noise.

The wintertime concentrations for CO and SO_2 were averaged for all locations for each year (Figure 4), revealing three significant results: (1) three phases are evident for CO: a sharp decrease in average concentration from 1996 to 1998, followed by a steady phase, and another decrease by 2006–2007, (2) both average and spatial variability (indicated by lower error bars) in CO among stations separated by 100s of kilometers were lower at the end of the period, (3) in the 2000s, SO_2 concentration decreased from 2004 to 2005 and later from 2008 to 2009. The different phases observed in the two topmost panels (Figure 4) were likely associated with the changes in policies of air pollution (until ~2005) and later by the widespread adoption of ethanol cars, discussed in more details by Andrade *et al.* (2015) and Carvalho *et al.* (2015). Here, we present some data of ethanol use (Figure 4(b) and (c)) to give context to the changes observed in the time series of pollutants. The decrease in concentrations of CO and SO_2 after 2004 was likely influenced by the increasing number ethanol fueled cars, which pollute less CO and SO_2, licensed each year (data for whole country), coupled with an increase in ethanol production in the state of

São Paulo – which is responsible for more than 60% of ethanol production in Brazil (Goldemberg, 2007). Overall, the ethanol fleet increased from 20% in 2005 to 60% in 2011. The true causation, however, of the reductions observed in the data goes beyond the scope of this paper. To accomplish this, a detailed analysis of sources of pollutants over temporal and spatial scales is required.

The wavelet decomposition revealed the strongest and statistically significant harmonics or temporal scales from 1996 to 2011 (SO_2 measured at Congonhas airport, Figure 5, and at São Caetano do Sul, Figure 6. Other examples were included in Figures S1–S12.). As expected, the annual cycle is strong and significant, between 256 and 512 days of duration. Because the series are composed of daily averages, the influence of the diurnal cycle is not shown. However, the contours between 4 and 8 days are present from the start of the period until ~2006–2007. This indicates that the concentrations are likely modulated in those scales by precipitation, horizontal advection, stability of the ABL, the solar radiation at the surface, or other effects. The results in Figures S7–S12 show

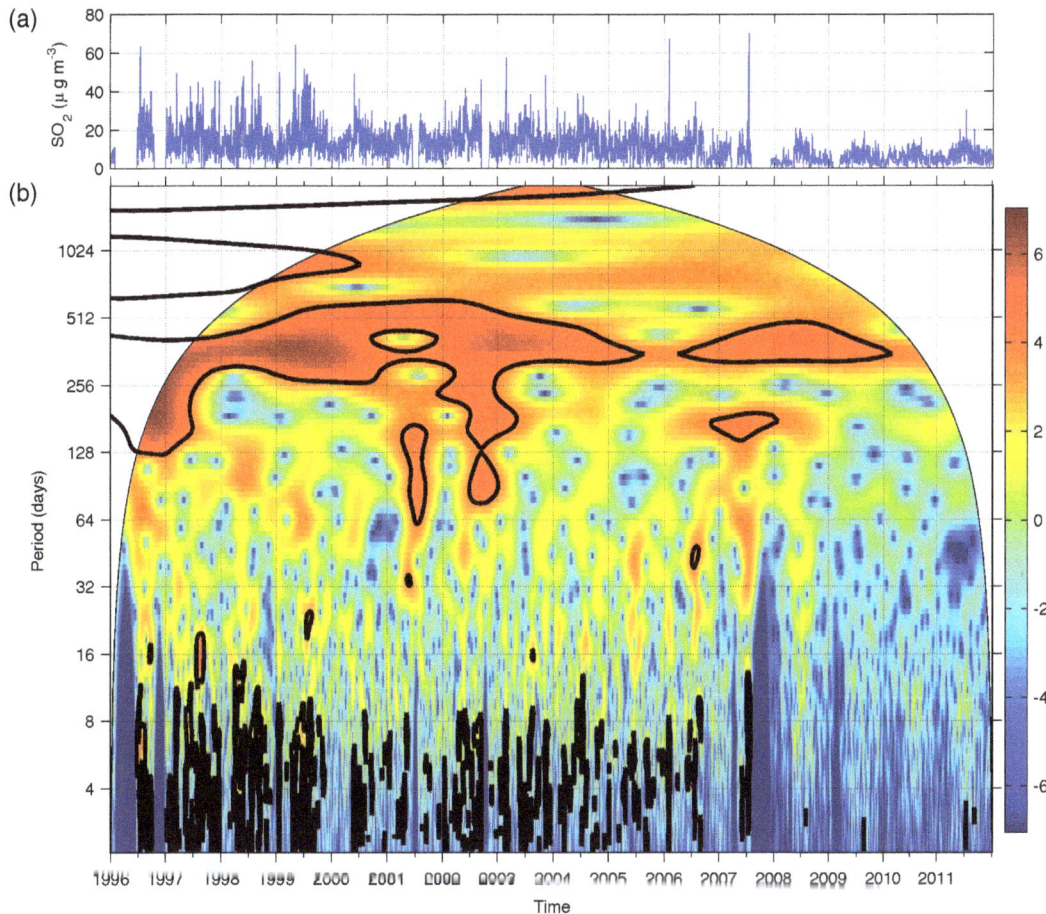

Figure 6. (a) Time series of the concentration of SO_2 ($\mu g\,m^{-3}$) for the station of São Caetano do Sul and (b) wavelet decomposition for the time series of SO_2. The color scale is proportional to the series variance while bold contours enclose regions where the wavelet power is statistically significant, when compared to a random noise.

a mixture of trends for some pollutants and cities. While the 4–8 days harmonic became less prevalent in Figures S7–S9, it was still strong and frequent for PM_{10} (Figures S10–S12), modulating the high variance observed until 2011. The same result was found in another study for the city of Rio de Janeiro (Zeri *et al.*, 2011). This variability was associated with the passage of frontal systems, which have a similar duration when they occur in the southeastern region of Brazil. While the dataset for Rio de Janeiro limited to 2 years, the longer time series of pollutants presented here made it possible to register a sudden change in the influence of this harmonic of 4–8 days of duration, becoming non-statistically significant after 2006–2007. The disappearance of this harmonic suggests a much weaker influence of the meteorological systems that reach the region, at least in time scales of several days, beyond the daily cycles. The daily cycle would still be influenced by the evolution of air temperature, humidity and wind speed, cycles influence by solar energy as well as by anthropogenic factors, such as traffic and industrial activity. In addition, the temporal resolution of this analysis (daily averages) makes it impossible to infer on short-lived spikes in concentrations (hours), beyond the acceptable limits determined by air quality agencies.

The patterns identified in the wavelet analysis are a result of the modulation of the variance by the harmonics. When the variance is reduced, the modulation loses significance and the patterns disappear from the wavelet plots. The reduction in both daily averages and variances of SO_2 and CO concentrations is obvious from the plots in Figures S1 and S4, with the exception of SO_2 in Cubatão panels S4(c) and S4(d). The sources of pollutants differ between locations, causing great differences in averages and variances between stations separated by a few kilometers. For example, São Caetano do Sul and Congonhas are separated by ~15 km, but the variance observed in the signal of SO_2 reduced from 2005 to 2006, in Congonhas, and later from 2007 to 2008, in São Caetano do Sul. As a result, the modulation with periods of 4–8 days identified by the wavelet lasted longer in São Caetano do Sul. Thus, the variance in the signals is a result of local sources and not induced by the harmonics. Finally, the harmonic of 4–8 days could be associated with low-pressure systems, as evidenced by the number of frontal systems reaching the state during July in Figures 2 and 3, or high pressure systems, leading to lower temperatures and reduced turbulent mixing, increasing concentrations of pollutants. A more detailed analysis using continuous meteorological data near the stations should make a clear distinction between both influences.

Low levels of pollutants should be a target of public policies so that the MASP, and cities in general, become more resilient to the influence of meteorological systems on air quality. The consequence of ideal public policies is a reduction in the vulnerability of city dwellers to the presence of meteorological systems that disturb the lower atmosphere for days, trapping pollutants during cold inversions or spreading – by turbulence – dust deposited over the ground. Wavelet analysis is a helpful tool that identifies the harmonics in time series of pollutants. Future work using finer temporal resolutions (hours) should explore other harmonics associated with human or industrial activity, such as patterns of traffic of cars or industrial activity.

Acknowledgements

The authors are grateful to CETESB for sharing the dataset of pollutants; data can be accessed at http://ar.cetesb.sp. gov.br/qualar/. The authors acknowledge the helpful comments from three anonymous reviewers as well as the assistance of C. J. Bernacchi with the language.

Supporting information

The following supporting information is available:

Figure S1. Time series of concentrations of CO (ppm). The standard of 9 ppm (maximum 8-h moving average) is marked with the dashed line.

Figure S2. Time series of concentrations of PM_{10} ($\mu g\, m^{-3}$). The standard of 150 $\mu g\, m^{-3}$ (24-h mean) is marked with the dashed line.

Figure S3. Time series of concentrations of O_3 ($\mu g\, m^{-3}$). The standard of 100 $\mu g\, m^{-3}$ (maximum 8-h moving average) is marked with the dashed line.

Figure S4. Time series of concentrations of SO_2 ($\mu g\, m^{-3}$). The standard of 20 $\mu g\, m^{-3}$ (24-h mean) is marked with the dashed line.

Figure S5. Evolution of monthly averages of PM_{10} ($\mu g\, m^{-3}$) from 1996 to 2011. Top: July and bottom: January. Bars denote the number of frontal systems reaching the coast (deep blue) and the countryside (light blue).

Figure S6. Evolution of monthly averages of O_3 ($\mu g\, m^{-3}$) from 1996 to 2011. Top: July and bottom: January. Bars denote the number of frontal systems reaching the coast (deep blue) and the countryside (light blue).

Figure S7. Top: time series of the concentration of CO (ppm) for the station Congonhas; bottom: wavelet decomposition for the time series. The color scale is proportional to the series variance while bold contours enclose regions where the wavelet power is statistically significant, when compared to a random noise.

Figure S8. Top: time series of the concentration of CO (ppm) for the station São Caetano do Sul; bottom: wavelet decomposition for the time series. The color scale is proportional to the series variance while bold contours enclose regions where the wavelet

power is statistically significant, when compared to a random noise.

Figure S9. Top: time series of the concentration of PM_{10} ($\mu g\, m^{-3}$) for the station Congonhas; bottom: wavelet decomposition for the time series. The color scale is proportional to the series variance while bold contours enclose regions where the wavelet power is statistically significant, when compared to a random noise.

Figure S10. Top: time series of the concentration of PM_{10} ($\mu g\, m^{-3}$) for the station São Caetano do Sul; bottom: wavelet decomposition for the time series. The color scale is proportional to the series variance while bold contours enclose regions where the wavelet power is statistically significant, when compared to a random noise.

Figure S11. Top: time series of the concentration of PM_{10} ($\mu g\, m^{-3}$) for the station Cubatão; bottom: wavelet decomposition for the time series. The color scale is proportional to the series variance while bold contours enclose regions where the wavelet power is statistically significant, when compared to a random noise.

Figure S12. Top: time series of the concentration of SO_2 ($\mu g\, m^{-3}$) for the station Cubatão; bottom: wavelet decomposition for the time series. The color scale is proportional to the series variance while bold contours enclose regions where the wavelet power is statistically significant, when compared to a random noise.

References

Andrade MF, Ynoue RY, Freitas ED, Todesco E, Vara Vela A, Ibarra S, Martins LD, Martins JA, Carvalho VSB. 2015. Air quality forecasting system for Southeastern Brazil. *Frontiers in Environmental Science* **3**: 9, doi: 10.3389/fenvs.2015.00009

Angevine WM, Grimsdell AW, Hartten LM, Delany AC. 1998. The flatland boundary layer experiments. *Bulletin of the American Meteorological Society* **79**: 419–431, doi: 10.1175/1520-0477(1998) 079<0419:TFBLE>2.0.CO;2.

Barbaro E, de Arellano JV-G, Ouwersloot HG, Schröter JS, Donovan DP, Krol MC. 2014. Aerosols in the convective boundary layer: shortwave radiation effects on the coupled land-atmosphere system. *Journal of Geophysical Research, [Atmospheres]* **119**: 5845–5863, doi: 10.1002/2013JD021237.

Bourotte C, Forti M-C, Taniguchi S, Bícego MC, Lotufo PA. 2005. A wintertime study of PAHs in fine and coarse aerosols in São Paulo city, Brazil. *Atmospheric Environment* **39**: 3799–3811, doi: 10.1016/j.atmosenv.2005.02.054.

Cançado JED, Saldiva PHN, Pereira LAA, Lara LBLS, Artaxo P, Martinelli LA, Arbex MA, Zanobetti A, Braga ALF. 2006. The impact of sugar cane-burning emissions on the respiratory system of children and the elderly. *Environmental Health Perspectives* **114**: 725–729.

Carvalho VSB, Freitas ED, Mazzoli CR, Andrade MF. 2012. Avaliação da influência de condições meteorológicas na ocorrência e manutenção de um episódio prolongado com altas concentrações de ozônio sobre a região metropolitana de São Paulo. *Revista Brasileira de Meteorologia* **27**: 463–474, doi: 10.1590/S0102-7786201200040 0009.

Carvalho VSB, Freitas ED, Martins LD, Martins JA, Mazzoli CR, Andrade MF. 2015. Air quality status and trends over the metropolitan area of São Paulo, Brazil as a result of emission control policies. *Environmental Science & Policy* **47**: 68–79, doi: 10.1016/j.envsci. 2014.11.001.

Castanho ADA, Artaxo P. 2001. Wintertime and summertime São Paulo aerosol source apportionment study. *Atmospheric Environment* **35**: 4889–4902, doi: 10.1016/s1352-2310(01)00357-0.

Climanálise. 2005. Produtos Climanálise INPE/CPTEC. http://www. cptec.inpe.br/products/climanalise/ (accessed 3 March 2011).

Collineau S, Brunet Y. 1993. Detection of turbulent coherent motions in a forest canopy. Part 1. wavelet analysis. *Boundary-Layer Meteorology* **65**: 357–379.

Daubechies I. 1992. *Ten Lectures on Wavelets*, Vol. **61**. Society for Industrial and Applied Mathematics: Philadelphia, PA, 377 pp.

Fajersztajn L, Veras M, Barrozo LV, Saldiva P. 2013. Air pollution: a potentially modifiable risk factor for lung cancer. *Nature Reviews Cancer* **13**: 674–678, doi: 10.1038/nrc3572.

Freitas E, Rozoff C, Cotton W, Dias PS. 2007. Interactions of an urban heat island and sea-breeze circulations during winter over the metropolitan area of São Paulo, Brazil. *Boundary-Layer Meteorology* **122**: 43–65, doi: 10.1007/s10546-006-9091-3

Godoy MLDP, Godoy JM, Roldão LA, Soluri DS, Donagemma RA. 2009. Coarse and fine aerosol source apportionment in Rio de Janeiro, Brazil. *Atmospheric Environment* **43**: 2366–2374, doi: 10.1016/j.atmosenv.2008.12.046.

Goldemberg J. 2007. Ethanol for a sustainable energy future. *Science* **315**: 808–810.

Gonçalves FLT, Carvalho LMV, Conde FC, Latorre M, Saldiva PHN, Braga ALF. 2005. The effects of air pollution and meteorological parameters on respiratory morbidity during the summer in Sao Paulo City. *Environment International* **31**: 343–349.

deLeon AP, Anderson HR, Bland JM, Strachan DP, Bower J. 1996. Effects of air pollution on daily hospital admissions for respiratory disease in London between 1987–88 and 1991–92. *Journal of Epidemiology and Community Health* **50**: S63–S70.

Martins MHRB, Anazia R, Guardani MLG, Lacava CIV, Romano J, Silva SR. 2004. Evolution of air quality in the Sao Paulo Metropolitan Area and its relation with public policies. *International Journal of Environment and Pollution* **22**: 430–440, doi: 10.1504/IJEP.2004.005679

de Miranda RM, de Fátima Andrade M, Worobiec A, Grieken RV. 2002. Characterisation of aerosol particles in the São Paulo Metropolitan Area. *Atmospheric Environment* **36**. 345–352, doi: 10.1016/s1352-2310(01)00363-6

Oliveira AP, Bornstein RD, Soares J. 2003. Annual and diurnal wind patterns in the city of São Paulo. *Water, Air and Soil Pollution: Focus* **3**: 3–15, doi: 10.1023/a:1026090103764

Sá LDA, Sambatti SBM, Galvao GP. 1998. Applying the Morlet wavelet in a study of variability of the level of Paraguay River at Ladario, MS. *Pesquisa Agropecuária Brasileira* **33**: 1775–1785.

Salvo A, Geiger FM. 2014. Reduction in local ozone levels in urban Sao Paulo due to a shift from ethanol to gasoline use. *Nature Geoscience* **7**: 450–458, doi: 10.1038/Ngeo2144

Samet JM, Dominici F, Curriero FC, Coursac I, Zeger SL. 2000. Fine particulate air pollution and mortality in 20 US Cities, 1987–1994. *New England Journal of Medicine* **343**: 1742–1749, doi: 10.1056/Nejm200012143432401

Schwartz J. 1996. Air pollution and hospital admissions for respiratory disease. *Epidemiology* **7**: 20–28.

Sharma VP, Arora HC, Gupta RK. 1983. Atmospheric pollution studies at Kanpur – suspended particulate matter. *Atmospheric Environment* **17**: 1307–1313, doi: 10.1016/0004-6981(83)90405-5

Silva Dias MF, Machado AJ. 1997. The role of local circulations in summertime convective development and nocturnal fog in São Paulo, Brazil. *Boundary-Layer Meteorology* **82**: 135–157, doi: 10.1023/A:1000241602661

Terradellas E, Soler MR, Ferreres E, Bravo M. 2005. Analysis of oscillations in the stable atmospheric boundary layer using wavelet methods. *Boundary-Layer Meteorology* **114**: 489–518, doi: 10.1007/S10546-004-1293-Y

Torrence C, Compo GP. 1998. A practical guide to wavelet analysis. *Bulletin of the American Meteorological Society* **79**: 61–78, doi: 10.1175/1520-0477(1998)079<0061:Apgtwa>2.0.Co;2

Zeri M, Oliveira JF, Lyra GB. 2011. Spatiotemporal analysis of particulate matter, sulfur dioxide and carbon monoxide concentrations over the city of Rio de Janeiro, Brazil. *Meteorology and Atmospheric Physics* **113**: 139–152, doi: 10.1007/S00703-011-0153-9

23

Improvement of land surface temperature simulation over the Tibetan Plateau and the associated impact on circulation in East Asia

Haifeng Zhuo,[1,2] Yimin Liu[1]* and Jiming Jin[3,4]

[1] State Key Laboratory of Numerical Modeling for Atmospheric Sciences and Geophysical Fluid Dynamics (LASG), Institute of Atmospheric Physics, Chinese Academy of Sciences, Beijing, China
[2] University of Chinese Academy of Sciences, Beijing, China
[3] College of Water Resources and Architectural Engineering, Northwest A&F University, Yangling, China
[4] Department of Watershed Sciences and Plants, Soils and Climate, Utah State University, Logan, UT, USA

*Correspondence to:
Y. Liu, No. 40, Huayanli, Institute of Atmospheric Physics, Chinese Academy of Sciences, Chaoyang District, Beijing 100029, China.
E-mail: lym@lasg.iap.ac.cn

Abstract

A new sensible-heat (SH) parameterization was used in the Weather Research and Forecasting model to improve the vertical heat transfer simulation in arid regions. With this new scheme, the simulated SH decreased, resulting in a 2 °C reduction of the cold bias in the land surface temperature over the Tibetan Plateau. The weakened SH led to anticyclonic circulation and cyclonical flow changes in the northern East China due to Rossby wave propagation. The summer-rainfall in East Asian is marginally improved. It is suggested that the SH simulation over the Plateau is a land–atmosphere coupling issue that is important to the dynamical downscaling in East Asian.

Keywords: Tibetan Plateau; land surface temperature; East Asian summer monsoon; WRF; CLM

1. Introduction

Owing to the unique geography of the Tibetan Plateau (TP) and land–atmosphere interaction characteristics, all of the global climate models underestimate surface air temperature in the TP, and most overestimate its precipitation (Hao *et al.*, 2013). Most of the reanalysis datasets also tend to underestimate temperature in the TP (You *et al.*, 2010; Wang and Zeng, 2012). Land surfaces form the lower boundary conditions of the atmosphere, which essentially dominate local energy and water transfer, and can affect local, regional, and global climate processes. Proper description of land surface processes in earth system models is a popular research topic.

Arid and semi-arid regions constitute approximately one-fourth of the total global land area, and account for approximately 40% of the land area in China (Hu and Zhang, 2001). They are mostly distributed in Northwest China and in the TP. Because vegetation coverage and rainfall are extremely rare in these regions, the land–atmosphere interaction process is dominated by heat transfer processes. However in the arid and semi-arid regions, surface temperature and energy are poorly simulated.

The unique solar radiation and energy transfer processes in the TP differ significantly from those in regions of lower altitude (Smith and Shi, 1992; Zhang *et al.*, 2014). The development of the turbulent transfer theory in recent years has allowed scientists to conduct numerous land surface parameterization field experiments in arid and semi-arid regions (Zeng *et al.*, 1998; Yang *et al.*, 2008; Chen *et al.*, 2010). In such studies, the surface energy transfer process was investigated, and several new parameterization schemes of land surface models were developed (Zeng *et al.*, 2012; Wang *et al.*, 2014).

Without considering vegetation, the emissivity ε of bare soil in Community Land Model (CLM) is set to a constant at approximately 0.96. Therefore, in arid and semi-arid regions, the main factors affecting the calculation of the land–atmosphere energy balance are surface albedo α, aerodynamic roughness length z_{0m}, thermal roughness length z_{0h}, soil heat capacity c_v, and soil thermal conductivity K_{soil}. After conducting more than 10 years of observation at three stations in alpine and desert areas including Gobi, Huang *et al.* (2005) assigned a surface albedo value of 0.255 ± 0.021 to the Dunhuang area of Gobi. Zhou *et al.* (2012) assigned a momentum roughness length z_{0m} of 0.61 ± 0.02 mm in Dunhuang, and an average thermal roughness length z_{0h} of approximately 0.05 mm during the daytime. Chen *et al.* (2009) used these parameters to evaluate the Biosphere–Atmosphere Transfer Scheme (BATS) and showed that the cold bias in land surface temperature (LST) simulation has been improved in the daytime, and its diurnal cycle has been accurately simulated. In addition, some studies have shown that vegetation degeneration in northwest arid areas not only affects local precipitation (Li and Xue, 2010) but also drive an anticyclone anomalies in the upper troposphere in the spring, and maintain to summer, at last affect the

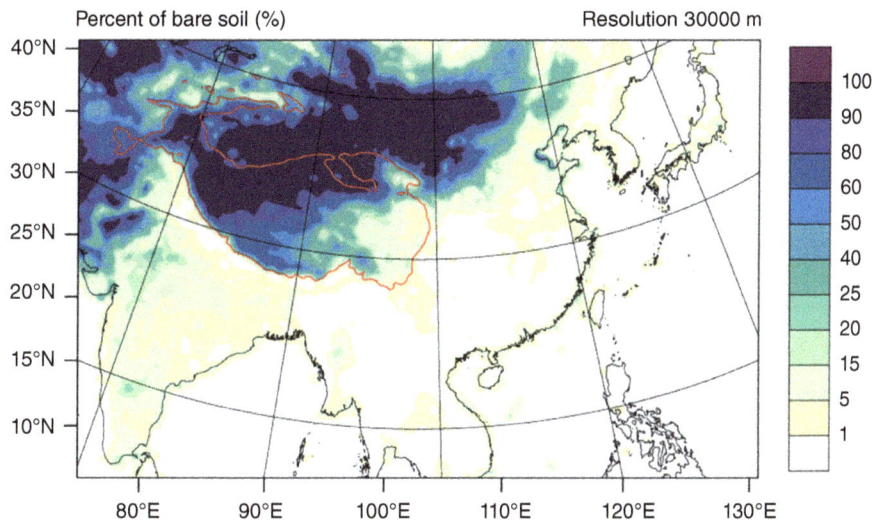

Figure 1. Model domain and the distribution of bare soil in the Community Land Model. The red solid line area indicates elevation >3000 m.

precipitation in the northeast part of TP through "silk road" wave train (Zhou and Huang, 2010; Huang *et al.*, 2011).

Moreover, many studies have derived surface sub-layer turbulent heat transfer parameterization for incorporating z_{0h} in past decades (Sheppard, 1958; Brutsaert, 1982; Zeng and Dickinson, 1998; Kanda *et al.*, 2007; Yang *et al.*, 2008). However, some problems remain (Hogue *et al.*, 2005; Jiménez *et al.*, 2011; Decker *et al.*, 2012). Many land surface models such as Noah land surface model (Noah LSM) and CLM overestimate the *SH* and thus underestimate *LST* during the dry season in arid areas (Hogue *et al.*, 2005). Chen *et al.* (2010) evaluated the parameterization of z_{0h} when simulating *LST* with six different schemes of the Noah LSM. They found that original Noah significantly underestimated *LST* and overestimated *SH* in the daytime. The offline run with revised parameterization schemes can give better results of both *LST* and *SH* owing to better simulation of the diurnal variation in z_{0h}.

Recently, through both theoretical analyses and data-model comparison, Zeng *et al.* (2012) used *in situ* observation and offline experiments at several stations in arid and semi-arid regions, including the Arizona desert and the TP, to further revise the coefficients of z_{0h} both in Noah and CLM. They also set a constraint of minimum friction velocity and take a prescribed soil texture. This measure significantly reduced the underestimation of *LST* in the daytime. Their new scheme has been successfully applied to the National Centers for Environmental Prediction Global Forecast System (NCEP-GFS) assimilation (Zheng *et al.*, 2012). However, it is unknown whether these improvements can be effectively applied in the land–atmosphere coupled model. Simulation is then applied over East Asia. These analyses will be used to examine the extent to which the new *SH* parameterization improves *LST* simulation. Moreover, we discuss the impacts of

land–atmosphere interaction on diabatic heating and local and downstream circulation.

This study helps us deeply understand the attribution of the land surface process of large-scale topography. Moreover, the discovery of direct and indirect effects of the land surface heat transfer process from TP to the downstream regions can be applied to improve the climate prediction of East Asia effectively.

2. Experiment design

2.1. Model and data

The Weather Research and Forecasting (WRF) model is currently the most widely used mesoscale model. Its latest WRFV3.6.1 (Skamarock *et al.*, 2008) coupled with CLM4 (Oleson *et al.*, 2013) was used in this study. The related physics includes MYNN surface layer and Mellor-Yamada Nakanishi and Niino Level 3 PBL. According to the model domain shown in Figure 1, the percentage of bare soil has gradually increased from Southeast China, with rich vegetation, to the Northwest China, which is sparsely vegetated. Particularly in the western plateau, most parts of the TP are sparsely vegetated.

We conducted two experiments: a control experiment (CTL) and a sensitivity experiment (SEN). Both used the same basic settings except that the new parameterization of *SH* was used in the SEN experiment. The central point of the model domain is (30°N, 103°E). The horizontal resolution was 30 km with 150 grid points from south to north and 240 grid points from west to east. Integration was from September 1, 2003, to August 31, 2010, and was continuous with the same initial condition at 00Z UTC. The forcing data used in this study were 6-hourly NCEP climate forecast system reanalysis (CFSR) products (Saha *et al.*, 2010). In addition, we used the daily site observation data of the Chinese Meteorological Administration (CMA) from 2003 to 2010;

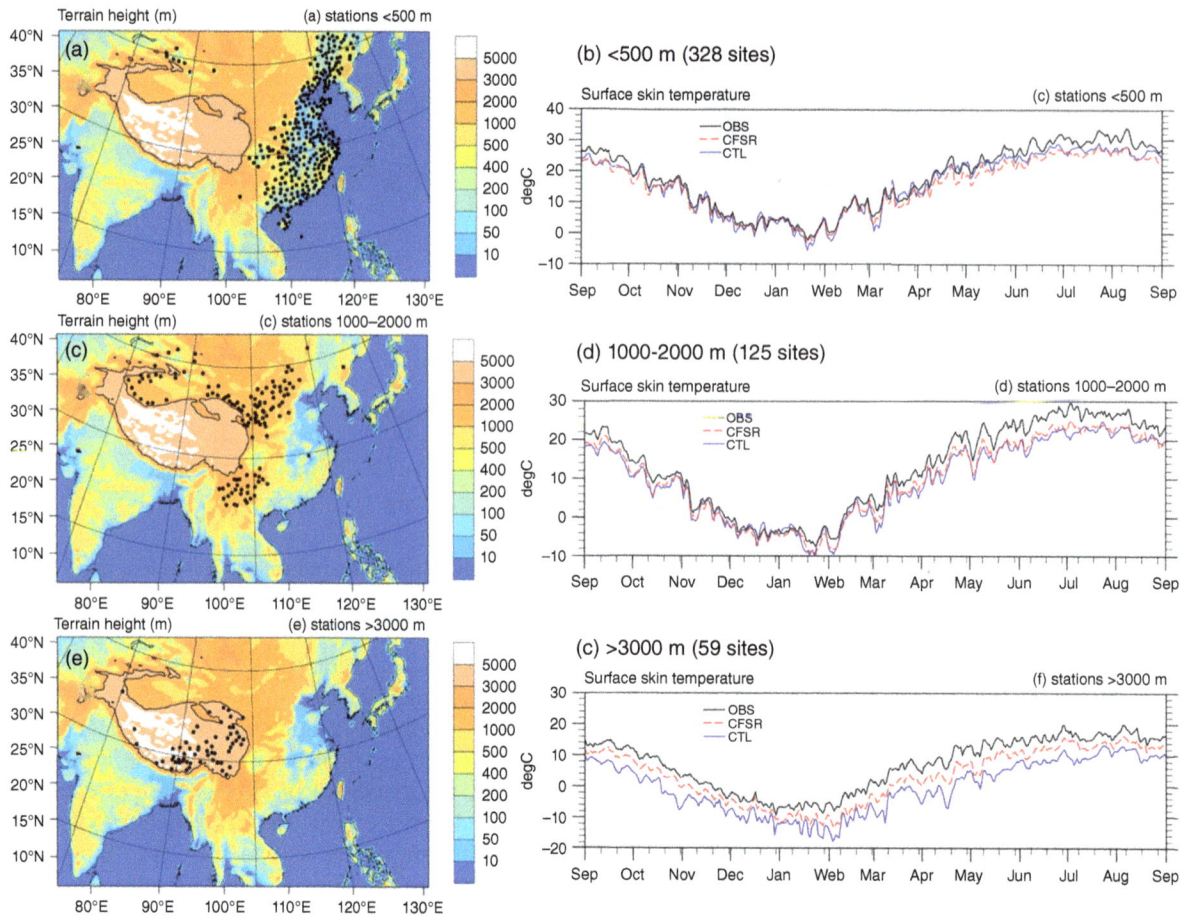

Figure 2. (a), (c), (e) Site distribution and corresponding land surface temperature time series of observation (black line), Climate Forecast System Reanalysis data (red dashed line), and Weather Research and Forecasting simulation (blue line) for (b) 0–500 m, (d) 1000–2000 m, and (f) >3000 m recorded September to August for 1-year integration.

the *LST* observation of the Moderate Resolution Imaging Spectroradiometer (MODIS) which recorded four times daily; and the Tropical Rainfall Measuring Mission (TRMM)-3B42 dataset.

2.2. Evaluation of the WRF model

In the present study, we compare the CMA station data against the corresponding model location in the CTL (Figure 2). The sites used for comparison are distributed in three different regions: elevation below 500 m (328 sites), at 1000–2000 m (125 sites), and higher than 3000 m (59 sites). The percentage of bare soil is larger at the stations located at higher elevations. In Figure 2, the left-hand side shows the terrain height and locations of the CMA sites, and the right-hand side shows averaged *LST*. The values of the CTL were obtained from the grid point in closest proximity to the site location. Because each site's *LST* is based on actual terrain elevation and the model's elevation may not be the same, we adjusted the temperature by using temperature vertical lapse rate of ~6 °C km^{-1}. The results show that the cold bias of *LST* gradually increased with an increase in altitude. The cold bias was obviously greater for the sites in the TP, with a root–mean–square deviation (RMSD) of ~7.6 °C and elevation higher than 3000 m,

than that in the eastern plain, with an RMSD of ~2.4 °C and elevation below 500 m. The original CLM model underestimates the *LST* in arid and semi-arid regions.

2.3. Model configuration with new scheme

Zeng and Dickinson (1998) previously reported a relationship between z_{0h} and z_{0m}. Recently, they revised the constants "a" and "b" from 0.13/0.45 to 0.36/0.5 and constrained the minimum friction velocity u_{*min} to 0.07 m in bare soil (Zeng *et al.*, 2012). The three parameters used in the *SH* of non-vegetated surfaces are z_{0h}, z_{0m}, and minimum friction velocity u_{*min}:

$$\ln\left(\frac{z_{0m}}{z_{0h}}\right) = a\left(\frac{u_* z_{0m}}{\nu}\right)^b \tag{1}$$

$$u_{*min} = 0.07\frac{\rho_0}{\rho}\left(\frac{z_{0m}}{z_{0g}}\right)^{0.18} \tag{2}$$

where $\nu = 1.5 \times 10^{-5}$ m^2 s^{-1} refers to the molecular viscosity. The air density at sea level ρ_0 is 1.22 kg m^{-3}, and the roughness length of bare soil is 0.01 m. It is shown that the underestimation of *LST* is significantly improved in the daytime (Zeng *et al.*, 2012; Wang *et al.*, 2014).

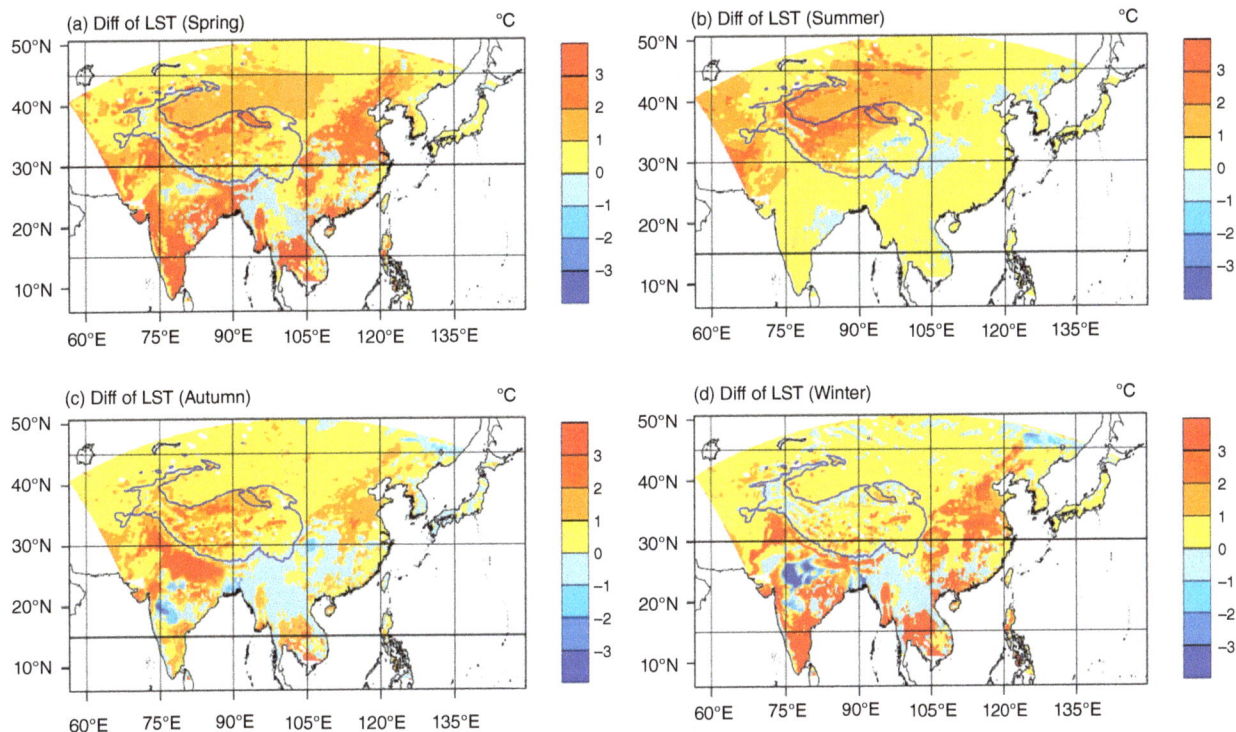

Figure 3. Differences in seasonal mean land surface temperature determined by subtracting the results of the control experiment from those of the sensitivity experiment (2003–2010): (a) spring, (b) summer, (c) autumn, and (d) winter.

3. Results

3.1. LST simulation

When a two-way feedback of land–atmosphere interaction was considered, the LST showed an obvious improvement in all of the arid and semi-arid regions, particularly in the TP and the Taklamakan Desert, but showed variations among seasons. The TP has the largest impacts in summer because the thermal forcing in the weak wind plays a more important role than the heating in strong wind in other seasons (Wu et al., 2007). The surface sensible heating over the Plateau pumps moisture from oceans so contributes the formation and variation of the Asian monsoon (Wu et al., 2012a, 2012b; Liu et al., 2102). The maximum LST increase, 3 °C, occurred in summer over the western TP; otherwise, the LST increase in the Taklamakan Desert was smaller than that in the TP. The changes in other regions indicate responses to the interaction in arid regions (Figure 3).

Figure 4 shows the differences in SH between SEN and CTL. The SH decreased in arid and semi-arid regions when using the new scheme. Of the four seasons, the SH decrease was most obvious in summer. Moreover, the surface wind changed with the SH; wind divergence was noted near the TP. The impact on East Asia was strong in summer. The cyclonic wind deviation was located in Northeast China, although the anticyclonic wind deviation occurred in the South China. Because thermal forcing had the strongest impacts on circulation in summer, the following analyses and discussions focus on summer.

3.2. Impact on the circulation and the East Asia climate in summer

To determine the causes of the cyclonic deviation of surface wind in summer, we analyzed the thermal heating effects of the TP on the atmospheric circulation. In summer, the cumulus convection is strong at the south slope and southern edge of the plateau; the maximum diabatic heating is $10\,\mathrm{K\,day^{-1}}$ at the south slope (Figure 5(a)). The updraft airflow is located at both sides of the TP due to the SH air pump effect (Wu et al., 2007). The flow on the south slope is considerably stronger owing to moisture physics feedback. However, the pumping effect was weakened after the application of the new scheme. The upward movement and the diabatic heating also became weaker on both north and south of the TP (Figure 5(b)).

The new scheme also affects the summer circulation and rainfall in East China. Figure 5(c) shows a comparison of CTL precipitation and TRMM observation, and Figure 5(d) shows the difference in precipitation and 700 hPa wind between SEN and CTL. The East Asia Summer Monsoon rain belt is underestimated by approximately $4\,\mathrm{mm\,day^{-1}}$ in East–Central China and is overestimated at the same rate in the eastern TP and the southeastern coast of China (Figure 5(c)). The new scheme improves the precipitation in most parts of East Asia. The underestimated rainfall in Northeast and East–Central China showed marginal improvement, as did the rainfall overestimated in the eastern TP, the Hetao region in Mongolia, and the southeast coast of China in summer. This result occurred because the new scheme obviously reduces the SH in the western and

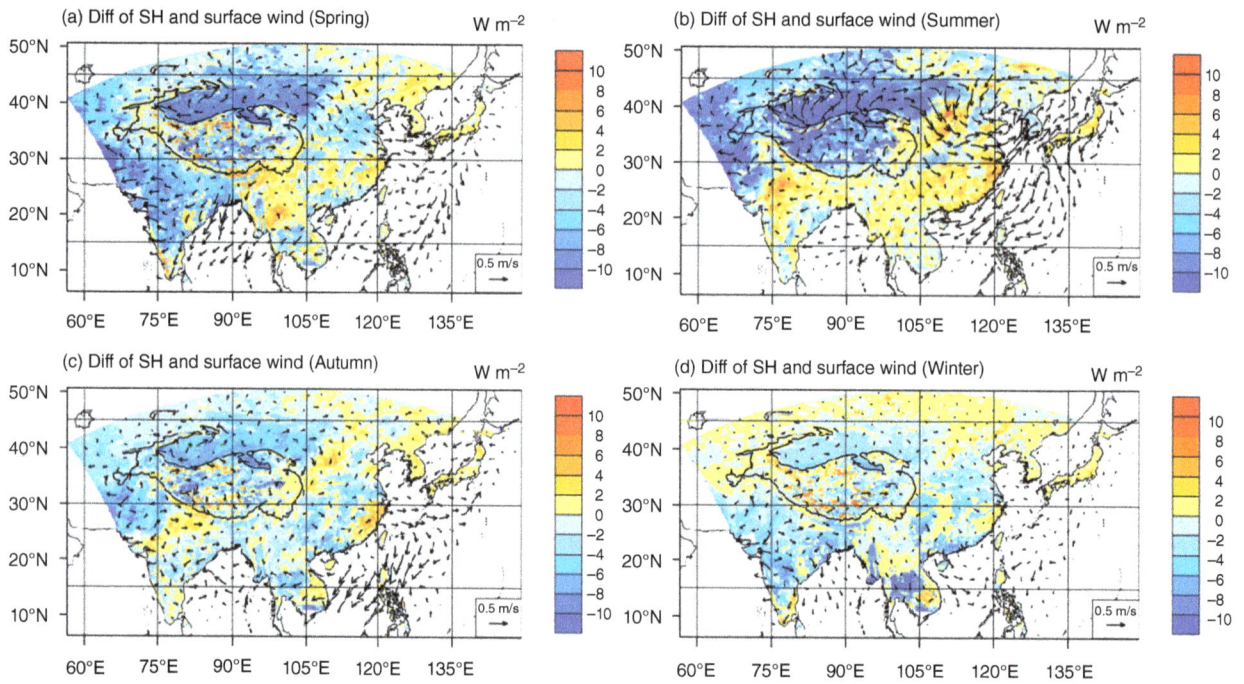

Figure 4. Comparison of seasonal mean surface sensible heat (shading, units: wm^{-2}) and surface wind (vector, units: m s^{-1}), determined by subtracting the results of the control experiment from those of the sensitivity experiment (2003–2010): (a) spring, (b) summer, (c) autumn, and (d) winter.

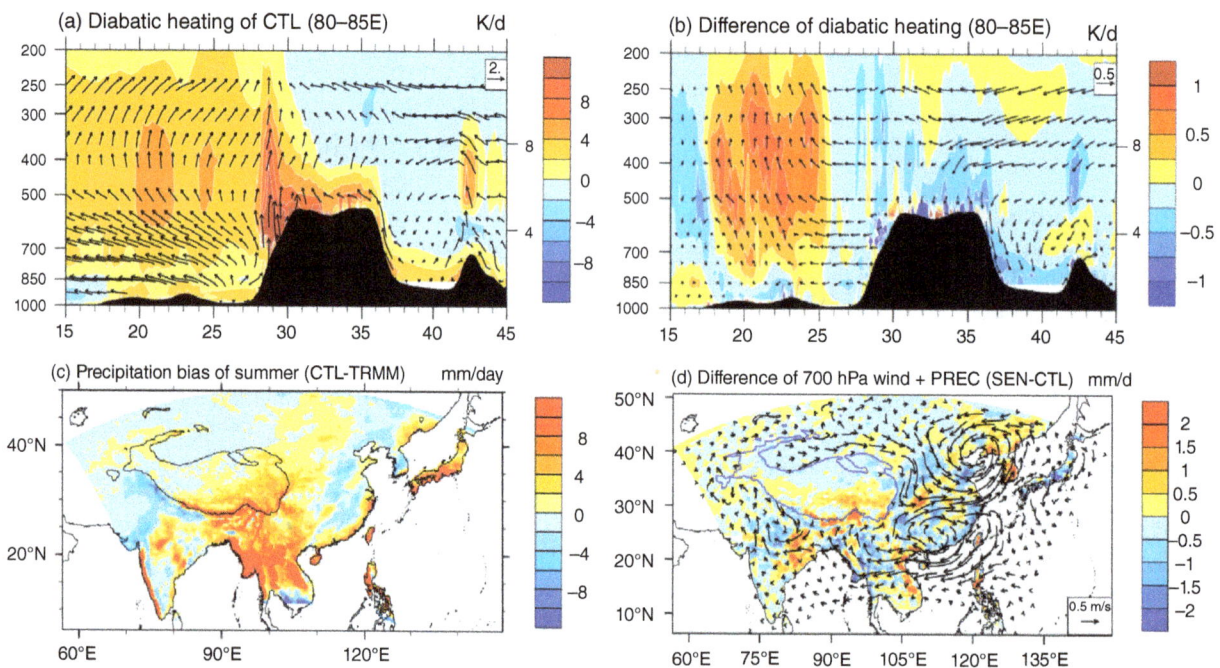

Figure 5. Diabatic heating, wind, and precipitation in summer. (a) 80°–85°E mean diabatic heating and wind in the control experiment (CTL); (b) 80°–85°E mean difference of diabatic heating and wind determined by subtracting the results of CTL from those of the sensitivity experiment (SEN); (c) precipitation bias of CTL and the Tropical Rainfall Measuring Mission (TRMM); (d) difference in precipitation and 700 hPa wind (SEN minus CTL).

central TP, where the ground surface is dominated by bare soil. Thus, the *SH* air pump effect of the TP is weakened, and the diabatic heating and the updraft airflow are reduced in both north and south sides of the TP. The weakened surface heating resulted in divergence and anticyclonic circulation close to the surface near the TP and cyclonical flow change in the northern

part of East China due to Rossby wave propagation. The southeasterly (westerly) winds of the cyclone led to convergence over Northeast (East–Central) China and contributed to the *in situ* rainfall increase. A compensatory anticyclone developed in South China, and its northwestern flow reduced the rainfall along the southeast coast.

In summary, the new *SH* scheme shows improvement in *LST* simulations in the TP and precipitation simulations in parts of East China in the summer monsoon season. On the other hand, soil moisture (Hong *et al.*, 2009; Moufouma-Okia and Rowell, 2010; Jaeger and Seneviratne, 2011) also play a role in the simulation in the regional models.

4. Summary

This study investigated the influence of a new *SH* parameterization scheme for bare soil in the land–atmosphere coupling system of WRF. The results show that the overestimation of the *SH* and underestimation of *LST* over arid and semi-arid regions are improved by using the new scheme. The new scheme significantly improves the *LST* simulation over most of the western and central TP. The average cold bias of the *LST* was reduced by approximately 2 °C in the TP. The decrease in *SH* affects the vertical circulation near the TP. The *SH* air pump effect in the TP and the updraft motion of airflow around the plateau were also weakened. The diabatic heating over both the southern and northern slopes of the TP was also reduced.

Moreover, the decrease in *SH* affects the circulation and precipitation in East Asia. The new scheme led to divergence with the development of an anticyclone over the TP and a cyclone and anticyclone over Northeast and South China, respectively. The rain belt simulations were thus improved. The cyclonic circulation in Northeast China led to more precipitation in Northeastern and Central China. The anticyclonic circulation in South China decreased the overestimated rainfall in the southeastern coast. It is suggested that the SH simulation over the Plateau is a land–atmosphere coupled issue that is important to climate dynamical downscaling, as well as the climate prediction and weather forecasting in East Asian. The influence of the new scheme on global climate models deserves future investigation.

Acknowledgements

This work is supported by the National Natural Science Foundation of China (Grant No. 91437219) and the Third Tibetan Plateau Scientific Experiment (Grant No. GYHY201406001).

References

Brutsaert WH. 1982. *Evaporation into the Atmosphere: Theory, History, and Applications.* D. Reidel Publishing Co. Dordrecht. Holland.

Chen W, Zhu D, Liu H, Sun SF. 2009. Land–air interaction over arid/semi-arid areas in China and its impact on the East Asian summer monsoon. Part I: calibration of the land surface model (BATS) using multicriteria methods. *Advances in Atmospheric Sciences* 26(6): 1088–1098.

Chen YY, Yang K, Zhou D, Qin J, Guo X. 2010. Improving the Noah land surface model in arid regions with an appropriate parameterization of the thermal roughness length. *Journal of Hydrometeorology* 11: 995–1006.

Decker M, Brunke M, Wang Z, Sakaguchi K, Zeng XB, Bosilovich MG. 2012. Evaluation of the reanalysis products from GSFC, NCEP, and ECMWF using flux tower observations. *Journal of Climate* 25: 1916–1944.

Hao ZC, Ju Q, Jiang WJ, Zhu CJ. 2013. Characteristics and scenarios projection of climate change on the Tibetan Plateau. *The Scientific World Journal* 2013(7): 1903–1912. Article ID 129793, doi: 10.1155/2013/129793.

Hogue TS, Bastidas L, Gupta H, Sorooshian S, Mitchell K, Emmerich W. 2005. Evaluation and transferability of the Noah land surface model in semiarid environments. *Journal of Hydrometeorology* 6(1): 68–84.

Hong S, Lakshmi V, Small EE, Chen F, Tewari M, Manning KW. 2009. Effects of vegetation and soil moisture on the simulated land surface processes from the coupled WRF/Noah model. *Journal of Geophysical Research, [Atmospheres]* 114(D18): 3151–3157.

Hu Y, Zhang Q. 2001. Some issues of arid environment dynamics (in Chinese). *Advances in Earth Sciences* 1: 18–23.

Huang RH, Wei GA, Zhang Q, Gao XQ. 2005. The preliminary scientific achievements of the field experiment on Air–Land Interaction in the Arid Area of Northwest China (NWC-ALAIEX). In *Proceedings of the 4th CTWF International Workshop on the Land Surface Models and their Applications*, 15–18 November, Zhuhai, China.

Huang RH, Chen W, Zhang Q. 2011. *Land–Atmosphere Interaction over Arid Region of Northwest China and Its Impact on East Asian Climate Variability.* China Meteorological Press: Beijing (in Chinese).

Jaeger EB, Seneviratne SI. 2011. Impact of soil moisture–atmosphere coupling on european climate extremes and trends in a regional climate model. *Climate Dynamics* 36(9–10): 1919–1939.

Jiménez C, Prigent C, Mueller B, Seneviratne SI, McCabe MF, Wood EF, Rossow WB, Balsamo G, Betts AK, Dirmeyer PA, Fisher JB, Jung M, Kanamitsu M, Reichle RH, Reichstein M, Rodell M, Sheffield J, Tu K, Wang K. 2011. Global intercomparison of 12 land surface heat flux estimates. *Journal of Geophysical Research* 116: D02102, doi: 10.1029/2010JD014545.

Kanda M, Kanega M, Kawai T, Moriwaki R, Sugawara H. 2007. Roughness lengths for momentum and heat derived from outdoor urban scale models. *Journal of Applied Meteorology and Climatology* 46: 1067–1079.

Li Q, Xue Y. 2010. Simulated impacts of land cover change on summer climate in the Tibetan Plateau. *Environmental Research Letters* 5: 015102, doi: 10.1088/1748–9326/5/1/015102.

Liu YM, Wu GX, Hong JL, Dong BW, Duan AM, Bao Q, Zhou LJ. 2012. Revisiting Asian monsoon formation and change associated with Tibetan plateau forcing: ii. change. *Climate Dynamics* 39(5): 1183–1195.

Moufouma-Okia W, Rowell DP. 2010. Impact of soil moisture initialisation and lateral boundary conditions on regional climate model simulations of the west African monsoon. *Climate Dynamics* 35(1): 213–229.

Oleson KW, Dai YJ, Bonan G, Bosilovich M, Dickinson R, Dirmeyer P, Hoffman F, Houser P, Levis S, Niu GY, Thornton P, Vertenstein M, Yang ZL, Zeng XB. 2013. Technical Description of the Community Land Model (CLM), NCAR Technical Note. NCAR/TN-503+STR, National Center for Atmospheric Research, Boulder, CO.

Saha S, Moorthi S, Pan HL, Wu XR, Wang JD, Nadiga S, Tripp P, Kistler R, Woollen J, Behringer D, Liu HX, Stokes D, Grumbine R, Gayno G, Wang J, Hou YT, Chuang HY, Juang HMH, Sela J, Iredell M, Treadon R, Kleist D, Delst P van, Keyser D, Derber J, Ek M. 2010. The NCEP climate forecast system reanalysis. *Bulletin of the American Meteorological Society* 91(8): 1015–1057.

Sheppard PA. 1958. Transfer across the earth's surface and through the air above. *Quarterly Journal of the Royal Meteorological Society* 84: 205–224.

Shi L, Smith EA. 1992. Surface forcing of the infrared cooling profile over the Tibetan Plateau. Part II: cooling-rate variation over large-scale plateau domain during summer monsoon transition. *Journal of Atmospheric Science* 49: 823–844.

Skamarock W, Klemp J, Dudhia J, Gill D, Barker D, Duda M, Huang X, Wang W, Powers J. 2008. A description of the advanced research

WRF Version 3, National Center for Atmospheric Research Technical Note. NCAR/TN-475+STR: 113 p.

Smith EA, Shi L. 1992. Surface forcing of the infrared cooling profile over the Tibetan Plateau. Part I: influence of relative longwave radiative heating at high altitude. *Journal of Atmospheric Science* **49**: 805–822.

Wang AH, Zeng XB. 2012. Evaluation of multi-reanalysis products with in situ observations over the Tibetan Plateau. *Journal of Geophysical Research* **117**: D05102, doi: 10.1029/2011JD016553.

Wang AH, Barlage M, Zeng XB, Draper CS. 2014. Comparison of land skin temperature from a land model, remote sensing, and in situ measurement. *Journal of Geophysical Research* **119**: 3093–3106.

Wu GX, Liu YM, Zhang Q, Duan AM, Wang T, Wan RJ, Liu X, Li W, Wang ZZ, Liang XY. 2007. The influence of mechanical and thermal forcing by the Tibetan Plateau on Asian Climate. *Journal of Hydrometeorology* **8**: 770–789.

Wu GX, Liu YM, Dong BW, Liang XY, Duan AM, Bao Q, Yu JJ. 2012a. Revisiting Asian monsoon formation and change associated with Tibetan plateau forcing: I. Formation. *Climate Dynamics* **39**(5): 1169–1181.

Wu GX, Liu YM, He B, Bao Q, Duan AM, Jin FF. 2012b. Thermal controls on the Asian summer monsoon. *Scientific Reports* **2**(5): 404.

Yang K, Koike T, Ishikawa H, Kim J, Li X, Liu HZ, Liu SM, Ma YM, Wang JM. 2008. Turbulent flux transfer over bare-soil surfaces: characteristics and parameterization. *Journal of Applied Meteorology and Climatology* **47**: 276–290.

You QL, Kang SC, Pepin N, Flügel WA, Yan YP, Behrawan H, Huang J. 2010. Relationship between temperature trend magnitude, elevation and mean temperature in the Tibetan plateau from homogenized surface stations and reanalysis data. *Global & Planetary Change* **71**(1): 124–133.

Zeng XB, Dickinson RE. 1998. Effect of surface sublayer on surface skin temperature and fluxes. *Journal of Climate* **11**: 537–550.

Zeng XB, Wang ZZ, Wang AH. 2012. Surface skin temperature and the interplay between sensible and ground heat fluxes over arid regions. *Journal of Hydrometeorology* **13**(4): 1359.

Zhang BQ, Wu PT, Zhao X, Gao X. 2014. Spatiotemporal analysis of climate variability (1971–2010) in spring and summer on the loess plateau, china. *Hydrological Processes* **28**(4): 1689–1702.

Zheng W, Wei H, Wang Z, Zeng XB, Meng J, Ek M, Mitchell K, Derber J. 2012. Improvement of daytime land surface skin temperature over arid regions in the NCEP GFS model and its impact on satellite data assimilation. *Journal of Geophysical Research* **117**: D06117, doi: 10.1029/2011JD015901.

Zhou L, Huang RH. 2010. Interdecadal variability of summer rainfall in Northwest China and its possible causes. *International Journal of Climatology* **30**: 549–557.

Zhou D, Huang G, Ma YM. 2012. Summer heat transfer over a Gobi underlying surface in the arid region of Northwest China (in Chinese). *Transactions of Atmospheric Sciences* **35**(5): 541–549.

24

Assessing the suitability of statistical downscaling approaches for seasonal forecasting in Senegal

R. Manzanas*[iD]

Meteorology Group, Instituto de Física de Cantabria, CSIC-Universidad de Cantabria, Santander, Spain

*Correspondence to:
R. Manzanas, Meteorology
Group, Instituto de Física de
Cantabria, Edificio Juan Jordá,
Avenida de los Castros, s/n,
39005 Santander, Spain.
E-mail:
rmanzanas@ifca.unican.es

Abstract

This work tests the suitability of statistical downscaling (SD) approaches to generate local seasonal forecasts of daily maximum temperature and precipitation for a set of selected stations in Senegal for the July–August–September season during the period 1979–2000. Two-month lead raw daily maximum temperature and precipitation from the five models included in the ENSEMBLES seasonal hindcast are compared against the corresponding downscaled predictions, which are obtained by applying the analog technique based on two different types of predictors: the direct surface variables and a combination of appropriate upper-air variables. Beyond correcting the large biases of the low-resolution raw model outputs, SD is found to add noteworthy value in terms of forecast association (as measured by interannual correlation), providing thus suitable (i.e. calibrated) predictions at the local-scale needed for practical applications, which means a clear advantage for the end-users of seasonal forecasts over the area of study. Moreover, a recommendation on the adequacy of surface (large-scale) predictors for SD of maximum temperature (precipitation) is also given.

Keywords: statistical downscaling; seasonal forecasting; Senegal

1. Introduction

Given their low spatial resolution, the global seasonal forecasts provided by the current climate models need to be satisfactorily translated to the local-scale required for most practical applications (see, e.g. Hanssen-Bauer *et al.*, 2005). One option for this is statistical downscaling (SD), which is based on empirical/statistical relationships linking the global model simulations (predictors) with the local observations of the target predictand variable (e.g. daily maximum temperature and precipitation in this work). However, though SD has been widely used in climate change studies, only a few works have applied it for seasonal forecasting (see, e.g. Gutiérrez *et al.*, 2004; Landman *et al.*, 2009; Frías *et al.*, 2010; Min *et al.*, 2011; Wu *et al.*, 2012; Shao and Li, 2013; Manzanas *et al.*, 2017). Moreover, SD methods have been mostly implemented for extra-tropical regions since several problems still hinder their successful application in the tropics (Paeth *et al.*, 2011). First, since the local climate is largely driven by meso-scale processes, the statistical relationships between the local- and the large-scale are weaker than in the extra-tropics. Second, reliable observational networks for the local predictand data are often not available. As a result of these factors, most of the downscaling studies undertaken to-date for West Africa have relied on dynamical approaches, that is, regional climate models (van den Hurk and van Meijgaard, 2010; Giorgi *et al.*, 2012; Sylla *et al.*, 2012), even though the skill of global seasonal

forecasts these models are nested to is limited there (see, e.g. Manzanas *et al.*, 2014). Therefore, assessing the suitability of SD approaches for seasonal forecasting over West Africa, where the capacity to invest in regional climate models is limited and the strong interannual climate variations are crucial for various socio-economic sectors (Ndiaye, 2010), is of large interest.

With these considerations in mind, this work focuses on seasonal forecasts of average daily maximum temperature and precipitation for Senegal, a region for which high-quality observations were available. In particular, the potential added value of SD is assessed by comparing the downscaled results with the raw model predictions in terms of forecast association and accuracy. Moreover, this study also tests the suitability of two different types of predictors which may be used for SD: the model counterpart of the variable being predicted (i.e. low-resolution surface maximum temperature/precipitation for predicting local maximum temperature/precipitation) and a combination of appropriate upper-air variables which best describe the synoptic phenomena determining the interannual variability of the local predictands.

The paper is organized as follows: the data used are described in Section 2. Section 3 details the methodology applied. Results are presented and discussed through Section 4 and a summary of the main conclusions obtained is given in Section 5.

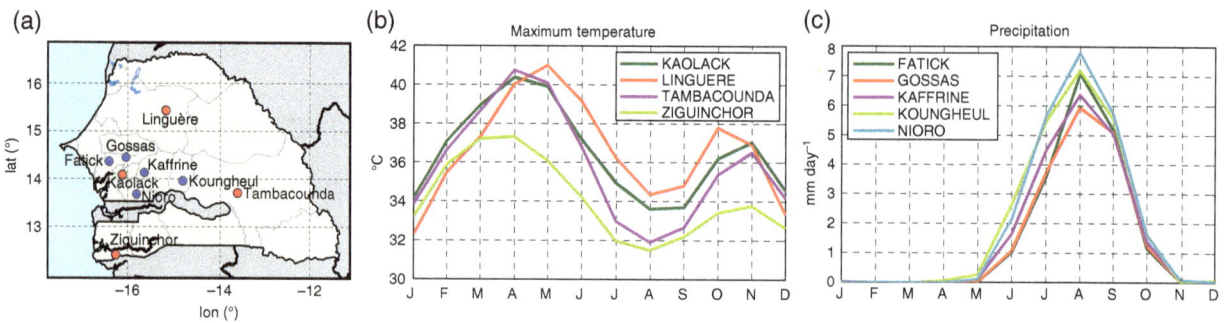

Figure 1. (a) Stations considered for maximum temperature (in red) and precipitation (in blue). (b) Annual cycles of maximum temperature for the period 1979–2000. (c): As (b), but for precipitation.

2. Data

2.1. Predictands

Daily observed maximum temperature (precipitation) from the Agence Nationale de la Météorologie du Sénégal for a set of 4 (5) stations was available for this work for the period 1979–2000 (panel (a) in Figure 1). These stations were quality-controlled by applying tests for detection of outliers and temporal inhomogeneities, and they presented less than a 2% of missing data for the period of study.

Panels (b) and (c) in Figure 1 show the observed annual cycle in these stations for maximum temperature and precipitation, respectively. Whereas maximum temperature presents a bimodal distribution with the first (second) peak around April–May (October–November) and the lowest values in July–August–September (JAS hereafter), precipitation is mainly conditioned by the seasonal migration of the inter-tropical convergence zone (Sultan and Janicot, 2003) and all the stations exhibit a unique rainfall peak in JAS. As the interannual variations of JAS precipitation are key for local agriculture (see, e.g. Wade *et al.*, 2015), only this season was considered in this work.

2.2. Predictors

Daily predictors from the ERA-Interim reanalysis (Dee *et al.*, 2011) were used as catalog for the search of analogs (see Section 3). Seasonal forecasts were obtained from the five models contributing to the ENSEMBLES seasonal hindcast (see Table 1). Note that, although the ENSEMBLES models are several years older than state-of-the-art seasonal forecasting systems, they form the most homogeneous and comprehensive multimodel ensemble publicly available to-date. Each of these models ran an ensemble of nine members which were produced by perturbing the observed state of the atmosphere and the ocean four times a year (the first of February, May, August and November), providing daily data for 7 month-long retrospective runs (see Weisheimer *et al.*, 2009, for further details about the experiment). Therefore, for JAS, only 2-month lead predictions were available.

Table 1. Main components of the five atmosphere–ocean coupled models contributing to the ENSEMBLES multimodel seasonal hindcast.

Centre	Atmospheric model and resolution	Ocean model and resolution
ECMWF	IFS CY31R1 (T159/L62)	HOPE (0.3–1.4°/L29)
UKMO	HadGEM2-A (N96/L38)	HadGEM2-O (0.33–1.0°/L20)
IFM-GEOMAR	ECHAM5 (T63/L31)	MPI-OM1 (1.5°/L40)
CMCC-INGV	ECHAM5 (T63/L19)	OPA8.2 (2.0°/L31)
MF	ARPEGE4.6 (T63)	OPA8.2 (2.0°/L31)

Table 2. Potential predictors considered for this work.

Code	Name	Level	Units
T	Temperature	850, 500, 200, 50 hPa	K
Z	Geopotential	850, 500, 200, 50 hPa	$m^2 s^{-2}$
U	Zonal wind	850, 500, 200, 50 hPa	$m s^{-1}$
V	Meridional wind	850, 500, 200, 50 hPa	$m s^{-1}$
Q	Specific humidity	850, 500, 200, 50 hPa	$g kg^{-1}$

Only predictor variables which were available for both ERA-Interim and the ENSEMBLES models were considered (see Table 2). All of them were re-gridded onto the same 2° regular grid covering the domain encompassed by (20–10°W) and (10–18°N).

3. Methodology

The popular non-parametric analog technique (Lorenz, 1963, 1969) assumes that similar (or analog) atmospheric configurations (e.g. a set of predictors defined over the aforementioned domain) lead to similar meteorological outcomes (local maximum temperature/ precipitation in this work). Here, a deterministic version of the technique which considers only the closest analog (Zorita *et al.*, 1995; Cubasch *et al.*, 1996) is applied. Therefore, for each daily atmospheric configuration simulated by the ENSEMBLES models, the corresponding local downscaled forecast is given as the observations corresponding to the most similar atmospheric configuration found in ERA-Interim. Similarity between atmospheric configurations is measured in terms of the Euclidean norm, which has been shown to perform satisfactorily in most cases (Matulla *et al.*, 2008). The same method has been already used for SD of seasonal forecasts in previous studies (see, e.g. Frías *et al.*, 2010; Manzanas *et al.*, 2017).

To avoid over-fitting, a *k*-fold cross-validation approach (Gutiérrez *et al.*, 2013) was followed, with $k = 4$ non-overlapping test periods, covering the full period of study 1979–2000. Finally, note that SD is performed on a daily basis, thus providing 2-month lead daily downscaled time-series.

Two types of different model predictors were used: the direct surface (SF) variables to be downscaled and a combination of upper-air (UA) variables accounting for the most relevant synoptic phenomena determining the local climate. Whereas the latter approach is the most common for SD in perfect prognosis (see, e.g. Wilby *et al.*, 2004), the utility of the former, which may be highly beneficial since no predictor screening is required, has been rarely tested to-date (see, e.g. Turco *et al.*, 2011).

For the UA case, a step-wise-like algorithm was used to find the optimum combination of predictor variables for each target predictand. Starting from a single predictor taken at random, in each iteration the algorithm performed the SD for all combinations resulting from including/excluding one extra variable (among those shown in Table 2), the downscaled results were validated against observations and the best combination was retained for the next iteration only if a relative improvement of a 1% was reached. Such an improvement was measured in terms of interannual correlation with observations, which is the basis of skill in seasonal forecasting. The optimum UA–predictor combination obtained from this automatic screening for maximum temperature (precipitation) was Z500-T850 (T500-Q850-U850), which account for thermodynamic- and circulation-related processes. For these UA predictor combinations, the leading principal components (PCs, see Preisendorfer, 1988) explaining the 95% of the entire predictor variance were considered (5 for the case of maximum temperature and 18 for precipitation). PCs were obtained, both for the reanalysis and for the seasonal forecasts, by projecting the corresponding standardized fields onto the empirical orthogonal functions obtained from the reanalysis, which were computed simultaneously on all predictor variables.

For each ENSEMBLES model, SD was independently applied to each of the nine available members, obtaining nine daily downscaled time-series. The multimodel ensemble mean (MM henceforth) was calculated by averaging the 45 (5 models × 9 members) available members, thus giving equal weights to all models and members.

4. Results and discussion

The 2-month lead daily downscaled predictions obtained for JAS for the period 1979–2000 were yearly aggregated and validated against the corresponding observations in terms of interannual correlation and mean absolute error (MAE), which account for different aspects of forecast quality: association and accuracy, respectively.

Panels (a) and (b) ((c) and (d)) of Figure 2 show the results obtained for maximum temperature (precipitation). For brevity, only the MM is shown since it was found to outperform the individual models in most of cases, which is in agreement with previous studies (see, e.g. Batté and Déqué, 2011; Landman and Beraki, 2012; Manzanas *et al.*, 2014). For each station, the gray bar corresponds to the MM raw outputs (shown for benchmarking purposes), whereas the blue (green) bar displays the results from SD when considering SF- (UA-) predictors from the MM. This figure allows to assess both the potential added value of SD – by comparing the blue and green bars with the gray one – and the relative performance of SF- and UA-predictors – by comparing blue and green bars.

For maximum temperature, SD outperforms in all cases the correlation of the MM raw outputs (see panel (a)) when considering SF-predictors (especially in Ziguinchor), whereas no substantial improvements are obtained for UA-predictors. Moreover, as expected by construction (SD methods are calibrated towards the observed climate), SD allows to reduce the MAE of the MM raw outputs (see panel (b)), either when considering SF- or UA-predictors (similar results are obtained in both cases).

For precipitation, whereas SF-predictors do not yield any added value in terms of correlation, the use of UA-predictors allows for clearly improving the forecast association of the MM raw outputs (see panel (c)). In particular, whereas the MM exhibits nearly zero correlations in all stations, SD yields significant (at a 95% confidence level) values in all of them (this effect is especially notable in Fatick, where an improvement of about 0.6 correlation units is reached). An explanation for this might be in relation to the results found by Manzanas *et al.* (2017), who proved that the use of UA-predictors can provide an opportunity to improve model precipitation in those cases for which the large-scale is well simulated by the model. Note that, whilst providing a good representation of the large-scale, models can still forecast erroneous precipitation since this variable is strongly affected by local forcing such as small-scale processes and/or orography, which usually are not properly represented in the models. Furthermore, as for the case of maximum temperature, SD outperforms the MM raw outputs in terms of MAE in all stations (see panel (d)), with SF- and UA-predictors yielding similar results.

For completeness, Table 3 shows the results obtained for the five individual ENSEMBLES models in two illustrative stations; Ziguinchor and Fatick, respectively. For providing the best correlation improvements (see Figure 2), SF- (UA-predictors) are considered in the former (latter). It is clear from Figure 2 and Table 3 that, beyond correcting the distinct biases found for

JAS (MM)

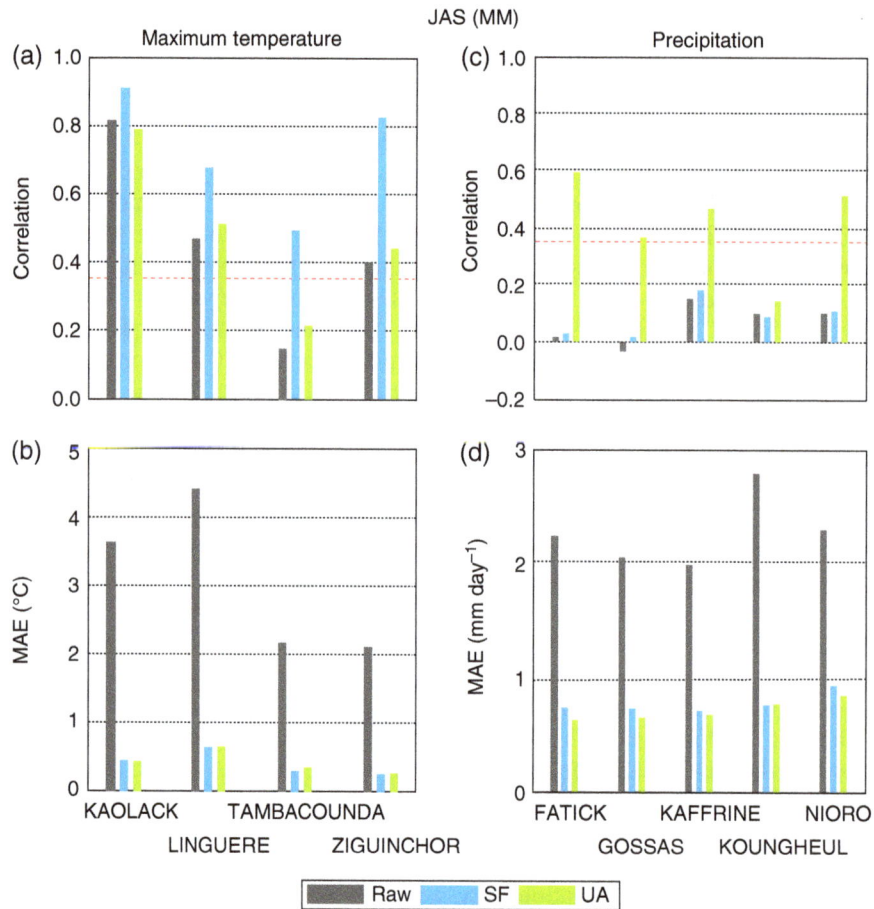

Figure 2. Results obtained for maximum temperature and precipitation in terms of interannual correlation (panels a and c) and MAE (panels b and d). For each station, the gray bar corresponds to the MM raw outputs (shown for benchmarking purposes), whereas the blue (green) bar shows the results from SD when considering SF- (UA-) predictor variables. Correlations above the red horizontal lines are statistically significant at a 95% confidence level, according to a Student's *t*-test.

the different models, SD can considerably improve the correlation attained by their raw outputs when using adequate predictors, providing thus more realistic local-scale climate information, which is needed for real user applications. At this point is important to highlight that simpler bias correction methods allow also to reduce the biases from the different models; however, differently to the SD method presented here, they can deteriorate forecast association (see, e.g. Manzanas *et al.*, 2014).

Although the results from this work mean a clear advantage for the end-users of seasonal climate forecasts in Senegal, it is important to note that they may be not extensible to other regions and/or seasons of interest, and further investigation is still needed to provide a more conclusive overview on the potential merits of SD in the context of seasonal forecasting. For instance, SD might be a beneficial option to compute climate impact indicators, which are sensitive to model biases (particularly those based on absolute thresholds, such as the number of heating or cooling degree days) and typically require working with properly calibrated daily data (Casanueva *et al.*, 2014). This kind of analysis is out of the scope of this paper, but might be matter of study in a future work.

5. Conclusions

This work assesses the suitability of different statistical downscaling (SD) approaches – which, as compared to dynamical downscaling, are computationally cheaper and do not require *a posteriori* correction since they directly incorporate observations into the method – to generate local seasonal forecasts of average daily maximum temperature and precipitation for a set of selected stations in Senegal for the July–August–September season during the period 1979–2000. To this, a nearest analog method is applied to surface (SF) and upper-air (UA) predictors from a number of global seasonal forecasting models (2-month lead predictions are considered). The daily downscaled predictions are yearly-aggregated and validated in terms of correlation and mean absolute error, which account for different aspects of forecast quality. The results obtained indicate that, beyond correcting the large biases of the different global forecasting models, SD adds noteworthy value to the low-resolution raw model outputs in terms of correlation. This clear advantage indicates that SD might be used by end-users in Senegal to obtain suitable (i.e. calibrated) forecasts at the local-scale needed some months ahead of the target season. Moreover,

Table 3. Results obtained for the five individual ENSEMBLES models for maximum temperature (precipitation) in Ziguinchor (Fatick), in terms of interannual correlation and MAE. For providing the best correlations improvements (see Figure 2), SF-(UA-predictors) were chosen.

		ECMWF		UKMO		IFM-GEOMAR		CMCC-INGV		MF	
		Raw	Downscaled	Raw	Downscaled	Raw	Downscaled	Raw	Downscaled	Raw	Downscaled
Ziguinchor	Correlation	−0.05	0.41	0.33	0.79	0.31	0.62	0.59	0.74	0.42	0.66
	MAE (°C)	3.14	0.30	3.47	0.23	3.33	0.35	5.09	0.26	3.68	0.31
Fatick	Correlation	−0.06	0.38	0.09	0.56	−0.17	0.52	−0.23	0.49	−0.24	0.53
	MAE (mm/day)	2.72	1.06	1.59	0.91	2.26	0.13	1.70	0.84	3.93	0.98

whereas UA variables are found to provide better results for SD of precipitation, simpler configurations relying exclusively on SF variables do perform better for maximum temperature. Note the convenience of the latter approach since no predictor screening is required, being thus a cost-effective and pragmatic choice.

Acknowledgements

This study was supported by the EU projects QWeCI and EUPORIAS, funded by the European Commission through the Seventh Framework Programme for Research under Grant Agreements 243964 and 308291, respectively. The author is grateful to the free distribution of the ECMWF ERA-Interim (http://www.ecmwf.int/en/research/climate-reanalysis/era-interim) data, as well as to the EU project ENSEMBLES, financed by the European Commission through the Sixth Framework Programme for Research under contract GOCE-CT-2003-505539, for the seasonal simulations provided and to the Agence Nationale de la Météorologie du Sénégal for the observational data.

References

Batté L, Déqué M. 2011. Seasonal predictions of precipitation over Africa using coupled ocean-atmosphere general circulation models: skill of the ENSEMBLES project multimodel ensemble forecasts. *Tellus A* **63**(2): 283–299. https://doi.org/10.1111/j.1600-0870.2010.00493.x.

Casanueva A, Frías MD, Herrera S, San-Martín D, Zaninovic K, Gutiérrez JM. 2014. Statistical downscaling of climate impact indices: testing the direct approach. *Climatic Change* **127**: 547–560. https://doi.org/10.1007/s10584-014-1270-5.

Cubasch U, von Storch H, Waszkewitz J, Zorita E. 1996. Estimates of climate change in Southern Europe derived from dynamical climate model output. *Climate Research* **7**(2): 129–149. https://doi.org/10.3354/cr007129.

Dee DP, Uppala SM, Simmons AJ, Berrisford P, Poli P, Kobayashi S, Andrae U, Balmaseda MA, Balsamo G, Bauer P, Bechtold P, Beljaars ACM, van de Berg L, Bidlot J, Bormann N, Delsol C, Dragani R, Fuentes M, Geer AJ, Haimberger L, Healy SB, Hersbach H, Holm EV, Isaksen L, Kallberg P, Koehler M, Matricardi M, McNally AP, Monge-Sanz BM, Morcrette JJ, Park BK, Peubey C, de Rosnay P, Tavolato C, Thepaut JN, Vitart F. 2011. The ERA-Interim reanalysis: configuration and performance of the data assimilation system. *Quarterly Journal of the Royal Meteorological Society* **137**(656, Part a): 553–597. https://doi.org/10.1002/qj.828.

Frías MD, Herrera S, Cofiño AS, Gutiérrez JM. 2010. Assessing the skill of precipitation and temperature seasonal forecasts in Spain: windows of opportunity related to ENSO events. *Journal of Climate* **23**(2): 209–220. https://doi.org/10.1175/2009JCLI2824.1.

Giorgi F, Coppola E, Solmon F, Mariotti L, Sylla MB, Bi X, Elguindi N, Diro GT, Nair V, Guiliani G, Turuncoglu UU, Cozzini S, Güttler I, O'Brien TA, Tawiik AB, Shalaby A, Zakey AS, Steiner AL, Stordal F, Sloan LC, Brankovic C. 2012. RegCM4: model description and preliminary tests over multiple CORDEX domains. *Climate Research* **52**: 7–29.

Gutiérrez JM, Cofiño AS, Cano R, Rodríguez MA. 2004. Clustering methods for statistical downscaling in short-range weather forecasts. *Monthly Weather Review* **132**(9): 2169–2183. https://doi.org/10.1175/1520-0493(2004)132h2169:CMFSDIi2.0.CO;2.

Gutiérrez JM, San-Martín D, Brands S, Manzanas R, Herrera S. 2013. Reassessing statistical downscaling techniques for their robust application under climate change conditions. *Journal of Climate* **26**(1): 171–188. https://doi.org/10.1175/JCLI-D-11-00687.1.

Hanssen-Bauer I, Achberger C, Benestad RE, Chen D, Frland EJ. 2005. Statistical downscaling of climate scenarios over Scandinavia. *Climate Research* **29**(3): 255–268.

van den Hurk BJJM, van Meijgaard E. 2010. Diagnosing land-atmosphere interaction from a regional climate model simulation over West Africa. *Journal of Hydrometeorology* **11**(2): 467–481.

Landman WA, Beraki A. 2012. Multi-model forecast skill for mid-summer rainfall over southern Africa. *International Journal of Climatology* **32**(2): 303–314.

Landman WA, Kgatuke MJ, Mbedzi M, Beraki A, Bartman A, Ad P. 2009. Performance comparison of some dynamical and empirical downscaling methods for South Africa from a seasonal climate modeling perspective. *International Journal of Climatology* **29**(11): 15351549. https://doi.org/10.1002/joc.1766.

Lorenz E. 1963. Deterministic non-periodic flow. *Journal of the Atmospheric Sciences* **20**(2): 130–141.

Lorenz E. 1969. Atmospheric predictability as revealed by naturally occurring analogues. *Journal of the Atmospheric Sciences* **26**(4): 636–646. https://doi.org/10.1175/1520-0469(1969)26h636:APARBNi2.0.CO;2.

Manzanas R, Frías MD, Cofiño AS, Gutiérrez JM. 2014. Validation of 40 year multimodel seasonal precipitation forecasts: the role of ENSO on the global skill. *Journal of Geophysical Research: Atmospheres* **119**(4): 1708–1719. https://doi.org/10.1002/2013JD020680.

Manzanas R, Lucero A, Weisheimer A, Gutiérrez JM. 2017. Can bias correction and statistical downscaling methods improve the skill of seasonal precipitation forecasts? *Climate Dynamics* 1–16. https://doi.org/10.1007/s00382-017-3668-z.

Matulla C, Zhang X, Wang X, Wang J, Zorita E, Wagner S, von Storch H. 2008. Influence of similarity measures on the performance of the analog method for downscaling daily precipitation. *Climate Dynamics* **30** (2–3): 133–144. https://doi.org/10.1007/s00382-007-0277-2.

Min YM, Kryjov VN, Oh JH. 2011. Probabilistic interpretation of regression-based downscaled seasonal ensemble predictions with the estimation of uncertainty. *Journal of Geophysical Research: Atmospheres* **116**(D8). https://doi.org/10.1029/2010JD015284.

Ndiaye O. 2010. The predictability of the Sahelian climate: seasonal Sahel rainfall and onset over Senegal, PhD Thesis. ProQuest Dissertations and Theses, Columbia University.

Paeth H, Hall NMJ, Gaertner MA, Domínguez Alonso M, Moumouni S, Polcher J, Ruti PM, Fink AH, Gosset M, Lebel T, Gaye AT, Rowell DP, Moufouma-Okia W, Jacob D, Rockel B, Giorgi F, Rummukainen M. 2011. Progress in regional downscaling of west African precipitation. *Atmospheric Science Letters* **12**(1, S1): 75–82. https://doi.org/10.1002/asl.306.

Preisendorfer R. 1988. *Principal Component Analysis in Meteorology and Oceanography*, 1st ed. Elsevier: Amsterdam.

Shao Q, Li M. 2013. An improved statistical analogue downscaling procedure for seasonal precipitation forecast. *Stochastic Environmental Research and Risk Assessment* **27**(4): 819–830. https://doi.org/10.1007/s00477-012-0610-0.

Sultan B, Janicot S. 2003. The west African monsoon dynamics. Part II: "pre-onset" and "onset" of the summer monsoon. *Journal of Climate* **16**(21): 3407–3427.

Sylla MB, Gaye AT, Jenkins GS. 2012. On the fine-scale topography regulating changes in atmospheric hydrological cycle and extreme rainfall over West Africa in a regional climate model projections. *International Journal of Geophysics* **2012**.

Turco M, Quintana-Seguí P, Llasat MC, Herrera S, Gutiérrez JM. 2011. Testing MOS precipitation downscaling for ENSEMBLES regional climate models over Spain. *Journal of Geophysical Research* **116**(D18). https://doi.org/10.1029/2011JD016166.

Wade M, Mignot J, Lazar A, Gaye AT, Carr M. 2015. On the spatial coherence of rainfall over the Saloum delta (Senegal) from seasonal to decadal time scales. *Frontiers in Earth Science* **3**(30). https://doi.org/10.3389/feart.2015.00030.

Weisheimer A, Doblas-Reyes FJ, Palmer TN, Alessandri A, Arribas A, Déqué M, Keenlyside N, MacVean M, Navarra A, Rogel P. 2009. ENSEMBLES: a new multi-model ensemble for seasonal-to-annual prediction. Skill and progress beyond DEMETER in forecasting tropical pacific SSTs. *Geophysical Research Letters* **36**. https://doi.org/10.1029/ 2009GL040896.

Wilby RL, Charles S, Zorita E, Timbal B, Whetton P, Mearns L. 2004. Guidelines for use of climate scenarios developed from statistical downscaling methods. Technical report, IPCC-TGCIA.

Wu W, Liu Y, Ge M, Rostkier-Edelstein D, Descombes G, Kunin P, Warner T, Swerdlin S, Givati A, Hopson T, Yates D. 2012. Statistical downscaling of climate forecast system seasonal predictions for the southeastern Mediterranean. *Atmospheric Research* **118**: 346–356. https://doi.org/10. 1016/j.atmosres.2012.07.019.

Zorita E, Hughes JP, Lettemaier DP, von Storch H. 1995. Stochastic characterization of regional circulation patterns for climate model diagnosis and estimation of local precipitation. *Journal of Climate* **8**(5): 1023–1042. https://doi.org/10.1175/1520-0442(1995)008h1023:SCORCPi2.0.CO;2.

25

On horizontal distribution of vertical gradient of atmospheric refractivity

Martin Grabner,[1]*[iD] Pavel Pechac[2][iD] and Pavel Valtr[2]

[1]Department of Frequency Engineering, Czech Metrology Institute, Brno, Czech Republic
[2]Department of Electromagnetic Field, Faculty of Electrical Engineering, Czech Technical University in Prague, Czech Republic

*Correspondence to:
M. Grabner, Department of
Frequency Engineering, Czech
Metrology Institute, Okruzni 31,
638 00 Brno, Czech Republic.
E-mail: mgrabner@cmi.cz

Abstract

The horizontal distribution of the vertical gradient of atmospheric refractivity is important for the assessment of the propagation of electromagnetic waves on nearly horizontal paths. A 5-year set of meteorological data, obtained from the ERA-Interim numerical weather product, has been used to analyze this distribution statistically. Vertical gradient maps of the Central European region have been processed to derive empirical probability distributions of the difference between local point gradient values at two locations. The difference of the effective (path-averaged) gradient along the path and the local gradient is shown to be statistically distributed so that the quantiles increase linearly with distance in the interval from 100 to 1000 km. Finally, a spatial correlation function is obtained and described by an exponential model with correlation distances determined in the range of 400–700 km.

Keywords: atmospheric refractivity; vertical gradient; electromagnetic propagation

1. Introduction

The atmosphere, as a dielectric medium, can be characterized by the spatial and temporal distribution of the refractive index of air that is dependent on pressure, temperature and humidity. Since the atmosphere is predominantly horizontally stratified so is the refractive index and it is the vertical component of the gradient which is the most important parameter describing the inhomogeneity of the atmosphere as a dielectric medium.

Spatial distribution of atmospheric refractivity is known to affect the propagation of electromagnetic (EM) waves in the atmosphere (Kerr, 1987). In particular, propagation on nearly horizontal paths is directly affected by the vertical gradient of refractivity due to the effect of the bending of EM waves. Statistics of the gradient obtained from long-term atmospheric measurements (e.g. Hall and Comer, 1969; Akiyama, 1977; Grabner and Kvicera, 2006) are used to predict statistical propagation parameters (ITU, 2015).

In frequency bands above 1 GHz that are widely utilized in terrestrial communication systems, typical path lengths are usually limited to less than 100 km. In this case, the vertical gradient of refractivity is often assumed to be constant along the whole propagation path and single-point statistics of the gradient are applied to assess the long-term link performance (ITU, 2015). On the other hand, there are applications such as unmanned aerial vehicle systems or low elevation links (Vanhoenacker-Janvier *et al.*, 2013) with transmitters in higher altitudes operating on significantly longer paths. Here the knowledge about the horizontal variation of the vertical gradient is needed in order to predict

propagation on such paths accurately. An appropriate statistical modeling of horizontal spatial distribution of the vertical gradient can be applied in remote sensing too (Jicha *et al.*, 2013).

In this study, the two-dimensional spatial distribution of the vertical gradient of atmospheric refractivity is analyzed based on 5-year meteorological data. The horizontal variation of the gradient is described by statistics of gradient difference and path-average gradient along the path. Finally, the spatial correlation characteristics of the gradient are derived and modeled.

2. Input data and processing

Numerical weather products ERA-40 and, recently, ERA-Interim (ERAI) (Dee *et al.*, 2011) by the European Centre for Medium-Range Weather Forecasts (ECMWF) are regularly used in ITU-R SG 9 as a source of long-term and global meteorological data for the propagation modeling purposes. The advantages of the ERAI dataset are twofold: (1) ERAI dataset is carefully checked by ECMWF and is free of serious errors and (2) it provides global coverage with sufficient spatial resolution 0.75° and temporal resolution 6 h. The ERAI data has been processed recently in order to develop global statistical maps of atmospheric refractivity parameters (Grabner *et al.*, 2014). Spatial correlation characteristics of the vertical refractivity gradient have been initially studied in (Grabner *et al.*, 2015). The vertical profiles of atmospheric refractivity were extracted from the ERAI database of meteorological parameters and vertical gradients were calculated by linear regression.

The following parameters were extracted from the ECMWF ERAI database with a time resolution of 6 h:

model level temperature T (K) in 12 model levels corresponding to the lowest part of the atmosphere (levels 49–60), model level specific humidity q (kg kg^{-1}) in the same 12 model levels, and surface pressure p_s (Pa). Additional time invariant parameters were also obtained, such as surface geopotential Φ_s (m^2 s^{-1}).

The pressure values p_i at model level boundaries ($i = 1, \ldots, 61$) are derived from the surface pressure p_s. The heights h_i above the surface corresponding to a particular model levels with the pressure p_i are then obtained from the geopotential values Φ_i. Those are, in turn, derived from Φ_s using a recommended standard iterative procedure that is determined from the discrete analog of the hydrostatic differential equation (ECMWF, 2016).

The atmospheric refractivity N (N-unit) ($N = (n - 1)10^6$, n is the refractive index of air) for radio waves (0–300 GHz at least) is dependent on pressure p (hPa), temperature T (K) and partial water pressure e (hPa). The relationship is often expressed by the formula:

$$N = K_1 \left(\frac{p - e}{T} \right) + K_2 \frac{e}{T} + K_3 \frac{e}{T^2} \qquad (1)$$

where for the Rüeger best-average model (Rüeger, 2002), $K_1 = 77.689 \pm 0.0094, K_2 = 71.295 \pm 1.3, K_3 = (3.75463 \pm 0.0076)10^5$. Partial water pressure e is obtained from the specific humidity of air q by the relation:

$$e = p \frac{q \frac{R_{vap}}{R_{dry}}}{1 + \left(\frac{R_{vap}}{R_{dry}} - 1 \right) q} \qquad (2)$$

where $R_{dry} = 287.0597$ J kg^{-1}K^{-1} and $R_{vap} = 461.5250$ J kg^{-1}K^{-1} are the gas constants of dry air and water vapor.

The vertical gradient of refractivity G (N-units km^{-1}) is derived by means of the linear regression of refractivity values at heights below the specified limit h_{max}. Thus, the vertical profiles of refractivity values $N_i(h_i)$ for heights $h_i < h_{max}$ were approximated by a linear equation:

$$N(h) = Gh + N(0) \qquad (3)$$

where h denotes the height above the surface and $N(0)$ is an additive constant. In the following, $h_{max} = 100$ m above the ground is chosen because, while refractivity at altitudes higher than 100 m is also relevant, the focus on the lowest troposphere is motivated by the higher spatial variability of refractivity near the ground and by propagation being more affected by the refractivity of the lowest layer. The heights obtained from the model levels as described above are not equidistant. For example, in Prague they are about: 0, 23, 58, 107, 173, 258, ... m above the ground. In this case, the refractivity at 100 m is approximated first by linear interpolation (from heights 58 and 107 m) and then the gradient is obtained by a linear regression of refractivity at heights 0, 23, 58 and 100 m above the ground.

An example of the spatial distribution of the vertical gradient less than 100 m above the surface is shown in

Figure 1. The spatial distributions depicted correspond to the times 0000 and 0600 UTC. The standard value of the gradient $G_{100m} \approx -40$ N-units km^{-1} is present over most of continental Europe and North America but super-refractive gradients are present over some sea and coastal areas.

3. Gradient differential statistics

3.1. Point gradient

In order to demonstrate the spatial variability of the vertical gradient of refractivity, the gradient difference $dG = G(0) - G(d)$ between two locations separated by a distance d is analyzed. Figure 2 shows the cumulative distributions (CDs) of dG for distances from 0 to 1000 km. The empirical distributions were calculated using data from the years 2008–2012 considering paths with a central location in Prague. The other points are given by eight azimuthal angles (measured from the northerly direction) ranging from 0 to 315° and by distances from 0 to 1000 km. The maximal distance (1000 km) analyzed was chosen as a practical upper bound since: (1) correlation is low on greater distances, see below, and (2) in many propagation scenarios, the propagation path of wanted or unwanted signals is likely shorter than this limit.

Note that Figure 2 shows both CDs and complementary CDs (i.e. $1 - F(x)$, where $F(x) = P(X < x)$ is the usual CD) in order to clearly see both tails of the distribution. The difference dG, CD is slightly asymmetrical toward the negative differences. The gradient difference extremes are clearly increasing with the distance of the two locations.

3.2. Effective gradient

Considering propagation paths longer than approximately 20–50 km, point gradient is no longer the most suitable parameter for propagation estimation. Instead, the averaged propagation effect is better characterized by the *effective gradient* (Mojoli, 1980; ITU, 1996). The effective (or path-average) gradient is obtained as a mean value of the gradient along the propagation path:

$$G_e(d) = \frac{1}{d} \int_0^d G(x) \; dx \qquad (4)$$

The effective gradient is usually more suitable to describe propagation on longer paths than the point gradient because local ray curvature is related to the local gradient and overall ray bending along the path is approximately given by a summation of local ray bending along the whole path.

From the propagation point of view, the extremal gradients occurring only rarely have the most significant impact on the propagation and their incidence statistics are of practical importance for the effective design of radio links. In this respect, time percentages as low as 0.01% are relevant. The effective gradient exceeded for

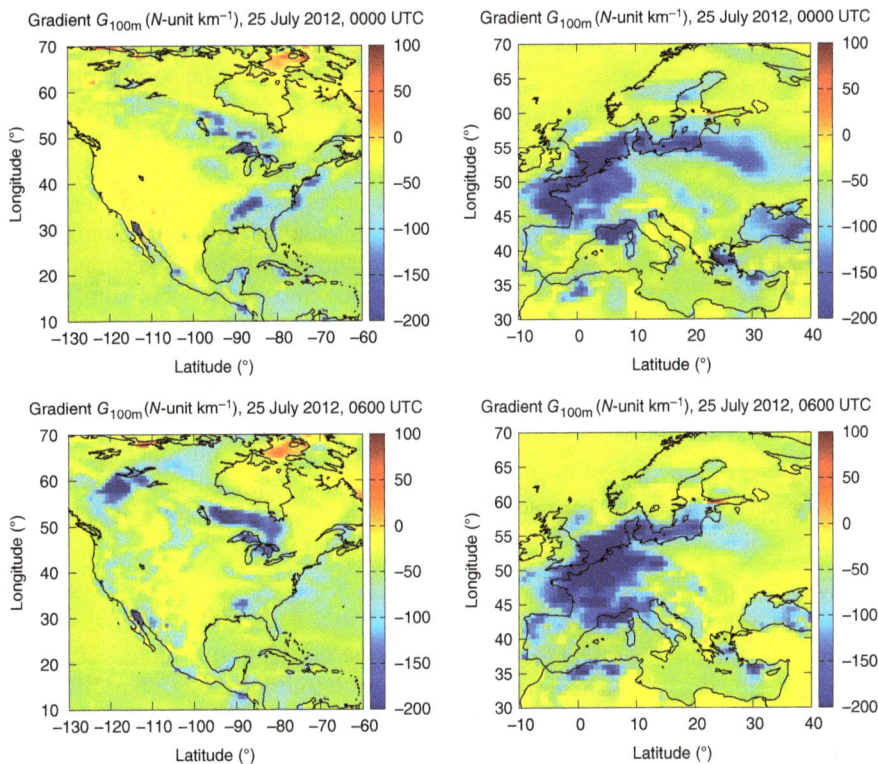

Figure 1. Spatial distribution of the vertical gradient of refractivity in North America and Europe on 25 July 2012.

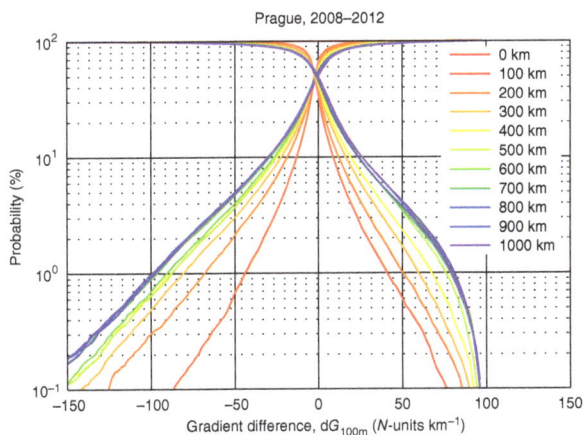

Figure 2. Cumulative distributions of gradient difference between two distant locations. The reference point is located in Prague.

0.01% of time is obtained from ERAI data and compared with the statistics provided by ITU-R P.530 (ITU, 2015). Figure 2 of ITU-R P.530 shows the dependence of the effective k-factor k_e exceeded 99.99% of the time on the path length for the continental temperate climate. Applying the known relation between the k-factor and the vertical gradient, the percentiles $k_{e99.99\%}$ are equivalent to the effective gradient percentiles according to the relation:

$$G_{e0.01\%} = 157 \left(1/k_{e99.99\%} - 1\right) \qquad (5)$$

ITU-R P.530 gives $G_{e0.01\%}(100) \approx 20\ N$-units km^{-1} for 100 km paths and $G_{e0.01\%}(200) \approx 0\ N$-units km^{-1}

for 200 km paths. On the other hand, the 0.01% percentiles $G_{e0.01\%}$ obtained from ERAI for the location of Prague decrease from about 30 to 0 N-units km^{-1} for distances ranging from 100 to 1000 km. Therefore, for path lengths above 100 km, ITU-R P.530 and ERAI agree sufficiently well.

There is a higher inconsistency between ITU-R P.530 and ERAI statistics for path lengths less than approximately 100 km. But in this region, the accuracy of the ERAI results should be naturally limited since the resolution 0.75° is not sufficient to analyze accurately the length scales below 80 km. (One must also take into account that the ITU-R P.530 statistics of k_e are originally derived indirectly from propagation measurements of a received signal level but the ERAI results come directly from meteorological data.) On the other hand, it is confirmed that G_e quantiles are approaching a constant value with further increasing path lengths.

Clearer path length dependence is revealed when analyzing the difference between the point gradient and the effective gradient. Figure 3 shows the CDs of the difference $dG_e = G(0) - G_e(d)$ for different path lengths d.

Both CDs and complementary CDs are depicted in Figure 3. The CD of the difference dG is highly symmetrical and with lower extremes than the point difference above which seems to be a logical consequence of the integral definition of G_e. The difference dG_e extremal values are clearly increasing with an increasing path length. For example, the 0.1% quantiles are from about −40 to −125 N-units km^{-1} for path lengths from 100 to 1000 km.

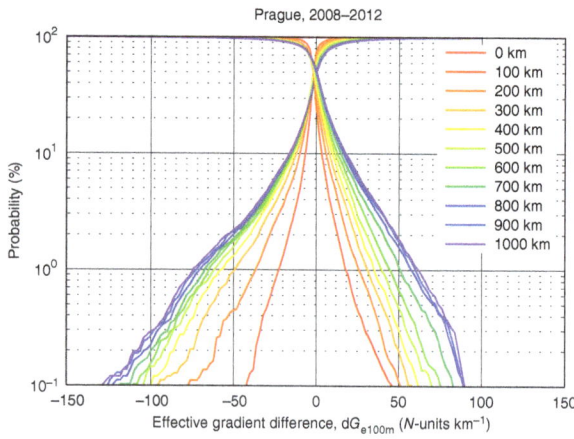

Figure 3. Cumulative distributions of the difference between point gradient and effective gradient along a path. The reference point is located in Prague.

While the results presented in this section could find their main application in EM wave propagation problems, the horizontal distribution of the vertical gradient has not been extensively studied so far. An important first step of the gradient structure modeling is presented in Section 4 by a derivation of a spatial correlation of the spatial random field (SRF).

4. Spatial random field characteristics of a gradient

Assuming the spatial distribution of the refractivity gradient as SRF (Christakos, 2005), its statistical parameters were extracted from the analyzed dataset. The continuous SRF is generally described by a joint probability density function (PDF) of the dimension m for any m, but for practical reasons only the case $m = 2$ is usually considered when the empirical PDF is being extracted from the data. Figure 4 shows the examples of two-dimensional joint PDFs of the vertical gradients at locations separated by a surface distance d from 50 to 950 km with the central location in Prague. Clearly, the degree of correlation between gradient values is decreasing with the horizontal distance. The single, as well as joint, PDFs of the gradient are not generally of the Gaussian type and are difficult to model analytically. [The single PDF of the gradient can be, e.g. expressed by the linear combination of three Gaussian distributions (Grabner and Kvicera, 2011).] Therefore, it is more convenient to use a correlation theory approach to model a horizontal structure of the gradient. The spatial correlation function (SCF) is defined as:

$$\rho(d) = C(x_1, x_2) / \sqrt{C(x_1, x_1) C(x_2, x_2)} \quad (6)$$

where x_1, x_2 denote different spatial locations separated by distance $d = |x_1 - x_2|$ and covariance C is obtained from:

$$C(x_1, x_2) = E\{G(x_1) G(x_2)\} \\ - E\{G(x_1)\} E\{G(x_2)\} \quad (7)$$

where $G(x)$ is the gradient at location x and $E\{\ \}$ denotes a mean value that is approximated by time averaging over the analyzed time period. Figure 5 shows the SCFs obtained for the central locations of Prague, London, New York and Denver (i.e. x_1 in Equations (6) and (7)) from the dataset (2008–2012). The other locations x_2 were considered to be at distances d (km) from the central location with an azimuth angle ϕ from 0 to 315°. The correlation is seen as being fairly isotropic in Prague and Denver, but some anisotropy is evident in New York. Such anisotropy could be caused by the proximity of the ocean–land transition. However, the results from London show that it is not straightforward to relate the anisotropy to this effect in general. The *mean* correlation function averaged over azimuth angles, also depicted in Figure 5, decreases with distance in a way that can be described by the following exponential model:

$$\rho(d) = exp(-d/d_c) \quad (8)$$

with the correlation distances $d_c = 631, 424, 430, 674$ km for all four locations, respectively. These correlation distances were obtained by a least-square fitting of the model Equation (8) to the average (averaging over azimuths) correlation functions as shown in Figure 5. The root mean square error of the fitted model values f_i with respect to the obtained average correlations $\rho_i = \rho(d_i)$ at distances, $d_i = 50i$, $i = 1, \dots, N$, $N = 20$, defined by:

$$\sigma = \sqrt{\frac{1}{N} \sum_{i=1}^{N} (f_i - \rho_i)^2} \quad (9)$$

is equal to 0.033, 0.038, 0.030 and 0.058, respectively. Note that the correlation distance d_c is expected to be a site-specific parameter dependent on local climatic and orographic conditions.

The SCF obtained is the simplest way to model spatial random distribution of the gradient. Using SCF allows, in principle, to model a *joint* PDF of a gradient or generate a SRF of the gradient from the SCF and from a *single point* PDF obtained from the local measurement.

5. Discussion and conclusions

The results presented show that on the horizontal distances within the interval (100, 1000 km), the difference between the local vertical gradients of refractivity at different locations is statistically distributed, so that its quantiles increase with the distance. The effective gradient along the propagation path that better characterizes propagation on extended paths is shown to be different from the local gradient in such a manner that the quantiles of the difference dG_e increase with distance (or path length) almost linearly. These statistical results (CDs and joint PDFs) were determined for the climate region of Central Europe and they are therefore site-specific to some extent. Nevertheless, the

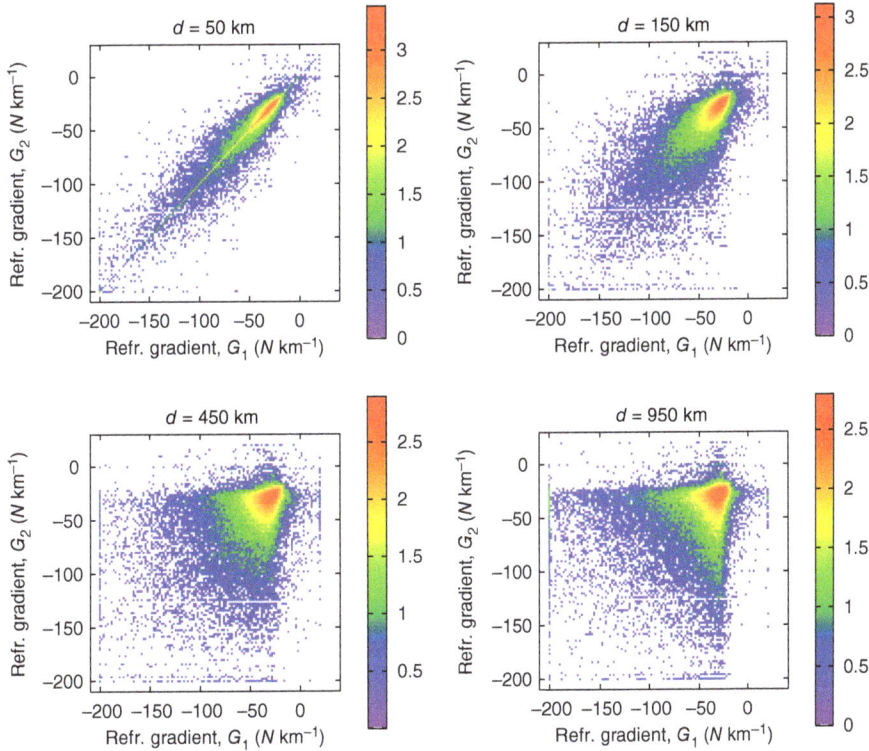

Figure 4. Empirical joint probability density (histogram) of the vertical gradients at locations separated by surface distance d (km). Numerical scale. $\sim log_{10}(N_{bin})$, where N_{bin} is the number of values in the bin. Central location in Prague, during 2008–2012.

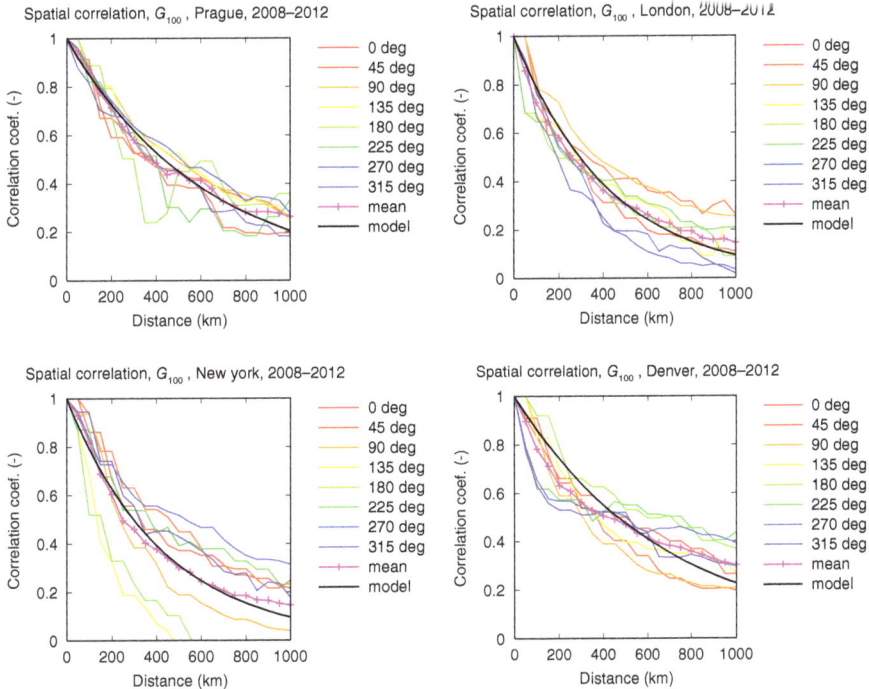

Figure 5. SCF of the vertical refractivity gradient in Prague, London, New York and Denver.

qualitative conclusions presented were determined to be valid also for other inland locations.

The SCF obtained is well-approximated by the exponential model Equation (8) with correlation distances between about 400 and 700 km depending on the location. It has been demonstrated that path direction may also influence the correlation distance at particular

areas due to anisotropy of the gradient horizontal distribution.

Finally, propagation effects due to the horizontal distribution of refractivity gradient are not analyzed in this study. Interested readers can refer to the work of Valtr and Pechac (2005) who studied this problem using an analytical ray-tracing technique.

Acknowledgement

This work was financially supported by the Czech Science Foundation grant no. 14-01527S.

References

Akiyama T. 1977. Studies on the radio refractive index in the tropospheric atmosphere. *Review of the ECL* **25**: 79–95.

Christakos G. 2005. *Random Field Models in Earth Sciences*. Dover Publications: Mineola, NY.

Dee DP, Uppala SM, Simmons AJ, Berrisford P, Poli P, Kobayashi S, Andrae U, Balmaseda MA, Balsamo G, Bauer P, Bechtold P, Beljaars ACM, van de Berg L, Bidlot J, Bormann N, Delsol C, Dragani R, Fuentes M, Geer AJ, Haimberger L, Healy SB, Hersbach H, Hlm EV, Isaksen L, Kallberg P, Khler M, Matricardi M, McNally AP, Monge-Sanz BM, Morcrette J-J, Park B-K, Peubey C, de Rosnay P, Tavolato C, Thpaut J-N, Vitart F. 2011. The ERA-Interim reanalysis: configuration and performance of the data assimilation system. *Quarterly Journal of the Royal Meteorological Society* **137**(656): 553–597.

ECMWF. 2016. *IFS DOCUMENTATION Cy43r1, Part III: Dynamics and Numerical Procedures*. ECMWF: Reading, UK.

Grabner M, Kvicera V. 2006. Refractive index measurements in the lowest troposphere in the Czech Republic. *Journal of Atmospheric and Solar-Terrestrial Physics* **68**: 1334–1339.

Grabner M, Kvicera V. 2011. Atmospheric refraction and propagation in lower troposphere. In *Electromagnetic Waves*, Zhurbenko V (ed). INTECH: Rijeka, Croatia.

Grabner M, Kvicera V, Pechac P, Kvicera M, Valtr P, Martellucci A. 2014. World maps of atmospheric refractivity statistics. *IEEE Transactions on Antennas and Propagation* **62**: 3714–3722.

Grabner M, Pechac P, Valtr P. 2015. Spatial correlation of vertical gradient of refractivity on large scales. In *Proceedings of the 9th European Conference on Antennas and Propagation (EuCAP)*. Lisbon, Portugal, 1–4.

Hall MPM, Comer CM. 1969. Statistics of tropospheric radio-refractive-index soundings taken over a 3-year period in the United Kingdom. *Proceedings of the IEEE* **116**: 685–690.

ITU. 1996. *Handbook on Radiometeorology*, 1st ed. Radiocommunication Bureau: Geneva, Switzerland.

ITU 2015. Propagation Data and Prediction Methods Required for the Design of Terrestrial Line-of-sight Systems. Geneva, Switzerland, ITU-R Rec. P.530-16.

Jicha O, Pechac P, Kvicera V, Grabner M. 2013. Estimation of the radio refractivity gradient from diffraction loss measurements. *IEEE Transactions on Geoscience and Remote Sensing* **51**(1): 12–18.

Kerr E (ed). 1987. *Propagation of Short Radio Waves*. Peter Peregrinus: London.

Mojoli LF. 1980. A new approach to the visibility problems in line-of-sight hops (ICC 1979). *Telettra Review* **31**: 14–21.

Rüeger, JM. 2002. Refractive index formulae for radio waves. In *Proceedings of the FIG XXII International Congress*, Washington, DC.

Valtr P, Pechac P. 2005. The influence of horizontally variable refractive index height profile on radio horizon range. *IEEE Antennas and Wireless Propagation Letters* **4**: 489–491.

Vanhoenacker-Janvier D, Bouchard P, Braten LE, Fabbro V, Kourogiorgas C, Rogers D. 2013. Channel models for aeronautical and low elevation radio links. In *Proceedings of the 7th European Conference on Antennas and Propagation (EuCAP)*. Gothenburg, Sweden, 3184–3186.

26

Analysis of the breeze circulations in Eastern Amazon: an observational study

Michell Fontenelle Germano,[1]* Maria Isabel Vitorino,[1,2] Júlia Clarinda Paiva Cohen,[1,2] Gabriel Brito Costa,[3] Jefferson Inayan de Oliveira Souto,[1] Mayse Thais Correa Rebelo[3] and Adriano Marlisom Leão de Sousa[4]

[1] Federal University of Para (UFPa), Department of Meteorology, Belém, Brazil
[2] Federal University of Para (UFPa), Graduate Program in Environmental Sciences (PPGCA), Belém, Brazil
[3] Federal University of Western Para (UFOPa), Institute of Biodiversity and Forestry, Santarém, Brazil
[4] Federal Rural University of Amazonia (UFRA), Socio-Environmental Institute, Belém, Brazil

*Correspondence to:
M. F. Germano, Federal University of Pará (UFPa), Augusto Corrêa Street, No. 1, 66075-110 Belém, Pará, Brazil.
E-mail:
michellfgermano@gmail.com

Abstract

An observational analysis was conducted in five different cities in Eastern Amazonia, in order to detect the breeze circulations in the region. The frequency of wind direction, wind speed, and precipitation was analyzed along with estimated spatio-temporal rainfall through the Climate Prediction Center Morphing Technique (CMORPH). The results show different types of breezes that occur in these cities, with regular time from 0900–2100 UTC for SB (sea breeze), 0000–0900 UTC for LB (land breeze), and 1200–0000 UTC for RB (river breeze). The SB has been shown to be more frequent from September to November (SON), while the LB is more prominent from March to May (MAM). However, the RB highlights throughout the whole year in Belém. The hour of occurrence of the SB circulation and the precipitation along the coast has shown a relationship.

Keywords: sea breeze; river breeze; land breeze; local circulation; Amazon rainforest; Climate Prediction Center Morphing Technique

1. Introduction

The Eastern Amazon (Figure 1 for the geographical location of the target area in this study) observes a variety of mesoscale convective systems, among which we can highlight mesoscale convective complex (CCMs) and squall lines (SLs) (Kousky, 1980; Maddox, 1980; Cohen *et al.*, 1995; Vitorino *et al.*, 1997; Silva Dias *et al.*, 2004; Lu *et al.*, 2005; Fitzjarrald *et al.*, 2008; Ramos da Silva *et al.*, 2011; Cohen *et al.*, 2014; Matos and Cohen, 2015).

The SLs that form along the Amazon Atlantic coast has its genesis associated with the sea breeze (SB) circulation (Kousky, 1980). This convective system is responsible for 45% of the precipitation in Eastern Pará (Figure 1) and in the Amazon central region during the rainy season (Greco *et al.*, 1990; Garstang *et al.*, 1994; Cohen *et al.*, 1995). The diurnal cycle of rainfall is highly influenced by the breeze circulations, having its maximum pronounced when these circulations occur (Kousky, 1980; Janowiak *et al.*, 2005).

The river breeze (RB) circulation, as well as the SB and the land breeze (LB), is the result from the difference of the heat capacity of the continent and water. During the day, the temperature in the continent is greater than on the water, while at night we observe the opposite. On the hot side has a low pressure with the air flow coming from the relatively cooler side; this air coming to the hot side goes up and returns to the cooler side, thus generating a local circulation.

According to Kousky (1980), the diurnal variation of the rainfall in the neighborhood of the coastal area of Eastern Amazon experience a maximum rainfall at night until the early due to the low level convergence associated to the mean flow coming from the ocean and the continental flow directed toward the ocean (LB).

About 700 km inland from the Amazon Atlantic coast, there is the confluence of the Amazon and Tapajos rivers, where the diurnal variability of the rainfall was analyzed according to the distance of the river. Near the Amazon bigger rivers, the precipitation is predominantly nocturnal, while on the continent it takes place both at night, associated with the passage of the SLs, as in the afternoon due to the RB (Fitzjarrald *et al.*, 2008; Cohen *et al.*, 2014).

Although the breeze circulations are well known, there are few studies that define the average patterns of these circulations observationally. In that sense, this study aims to investigate the mesoscale atmospheric circulation, associated with the breeze circulations and its contribution in the diurnal cycle of precipitation.

2. Data and methodology

The study was conducted in Northeastern Pará region located in Northern Brazil using data from automatic weather stations installed in five different cities: Belém, Pará's state capital, located in (−1.4103°; −48.4383°); Soure, located in the Marajó's island

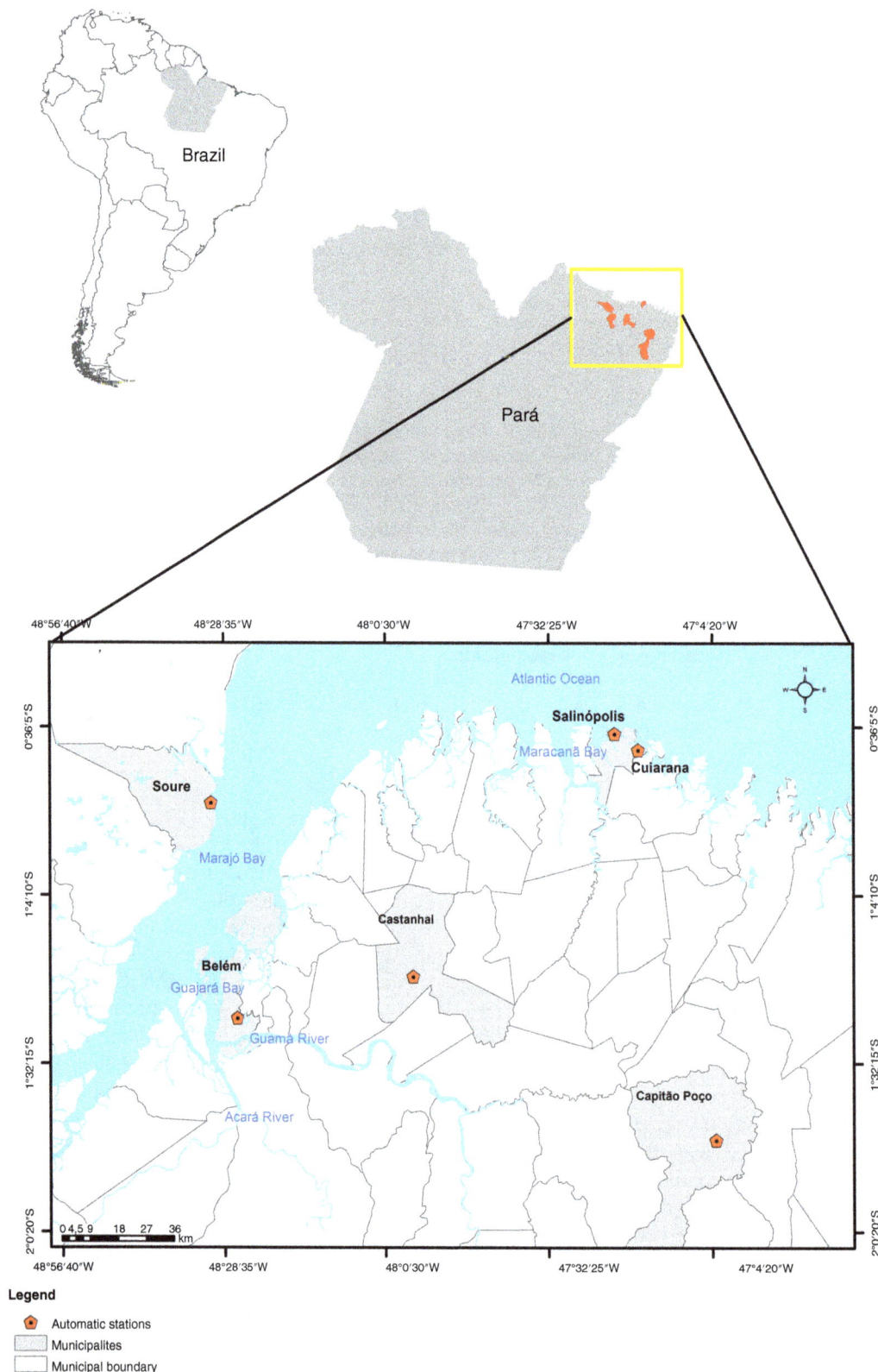

Figure 1. Map of the study area showing the location of the automatic weather stations.

(−0.8112°; −48.5158°); Salinópolis (−0.61868°; −47.3511°); Cuiarana (−0.6636°; −47.2842°); Castanhal (−1.2952°; −47.9281°), and Capitão Poço (−1.7489°; −47.0614°) as shown in Figure 1. It is worth emphasizing that each location has different surface characteristics, Belém (population: 1 0393 399)

is an urban metropolis; Soure (population: 23 001) is located on the western bank of the Marajó Bay – a large estuary of the Amazon and Tocantins rivers; Cuiarana and Salinópolis, both part of the same municipality (population: 37 421), are located on the coastline near the Atlantic Ocean; Castanhal (population: 173 149);

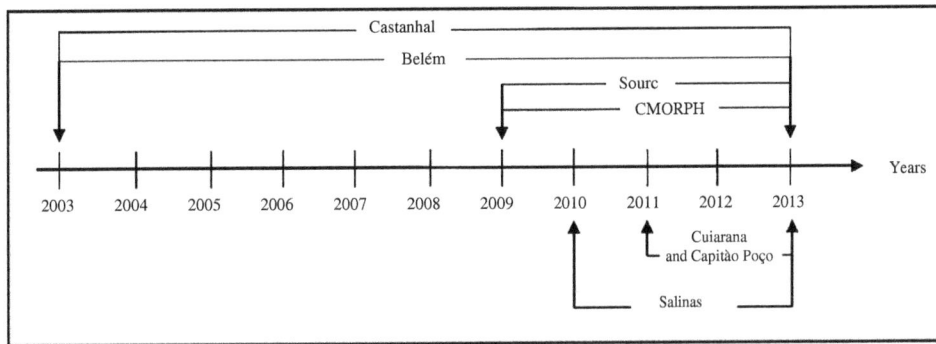

Figure 2. Description of the time series data for each location.

and Capitão Poço (population: 51 893) are located a little further inland.

The data that we have used in this study were precipitation (mm), wind speed (m s^{-1}), and direction (degrees). The data were retrieved from automatic weather stations located in each city. In Cuiarana, these data were collected and stored every 10 min in the period from 2011 to 2013. On the other hand, the data collected in Belém, Castanhal, Capitão Poço, Soure, and Salinópolis were collected every hour, for different periods (Figure 2): Belém and Castanhal (2003–2012); Soure (2009–2012); Cuiarana and Capitão Poço (2011–2013), and Salinópolis (2010–2013). After retrieving the data, hourly averages were computed. The hourly averages were determined for each city considered in this study. The months with more than 15 days missing were not computed and the days with no information were considered as NA (not available).

To calculate the hourly frequency of wind speed and direction, and precipitation we used the Metvurst package – Meteorological visualization utilities using R for science and teaching (Appelhans et al., 2013). To obtain the average variability of the circulation and precipitation we applied the calculation of the hourly frequency. After that, the data were divided into seasonal series.

In order to evaluate the diurnal cycle of precipitation associated to the breezes circulations with high spatio-temporal (0.25° lat/long – 3/3 h), we used the data from the Climate Prediction Center Morphing Technique (CMORPH), according to Joyce et al. (2004). The intervals considered in this study for the definition of the breeze circulations were: SB 0900–2100 UTC, LB 2100–0900 UTC, and RB 1200–0000 UTC. These criteria were defined from the observation of the average standards of breezes circulations, as shown in several studies (Haurwitz, 1947; Fitzjarrald et al., 2008; Teixeira, 2008).

3. Results and discussion

3.1. Average patterns of the local circulations

In general, the wind direction diurnal cycle presents a Northeastern (NE) main maximum for most of the locations, featuring the occurrence of the SB on the same direction of the trade winds. The SB is most frequent during the period from 1200 to 2100 UTC as shown in Figure 3(a). These local winds are associated with the mean flow of the trade winds that through displacement of the subtropical ridges can change the intensity of local circulation (Cavalcanti, 1982; Uvo, 1989, Baptista, 2003). In that sense, it is observed that the SB for Salinópolis and Cuiarana is much more pronounced when compared to other localities, due to its proximity to the Atlantic Ocean. We also notice that the wind speed (Figure 3(c)) generally increases with the occurrence of the SB. This is due to the fact that the SB is in the same quadrant of the mean flow – vectorially the SB is in the same quadrant as the mean flow, thereby increasing its intensity (Frizzola and Fisher, 1963; Atkinson, 1981; Costa and Lyra, 2012; Santos et al., 2013). In Castanhal and Capitão Poço, we observed an eastern pattern on the wind direction, having its most pronounced frequency during the hours from 1200 to 1800 UTC along with an increase in the wind speed, again due to the penetration of the SB.

The increase in the wind speed (Figure 3(c)) is directly related to the time when the air-sea temperature contrast is higher, forcing the inland penetration of the SB. The penetration of the SB is clearly represented by a slight shift in the wind direction (Figure 3(a)) from NE to E in Capitão Poço, Castanhal, and Soure. Cuiarana and Salinas did not have an apparent wind direction shift, suggesting that the SB penetrates in the same direction as the mean flow, as discussed before. However, the wind speed increase in Cuiarana and Salinas is as noticeable as in the other cities. Pearce (1955) states that the SB is at first slow, becoming more rapid as the heating continues; the onset of the SB well inland does not occur until a few hours after heating commences. This air-sea temperature contrast determines whether we have the formation of the SB or not; therefore, when we have large-scale weather systems such as the Intertropical Convergence Zone (ITCZ) over the region this contrast becomes smaller, weakening, and sometimes not allowing the formation of the SB. A further analysis of this seasonal difference will be discussed in Section on Seasonal Analysis of the Breeze Circulations.

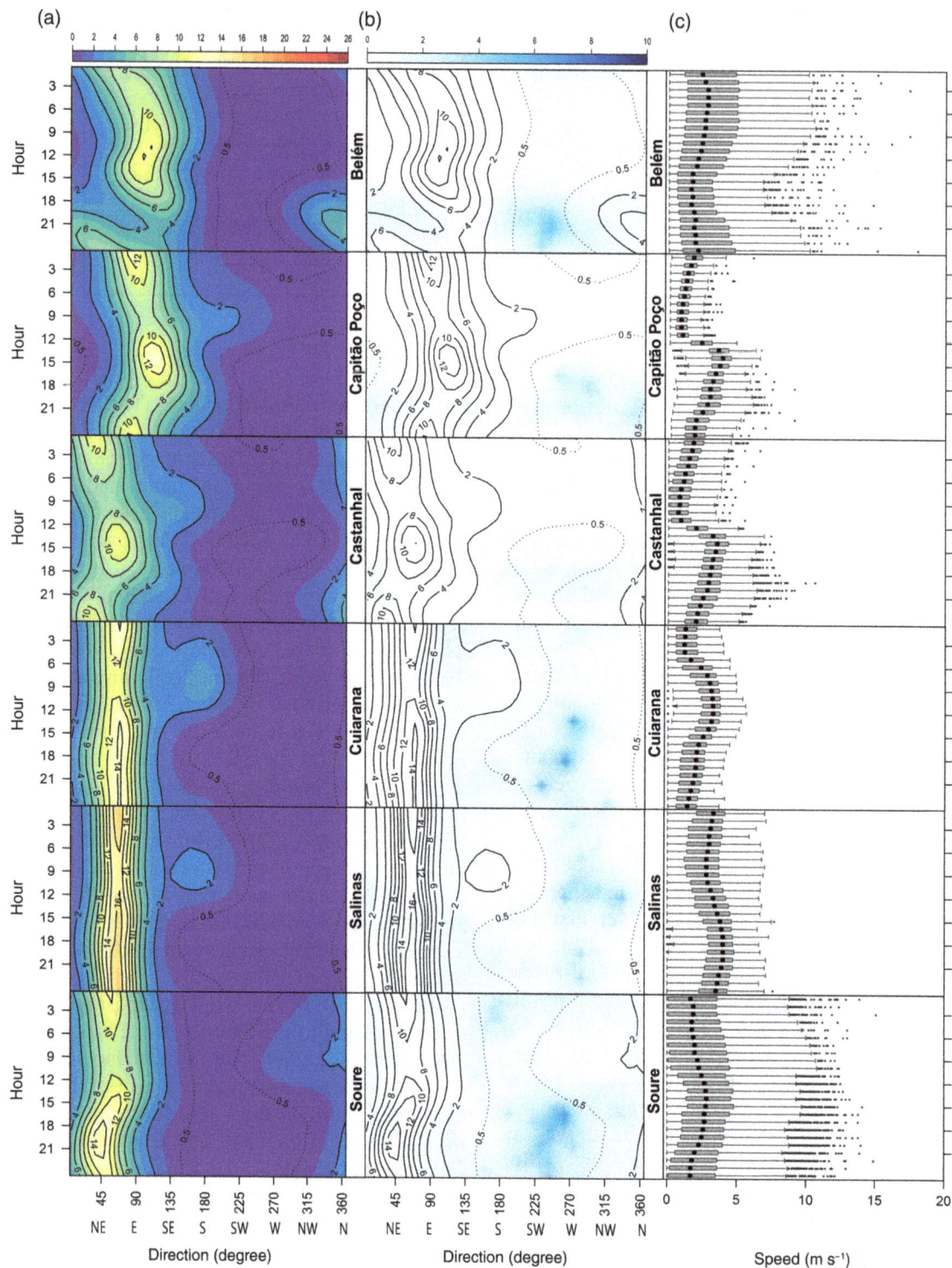

Figure 3. Hourly frequency: (a) wind direction (line, %); (b) wind direction (line, %) and precipitation (shaded, mm); and (c) the median for the wind speed (black points, m s^{-1}), with its respective quartiles (shaded gray bar, m s^{-1}); the outside bars represent upper and lower whiskers and outliers are represented by asterisks outside of the plot.

The LB was identified from a secondary maximum in the wind direction from Southeast (SE)/South (S) in Salinópolis, Cuiarana, Castanhal, Capitão Poço, and Soure. We also notice that the most frequent hour of this circulation is from 0300 to 0900 UTC. Although we also observed the LB in Soure, the direction of this circulation is different due to its geographical position. The LB in Soure is in a different direction in relation to other cities; in this case, the LB is from Northwest (NW)/West (W), but at the same time, as in the other cities. In general, we observe the lowest wind speed values (Figure 3(c)) at the same time when the LB is present. Therefore, the LB circulation is less intense when compared to the SB, since it is opposed to the direction of the mean flow (Atkinson, 1981; Teixeira, 2008).

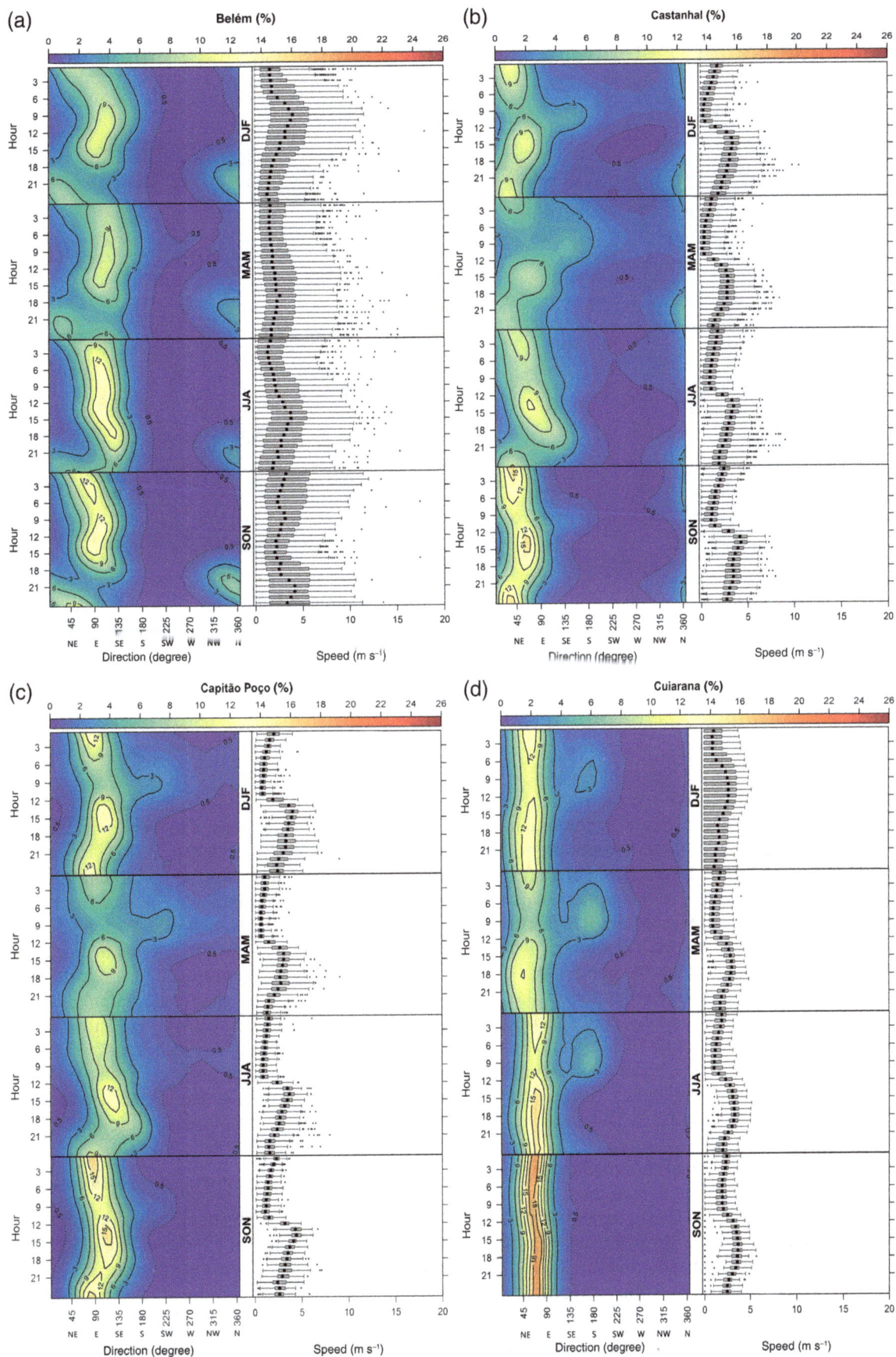

Figure 4. Seasonal hourly frequency of the wind direction (left panel) and speed (right panel) for Belém (a), Castanhal (b), Capitão Poço (c), Cuiarana (d), Salinópolis (e), and Soure (f).

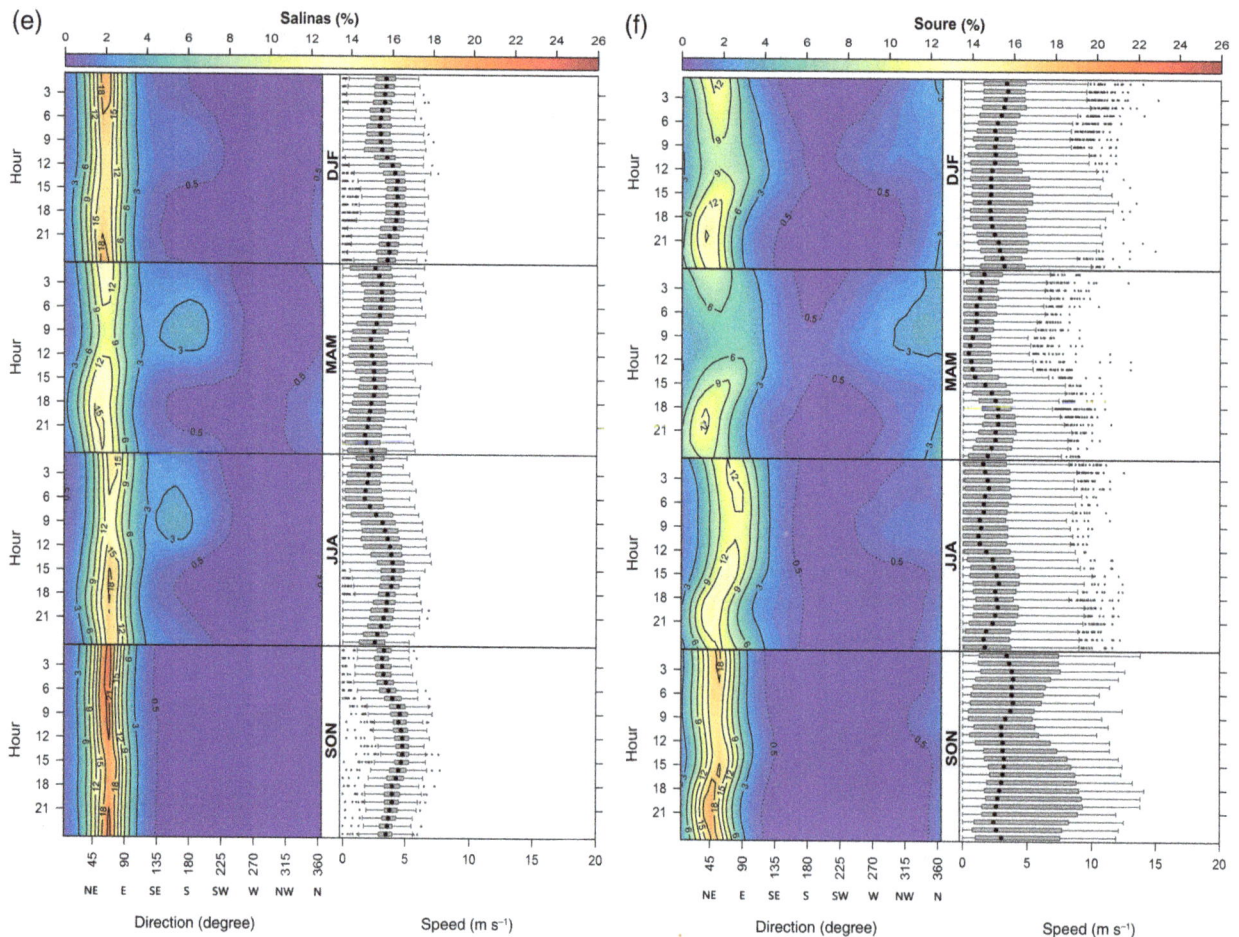

Figure 4. Continued.

In Belém, we cannot observe the LB signal, but we notice the presence of a secondary maximum in the wind direction from NW, with occurrence from 1500 to 0000 UTC, along with a decrease in the wind speed during the time of this secondary maximum. This may be associated with the presence of the RB due to the Marajó Bay proximity to the city. Therefore, this influence becomes much bigger than the oceanic effect on the SB. Other studies have shown the presence of the RB in Amazonian cities (Silva Dias *et al.*, 2004; Fitzjarrald *et al.*, 2008; Cohen et al, 2014; Santos *et al.*, 2014; Matos and Cohen, 2015); generally this temperature contrast between the forest and the rivers can reach more than 6 °C (Greco *et al.*, 1992; Oliveira and Fitzjarrald, 1993).

The precipitation (Figure 3(b)) occurs during the late afternoon and early evening, this result agrees with the results found by Kousky (1980), indicating that the largest rain volumes are in the afternoon, between 1700 and 2100 UTC, with the occurrence of rain also at night. The SB causes this nighttime maximum in local precipitation. According to Janowiak *et al.* (2005), with daytime heating the precipitation rapidly develops along the coast, having its most pronounced effects inland from the coast, producing a nocturnal maximum due to its propagation inland. The effects of the SB on the precipitation are more evident in locations that are farther

inland, such as Belém, Castanhal, Capitão Poço, and Soure, which all experience a higher frequency of nocturnal precipitation from 1500 to 2100 UTC, while Salinópolis and Cuiarana observe precipitation earlier in the day. These results indicate that the SB enters the coast, but its ascending branch which originates convection, is established within the continent. Furthermore, we can also observe the presence of a secondary maximum of precipitation in Soure and Salinópolis, during the early hours of the day from 0000 to 0300 UTC. These results converge again with Kousky (1980), who indicates that the rainfall peak at 0000–0300 UTC is associated to the passage of the SB front at Soure, when the convective cells are not totally developed.

3.2. Seasonal analysis of the breeze circulations

Figure 4 shows the seasonal variation of the wind direction for each city, which shows that the SB is more pronounced during the period of September–November (SON), while the LB is more pronounced during the period of March–May (MAM). These seasonal differences both in wind speed and direction in SON are directly connected with the intensification of the South Atlantic subtropical high pressure (Baptista, 2003), contributing to the inland penetration of the SB, which is perpendicular to the coast. According to Nobre and

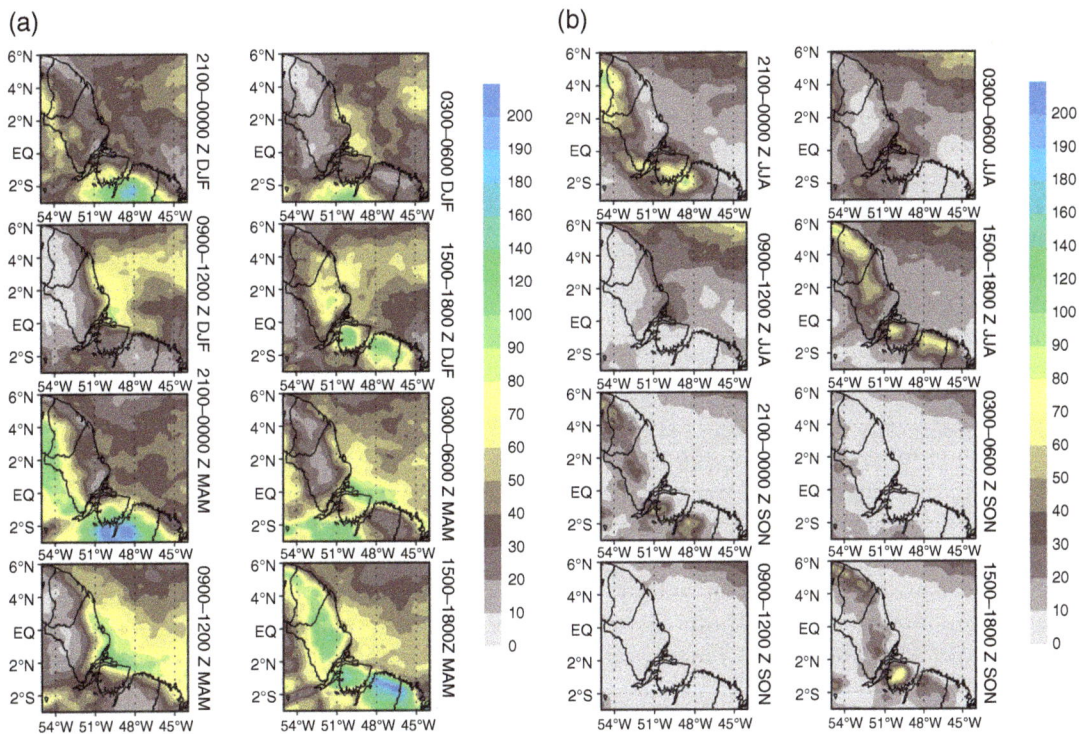

Figure 5. Precipitation patterns associated with the diurnal cycle in the intervals of 2100–0000 UTC, 0300–0600 UTC, 0900–1200 UTC, and 1500–1800 UTC for the seasonal periods of DJF, MAM, JJA, and SON, respectively in mm/3 h.

Shukla (1996), the ITCZ is most active over Northern and Northeastern Brazil during MAM. Due to the intense convection associated with the ITCZ during this period, the mean flow may suffer attenuation, reducing the intensity of the wind that enters the coast. Therefore, the LB shows greater intensity in this period due to the performance of large-scale weather systems such as the ITCZ.

It is observed that cities located closer to the coastline as Cuiarana and Salinópolis, has a predominance of winds from E/NE during most of the year, due to its geographical position, facilitating the SB initiation. The LB was more prominent in the period June–August (JJA) and MAM. On the other hand, in Soure it appears that an increased LB occurs during SON, whereas the LB becomes more evident during December–February (DJF) and MAM. As Belém, Castanhal, and Capitão Poço are located further inland, it is clear the SB is weaker than in the cities closer to the coastline. The LB in Castanhal and Capitão Poço are stronger during MAM, while we cannot identify the LB in Belém. In Belém, we also notice the occurrence of the RB during every month of the year, as well as the SB more intensely during JJA.

Figure 5(a) shows that the highest precipitation values are on the coast, in the period of 1500–1800 UTC. These maximums are related to the SB circulation. We can observe that during the hours of 0900–1200 UTC precipitation maximums are more localized on the ocean, inducing higher accumulates in located near the coastline as in Soure, Cuiarana, and Salinópolis. These results agree with Negri *et al.* (1994), who demonstrates

the interaction between the SB and LB causing maximum rainfall on the continent and ocean, respectively. During the months of MAM is more noticeable the influence of the LB in the rainfall over the ocean and the SB on the continent. As discussed previously, the LB is more frequent during the months of MAM, justifying the highest values found at the ocean in the period of 0900–2100 UTC. However, it should be noted that this pattern is repeated in the diurnal cycle in DJF when the ITCZ is located farther North. It is also possible to verify that the inland precipitation moves during the interval of 2100–0000 UTC, possibly associated with the displacement of the SB front and the squall lines (Kousky, 1980; Cohen *et al.*, 1995).

Figure 5(b) shows that the influence of the LB in the precipitation is not evident during the months of JJA and SON. However, we can still clearly see the influence of the SB during the period of 1500–1800 UTC both JJA as SON. In SON the maximum precipitation decrease, as these are the months dry period in the region, making it evident the effects of the SB in the precipitation since there are no large-scale weather systems in this period (Figueroa and Nobre, 1990; Rao *et al.*, 1996). It is noteworthy that the same propagation observed in Figure 5(a) is still observed in the months of SON and JJA. These results suggest that the SB circulation would be the main trigger in the formation of the Amazonian SLs, as hypothesized by Kousky (1980) and Cohen *et al.* (1995). However, it is known that the SBs usually occur every day, while the same does not occur with the SL, indicating that besides the SB there must be some other mechanism for the formation of the SLs.

4. Conclusion

The main goal of this study was to analyze the local circulations in Belém, Castanhal, Capitão Poço, Soure, Cuiarana, and Salinópolis and its influence on the diurnal cycle of precipitation. For this, we used rainfall data and horizontal wind (direction and speed) from the automatic stations located in these cities, as well as estimated rainfall through the CMORPH.

It was observed that the SB has a regular frequency in the NE/E quadrant. The LB was more intense in coastal regions during the hours of 0900–2100 UTC, moving in the same direction of the mean flow. On the other hand, the LB was observed only in Cuiarana, Salinópolis, Castanhal, and Capitão Poço in the quadrant SE/S, while in Soure, the LB was observed in the NW/N quadrant due to its geographical position. In general, the time of its performance takes place from 0000 to 0900 UTC. The SB was more intense during the SON, while the LB was observed more pronouncedly during MAM and JJA. The hourly rainfall was more related to the SB than the LB.

In contrast, Belém observed the presence of the RB in the NW/N quadrant, with time of occurrence from 1200 to 0000 UTC; this circulation has been regularly observed throughout the year. Although we can detect the signal of the SB in cities located farther inland, it is weaker compared to coastal regions.

The rainfall estimates through CMORPH confirm the influence of the SB circulation generating a cumulative rainfall during the period of 1500–1800 UTC on the coast and inland 2100–0000 UTC. Accumulated located more on the ocean due to the LB is also evidenced in the study, during 0900–1200 UTC.

Finally, despite the breezes are well studied, there are few studies that examined observationally the breezes circulations in the Amazon. Local circulations in the Amazon region contribute to the precipitation regime and are represented in the climatological normal, being relevant for the improvement of weather and climate models, and knowledge of the interactions between scales.

Acknowledgements

The authors gratefully acknowledge the financial support from PROPESP/UFPA for funding the scientific initiation scholarship and fees. The National Institute of Meteorology and the Large-Scale Biosphere-Atmosphere Experiment in Amazonia for the availability of meteorological data. This study was partially conducted during a visiting scholar period at University of Nevada – Reno, sponsored by the Capes Foundation within the Ministry of Education, Brazil.

References

Appelhans T, Sturman A, Zawar-Reza P. 2013. Synoptic and climatological controls of particulate matter pollution in a Southern Hemisphere coastal city. *International Journal of Climatology* 33(2): 463–479, doi: 10.1002/joc.3439.

Atkinson BW. 1981. *Meso-Scale Atmospheric Circulations*. Academic: San Diego, CA, 412 pp.

Baptista MC. 2003. Uma análise do campo de vento de superfície sobre o Oceano Atlântico Tropical e Sul usando dados do escaterômetro do ERS. MSc thesis, INPE, São José dos Campos, Brazil.

Cavalcanti IFA. 1982. Um estudo sobre interações entre sistemas de circulação de escala sinótica e circulações locais. MSc thesis. INPE-2494-TDL/097, São José dos Campos, Brazil.

Cohen JC, Silva Dias MA, Nobre CA. 1995. Environmental conditions associated with Amazonian squall lines: a case study. *Monthly Weather Review* 123(11): 3163–3174, doi: 10.1175/1520-0493(1995) 123<3163:ECAWAS>2.0.CO;2.

Cohen JCP, Fitzjarrald DR, D'Oliveira FAF, Saraiva I, Barbosa IRDS, Gandu AW, Kuhn PA. 2014. Radar-observed spatial and temporal rainfall variability near the Tapajós-Amazon confluence. *Revista Brasileira de Meteorologia* 29(SPE): 23–30, doi: 10.1590/0102-778620130058.

Costa GB, Lyra R. 2012. Análise dos padrões de vento no Estado de Alagoas. *Revista Brasileira de Meteorologia* 27(1): 31–38, doi: 10.1590/S0102-77862012000100004.

Figueroa SN, Nobre CA. 1990. Precipitation distribution over central and western tropical South America. *Climanalise* 5(6): 36–45.

Fitzjarrald DR, Sakai RK, Moraes OL, Cosme de Oliveira R, Acevedo OC, Czikowsky MJ, Beldini T. 2008. Spatial and temporal rainfall variability near the Amazon-Tapajós confluence. *Journal of Geophysical Research. Biogeosciences* 113(G1): G00B11, doi: 10.1029/2007JG000596.

Frizzola JA, Fisher EL. 1963. A series of sea breeze observations in the New York City area. *Journal of Applied Meteorology* 2(6): 722–739, doi: 10.1175/1520-0450(1963)002<0722:ASOSBO>2.0.CO;2.

Garstang M, Massie HL Jr, Halverson J, Greco S, Scala J. 1994. Amazon coastal squall lines. Part I. Structure and kinematics. *Monthly Weather Review* 122(4): 608–622, doi: 10.1175/1520-0493(1994)122<0608:ACSLPI>2.0.CO;2.

Greco S, Swap R, Garstang M, Ulanski S, Shipham M, Harriss RC, Talbot R, Andreae MO, Artaxo P. 1990. Rainfall and surface kinematic conditions over central Amazonia during ABLE 2B. *Journal of Geophysical Research. Atmospheres* 95(D10): 17001–17014, doi: 10.1029/JD095iD10p17001.

Greco S, Ulanski S, Garstang M, Houston S. 1992. Low-level nocturnal wind maximum over the central Amazon basin. *Boundary-Layer Meteorology* 58(1–2): 91–115, doi: 10.1007/BF00120753.

Haurwitz B. 1947. Comments on the sea-breeze circulation. *Journal of Meteorology* 4(1): 1–8, doi: 10.1175/1520-0469(1947)004<0001:COTSBC>2.0.CO;2.

Janowiak JE, Kousky VE, Joyce RJ. 2005. Diurnal cycle of precipitation determined from the CMORPH high spatial and temporal resolution global precipitation analyses. *Journal of Geophysical Research. Atmospheres* 110(D23): D23105, doi: 10.1029/2005JD006156.

Joyce RJ, Janowiak JE, Arkin PA, Xie P. 2004. CMORPH: a method that produces global precipitation estimates from passive microwave and infrared data at high spatial and temporal resolution. *Journal of Hydrometeorology* 5: 487–503, doi: 10.1175/1525-7541(2004)005<0487:CAMTPG>2.0.CO;2.

Kousky VE. 1980. Diurnal rainfall variation in northeast Brazil. *Monthly Weather Review* 108(4): 488–498.

Lu L, Denning AS, da Silva-Dias MA, da Silva-Dias P, Longo M, Freitas SR, Saatchi S. 2005. Mesoscale circulations and atmospheric CO_2 variations in the Tapajós Region, Pará, Brazil. *Journal of Geophysical Research. Atmospheres* 110(D21): D21102, doi: 10.1029/2004JD005757.

Maddox RA. 1980. Mesoscale convective complex. *Bulletin of the American Meteorological Society* 61: 1374–1387, doi: 10.1175/1520-0477(1980)061<1374:MCC>2.0.CO;2.

Matos AP, Cohen JCP. 2015. Circulação de brisa e a banda de precipitação na margem leste da baía de Marajó. *Ciência e Natura* 38: 21–27.

Negri AJ, Adler RF, Nelkin EJ, Huffman GJ. 1994. Regional rainfall climatologies derived from Special Sensor Microwave Imager (SSM/I) data. *Bulletin of the American Meteorological Society* 75(7): 1165–1182, doi: 10.1175/1520-0477(1994)075<1165:RRCDFS>2.0.CO;2.

Nobre P, Shukla J. 1996. Variations of sea surface temperature, wind stress, and rainfall over the tropical Atlantic and South America. *Journal of Climate* **9**(10): 2464–2479, doi: 10.1175/1520-0442(1996) 009<2464:VOSSTW>2.0.CO;2.

Oliveira AP, Fitzjarrald DR. 1993. The Amazon river breeze and the local boundary layer: I. Observations. *Boundary-Layer Meteorology* **63**(1): 141–162, doi: 10.1007/BF00705380.

Pearce RP. 1955. The calculation of a sea-breeze circulation in terms of the differential heating across the coastline. *Quarterly Journal of the Royal Meteorological Society* **81**(349): 351–381, doi: 10.1002/qj.49708134906.

Ramos da Silva R, Gandu AW, Sá LD, Dias MAS. 2011. Cloud streets and land–water interactions in the Amazon. *Biogeochemistry* **105**(1–3): 201–211, doi: 10.1007/s10533-011-9580-4.

Rao VB, Cavalcanti IF, Hada K. 1996. Annual variation of rainfall over Brazil and water vapor characteristics over South America. *Journal of Geophysical Research. Atmospheres* **101**(D21): 26539–26551, doi: 10.1029/96JD01936.

Santos SRQ, Vitorino MI, Braga CC, Campos TB, Santos AP. 2013. O efeito de brisas maritimas na Cidade de Belém-PA: utilizando análise em Multivariada (The effect of sea breeze over Belém-PA: using multivariate analysis). *Revista Brasileira de Geografia Física* **5**(5): 1110–1120.

Santos MJ, Silva Dias MA, Freitas ED. 2014. Influence of local circulations on wind, moisture, and precipitation close to Manaus City, Amazon Region, Brazil. *Journal of Geophysical Research. Atmospheres* **119**(23): 13,233–13,249, doi: 10.1002/2014JD021969.

Silva Dias MAF, Dias PS, Longo M, Fitzjarrald DR, Denning AS. 2004. River breeze circulation in eastern Amazonia: observations and modelling results. *Theoretical and Applied Climatology* **78**(1–3): 111–121, doi: 10.1007/s00704-004-0047-6.

Teixeira RFB. 2008. O fenômeno da brisa e sua relação com a chuva sobre Fortaleza-CE. *Revista Brasileira de Meteorologia* **23**(3): 282–291, doi: 10.1590/S0102-77862008000300003.

Uvo CRB. 1989. A Zona de Convergência Intertropical (ZCIT) e sua relação com a precipitação da Região Norte do Nordeste Brasileiro. MSc thesis. INPE, São José dos Campos, Brazil.

Vitorino MI, Silva MES, Alves JMB. 1997. Classificação de sistemas convectivos de mesoescala no setor norte do Nordeste brasileiro. *Revista Brasileira de Meteorologia* **12**(1): 21–32.

27

A 31-year trend of the hourly precipitation over South China and the underlying mechanisms

Shenming Fu,[1] Deshuai Li,[2] Jianhua Sun,[3] Dong Si,[4] Jian Ling[5],* and Fuyou Tian[4]

[1]Institute of Atmospheric Physics, Chinese Academy of Sciences, International Center for Climate and Environment Sciences, Beijing, China
[2]College of Atmospheric Sciences, Lanzhou University, China
[3]Institute of Atmospheric Physics, Chinese Academy of Sciences, Laboratory of Cloud–Precipitation Physics and Severe Storms, Beijing, China
[4]China Meteorological Administration, National Climate Center, Beijing, China
[5]Institute of Atmospheric Physics, Chinese Academy of Sciences, LASG, Beijing, China

*Correspondence to:
J. Ling, LASG, Institute of
Atmospheric Physics, Chinese
Academy of Sciences, 40
Huayanli, Chaoyang District,
Beijing 100029, China.
E-mail: lingjian@lasg.iap.ac.cn

Abstract

On the basis of a newly developed intensive hourly observational precipitation dataset, the precipitation trend of South China was investigated. Results indicate that the hourly precipitation over South China featured a significant increasing trend, particularly for the extreme precipitation category. The trend is mainly due to the increasing frequency of the precipitation events. A possible mechanism accounting for this increasing trend was proposed: the global warming may be the original forcing for the trend, through its modulation on activities of the western Pacific subtropical high, but the low-level vorticity is the most important direct trigger.

Keywords: hourly precipitation; vorticity budget; western Pacific subtropical high; rotated empirical orthogonal function (REOF)

1. Introduction

Heavy rainfall events are one of the most severe disastrous weathers worldwide (Tao, 1980; Karl and Knight, 1998; Zhou et al., 2008; Grimm and Tedeschi, 2009; Hitchens et al., 2013; Stevenson and Schumacher, 2014). Every year, the flash flood, debris flow and urban waterlogging triggered by heavy rainfall events result in substantial casualties, great economic losses and other large social consequences (Tao and Ding, 1981; Karl and Knight, 1998; Zhao et al., 2004; Ramos et al., 2014). Moreover, as a key component of the hydrological cycle, the precipitation is vital in determining the distribution of the water resource that participates in many physical, chemical and biogeological processes of the Earth system. Therefore, clarifying the trend of the precipitation and revealing the main mechanisms accounting for the variation of precipitation are very important in the recent climate research (Karl and Knight, 1998; Buffoni et al., 1999; Black, 2009; Chen et al., 2009; Yu and Li, 2012; Ramos et al., 2014).

The South China, which is under the influences of the East Asian summer monsoon, the Indian Monsoon, the western Pacific subtropical high, and the dynamical/thermodynamical effects of the Tibetan Plateau, is one of the most famous and important rainy regions in Asia (Tao, 1980; Zhao et al., 2004; Zhou et al., 2008). In recent years, serious floods caused by torrential rainfall events occurred frequently in South China (e.g. the devastating floods in 1982, 1994, 1998, 2005 and 2008). Furthermore, many studies (Qian and Lin, 2005; Zhai et al., 2005; Ren et al., 2006; Fischer et al., 2012) reveal that, under the background of the global climate change, the precipitation over South China show significant changes, which may render a potential increasing flood risk. Possible mechanisms accounting for the variation of precipitation were discussed, mainly focused on the El Niño Southern Oscillation (ENSO) and the Pacific Decadal Oscillation (Chan and Zhou, 2005), the tropical cyclone activities (Ren et al., 2006), the variation of the East Asian summer monsoon (Ding et al., 2008), the surface flux over the Indochina Peninsula and the South China Sea (Liang and Qian, 2009), the surface air temperature (Yu and Li, 2012), as well as the remote impacts of the Arctic Oscillation (Li and Leung, 2013). It should be noted that, although many factors were found to be capable of influencing/determining such variation over South China, thus far, no comparisons have been conducted among these factors. Moreover, most of the previous studies were based on sparse station observation with low temporal resolution (6-h or daily) and reanalysis data with coarse horizontal resolution ($2.5° \times 2.5°$), thus, rich features of the hourly precipitation cannot be revealed in detail. Therefore, this study intends to evaluate the trend of the hourly precipitation over South China more thoroughly using a new intensive hourly station observational rainfall dataset, and to determine the main factors accounting for the variation of hourly precipitation.

Data and methods are presented in the next section, main results are shown in Section 3, and finally a conclusion is provided in Section 4.

2. Data and methods

In this study, the hourly observational precipitation data at 2420 stations (Figure 1(a)) of the China Meteorological Administration (CMA) from 1982 to 2012 were used to determine the rainy region of South China in the warm season (from May to September) objectively, and also to calculate the trend of the hourly precipitation. Six-hourly European Centre for Medium-range Weather Forecasts Interim reanalysis (ERA-I) with a horizontal resolution of $0.75° × 0.75°$ (Simmons *et al.*, 2007) was used for evaluating the relative importance of various factors to the rainfall event. The Rotated Empirical Orthogonal Function (REOF) analysis (Kaiser, 1958), the correlation analysis and the vorticity budget are the main research methods. The Student's *t*-test and the Mann–Kendall trend test (Mann, 1945; Kendall, 1975) were used to verify the significance of the correlation and the trend, respectively. The vorticity budget equation (Kirk, 2003) used in this study is as follows:

$$\frac{\partial \zeta}{\partial t} = \text{HAV} + \text{VAV} + \text{TIL} + \text{STR} + \text{AF} \quad (1)$$

where ζ is the vertical vorticity, HAV represents the horizontal advection of vorticity, VAV denotes the vertical advection of vorticity, TIL stands for the titling effect, STR represents the stretching effect and AF denotes the advection of the planetary vorticity (refer Appendix A for detailed information).

3. Results

3.1. The trend of hourly precipitation over South China

In this study, the REOF analysis (Horel, 1982) was conducted to detect the similar localized modes of the hourly precipitation objectively. The data scarcity of each station that was used in the REOF analysis was confined to be less than 2%. Thus, out of the 2420 stations shown in Figure 1(a), only 1095 meet the above criteria, which distribute homogenously in the middle, east and south China (not shown). As reported by Kaiser (1958), eigenvalue separations were used to test the number of region divisions, and their corresponding maximum loading vectors were used to determine the climate divisions. A total of ten spatial patterns were determined over central eastern China, with their accumulated variance contribution exceeding 60% (Li *et al.*, 2015). The rainy area over South China can be well represented by the second leading mode of the REOF analysis (Figure 1(b)) which contributes 7.84% to the total variation of precipitation over China. The blue dashed rectangle in Figure 1(b) was defined as the area of the South China, which was used to calculate the averaged value for each potential factor that may influence the precipitation over South China, based on the ERA-I reanalysis data. This region contains 99 stations; therefore, the mean hourly precipitation over South China was defined as the averaged value of the hourly precipitation of all these stations.

As reported by Karl and Knight (1998), changes in precipitation amount are determined by the change in the frequency of occurrence of precipitation events (the frequency factor) and the change in the intensity of precipitation per event (the intensity factor). In order to evaluate the precipitation trend thoroughly, the hourly precipitation intensity (mm h^{-1}) was classified into ten categories. All the hourly precipitation observed in 99 stations over South China was firstly sorted from lowest to highest value and then equally divided into ten categories. That is, Category 1 represents the first 10 percentile of hourly precipitation intensity corresponding to the very light precipitation, and Category 10 stands for the last 10 percentile of hourly precipitation intensity corresponding to very intense precipitation events. The trends of these ten categories were calculated following Karl and Knight (1998), and an increasing trend was detected for all categories. The hourly precipitation in categories from five to ten accounts for up to 96.9% of the total precipitation amount during the warm season (Table 1). Therefore, the variation of the precipitation over South China was dominated by these categories, and their trends are discussed in detail (Figure 2(a)).

As Figure 2(a) illustrates, all the categories from five to ten featured an increasing trend, which is dominated by the frequency factor. This means that the increasing trend of the hourly precipitation over South China was mainly because of the growth in the number of the precipitation events, whereas the intensity of the precipitation event changed slightly, except for the extreme precipitation category. The time series of the extreme precipitation events over South China during the last 31 years are shown in Figure 2(b). It clearly shows the remarkable increasing trend exceeding the 95% confidence level. It is obvious that, the category with larger intensity of the precipitation has larger increasing trend. For the extreme precipitation category (the Category 10), which accounted for about 57.1% of the total precipitation amount and generally featured a mean precipitation rate of 14.73 mm h^{-1} (Table 1), has the largest increasing trend as well as its corresponding frequency and intensity factors. Moreover, the intensity factor account for about 10% of the total increasing trend (Figure 2(a)). It suggests that the extreme hourly precipitation events over South China featured a significant growth in their total number and a weak increasing in their intensity, both of which may increase the flood risk over South China remarkably.

3.2. The dominant factors for the hourly precipitation over South China

In order to determine the dominant factors accounting for the variation of precipitation, the factors proven to be capable of influencing/dominating the rainfall events over South China through detailed case studies (Tao, 1980; Zhao *et al.*, 2004; Xia *et al.*, 2006; Xia and

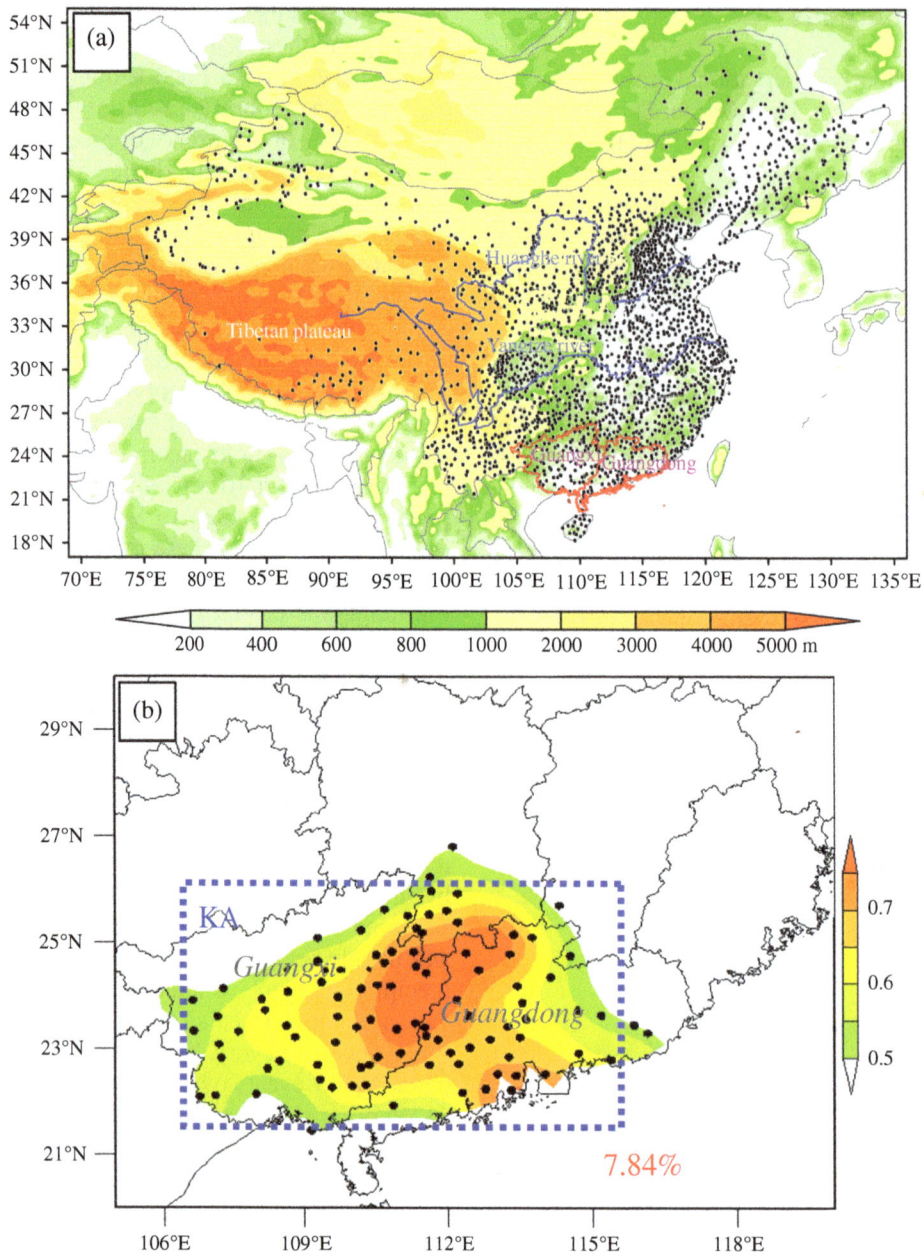

Figure 1. (a) Stations with hourly precipitation observation (black dots) and the terrain characteristics (shaded, units: m). (b) The forth leading mode derived from hourly precipitation during the period of 1982–2012 (the shaded illustrates the value of the loading vector). The blue rectangle represents the area of South China used in this study.

Table 1. The contribution rate (CR, %) of each category to the total precipitation amount in the warm season, and the averaged intensity (AI, mm h^{-1}) of each category.

Category	5	6	7	8	9	10	5–10
CR (%)	2.1	3.5	5.8	9.9	18.5	57.1	96.9
AI (mm h^{-1})	0.6	0.95	1.54	2.59	4.83	14.73	4.21

Zhao, 2009; Zhao and Wang, 2009; Fu *et al.*, 2010) had been evaluated using the 6-h ERA-I reanalysis data. Because the time interval of the ERA-I analysis was 6 h, the hourly precipitation at 0000, 0600, 1200 and 1800 UTC was used to calculate the correlation with each possible factor.

In this study, 20 factors suggested by previous studies were evaluated, including vertical velocity (VV), vorticity (VOR), vertical helicity (HEL), front/baroclinity (BAR) denoted by $|\nabla_h T|$, where ∇_h is the horizontal gradient operator and TEM is the temperature, divergence of the moisture transport (DMT), temperature (TEM), temperature advection (TA), baroclinic conversion term between the available potential energy and kinetic energy (Lorenz, 1955), convective instability, convective available potential energy, thickness between 925 and 500 hPa, specific humidity, entire atmosphere precipitable water, potential vorticity, surface pressure, zonal wind, meridional wind, horizontal wind velocity, surface latent heat flux and surface sensible heat flux. It should be noted that, all

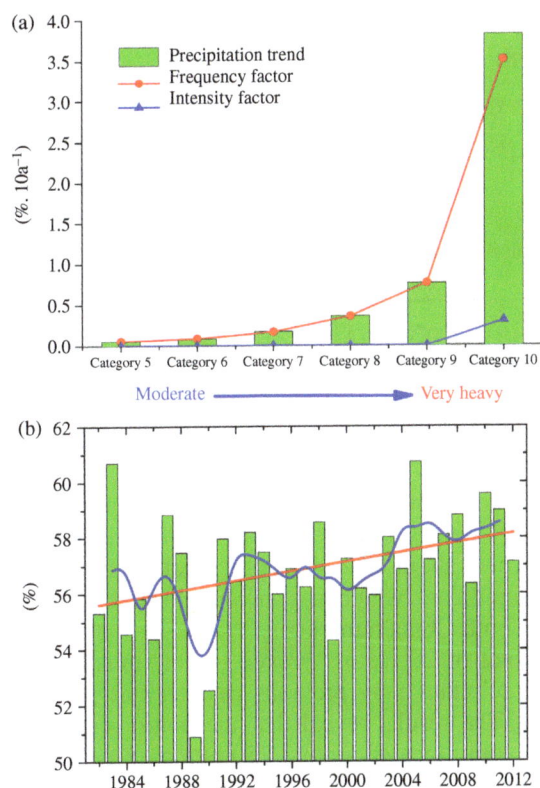

Figure 2. (a) Precipitation trend of each category (10^{-1}% year^{-1}), with their corresponding contribution from frequency and intensity factor respectively. (b) Time series of the contribution (%) of the hourly precipitation in Category 10 to the total precipitation over South China, where the blue curve is the smoothed curve (using a 9-point binomial filter), and the red line is its linear trend.

the three-dimensional variables were calculated at ten successive vertical levels from 925 to 500 hPa, and the convective instability were calculated between 925 and 500 hPa, 900 and 700 hPa, 700 and 500 hPa, 900 and 800 hPa, as well as 800 and 700 hPa, respectively.

The maximum correlation coefficient of each factor was identified (for the three-dimensional variable, the correlation coefficients is different for different level), and then sorted from large to small. In this study, only eight factors can be regarded as in good correlation with the hourly precipitation passing the significant test at 95% level. They are the seven factors shown in Figure 3, and the convective instability calculated between 900 and 700 hPa with a correlation coefficient of −0.215. All other factors featured correlation coefficients below 0.11, and thus were not discussed in this study. Among all the eight factors, VOR has the maximum correlation coefficient among the levels between 925 and 600 hPa (Figure. 3), whereas VV has the largest correlation at 550 hPa, slightly greater than that of VOR. Thus, overall, VOR was the most important factor that impacts the hourly precipitation over South China. As Figure 3 illustrates, DMT at 900 hPa ranks the third place (−0.245). DMT was in negative correlation with the precipitation below 750 hPa, implying the importance of low-level moisture convergence to the precipitation; whereas positive correlation

appeared above 750 hPa, indicating the importance of divergence at higher levels to the precipitation. In addition, the non-divergent level was mainly located around 750 hPa. The convective instability, HEL, TA, TEM and BAR rank from the forth to eighth place respectively. It reveals that convective unstable layer, positive helicity, warm TA, cold low-level temperature and strong baroclinity are generally conducive to the precipitation over South China.

The maximum correlation (0.442) between VOR and precipitation is at 850 hPa, therefore, the low-level vorticity, that is closely related to the shear lines (Zhao et al., 2004), the quasi-stationary front (Tao, 1980; Xia and Zhao, 2009), the mesoscale vortices (Tao, 1980; Fu et al., 2010) and the tropical disturbance/cyclones (Zhao et al., 2004) is vital in triggering the precipitation over South China. Because the variation of VOR is governed by the vorticity budget equation, the correlation coefficients between each vorticity budget term on the right hand side of the Equation (1) and VOR were evaluated. It indicates that STR was highly correlated with VOR, with a correlation coefficient up to 0.636, which means that the convergence was the dominant factor for the maintenance of positive vorticity at lower level. HAV ranks the second place (−0.259), implying HAV is the most detrimental factor consuming the positive vorticity. It should be noted that, the above results were also confirmed by many case studies (Zhao et al., 2004; Fu et al., 2010).

3.3. Mechanisms of the trend of hourly precipitation

In order to explore the hourly precipitation trend over South China, correlation between the 850-hPa VOR and hourly precipitation as well as the variance of the 850-hPa VOR were calculated during the warm season of each year (613 samples per year). As Figure 4(a) illustrates, every year, VOR was generally well correlated with the hourly precipitation, particularly for the year 1982 (0.511), 1994 (0.529), 1998 (0.521), 2002 (0.539), 2006 (0.515), 2008 (0.516) and 2009 (0.548), whereas 1991 features the smallest correlation coefficient (0.231), when a severe drought appeared over South China. The annual correlation coefficient between VOR and the hourly precipitation features an increasing trend (exceeds the 90% confidence level), which means that the low-level vorticity may become more important in the rainfall events over South China. The annual variance of VOR generally featured an increasing trend (Figure 4(b)), exceeding the 95% confidence level. This implies that, the weather systems directly triggered the rainfall events (e.g. the synoptical, subsynoptical, mesoscale weather systems, etc.) generally became more frequent and/or stronger. Therefore, the vorticity related perturbation circulations generally became more favorable for the rainfall events over South China. It might be the reason for the increasing trend of hourly precipitation there. Moreover, from Figure 4(a) and (b), it is obvious that, overall, the

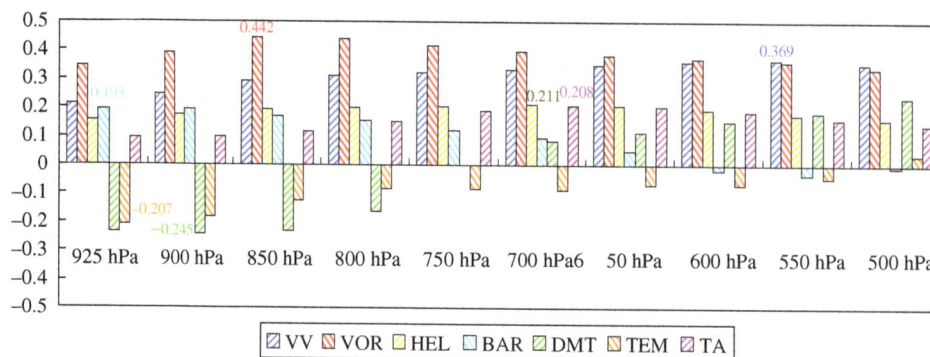

Figure 3. Correlation between the mean hourly precipitation and the seven factors averaged over South China at different levels. The maximum correlation coefficient of each factor is marked with corresponding color.

stronger the annual variance of VOR, the larger the correlation coefficient (their correlation coefficient is 0.48), which also confirms that, the vorticity related perturbation circulations were very important in triggering the rainfall events over South China.

From Figure 4(c), the longitude of westernmost ridge point of the western Pacific subtropical high features a decreasing trend (exceeds the 90% confidence level), implying that the subtropical high generally stretched more westward and thus influenced wider areas over South China. Therefore, the circulations associated with the subtropical high became more favorable for the divergence around 500 hPa over South China, which can be confirmed by the significant increasing trend (exceeding the 90% confidence level) of averaged divergence at 500 hPa over South China (Figure 4(d)). Because the non-divergent level was mainly located around 750 hPa (Figure 3), according to the continuity equation, the stronger divergence around 500 hPa favors stronger convergence at lower levels. This result can be confirmed by the correlation coefficient between the divergence at 500 and 850 hPa for 31 years with a value of −0.38 (above the 95% confidence level). Because STR dominated the variation of vorticity (Section 3.2), stronger lower-level convergence were more conducive to the occurrence/enhancement of positive vorticity related anomalous circulation, therefore, the low-level VOR's variance shows an increasing trend (Figure 4(b)), which may account for the increasing trend of the precipitation over South China (Figure 3).

4. Conclusion and discussion

In this study, on the basis of a new intensive hourly observational precipitation dataset during the warm seasons of 1982–2012, the rainy region over South China was determined objectively using REOF analysis. The trend of the hourly precipitation over South China was investigated in detail by calculating both frequency and intensity factors. A significant increasing trend was detected in the hourly precipitation over South China, particularly for the extreme precipitation category, which accounts for about 57.1% of the total precipitation. Generally, the increasing trend was

due to the growth in the number of the precipitation events; whereas for the extreme precipitation events, an increase of about 10% in the precipitation intensity was also significant, which may increase the risk of flooding over South China remarkably.

A total of 20 factors proven to determine/influence the precipitation by previous case studies were evaluated. The results reveal that the non-divergent level was mainly located around 750 hPa over South China and the low-level vorticity was the key factor influencing the hourly precipitation. The low-level shear lines, quasi-stationary front, mesoscale vortices and tropical disturbance/cyclones are vital in triggering the precipitation over South China directly. The vorticity budget was used to understand the variation of VOR, and STR which is determined by the divergence was found to be the dominant factor.

A possible mechanism accounting for the increasing trend of the hourly precipitation over South China is proposed. The western Pacific subtropical high stretched more westward during the last 31 years, which favored the divergence at higher levels over South China. Thus, low-level convergence was enhanced, which rendered an increasing trend of VOR in the lower level through term STR. This increasing trend in the low-level vorticity was corresponding to the increasing and/or intensifying of the weather systems that triggered the precipitation directly. Thus the precipitation over South China shows an increasing trend.

Recently, He and Zhou (2015) proposed that, the change in the zonal sea surface temperature gradient between the tropical Indian Ocean and the tropical western Pacific which is closely related to the global warming, dominated the variation of the intensity of the western Pacific subtropic high. Therefore, the global warming may be the original forcing for the increasing trend of hourly precipitation over South China through its modulation on activities of the western Pacific subtropical high.

Acknowledgements

The authors thank the European Centre for Medium-range Weather Forecasts and China Meteorological Administration

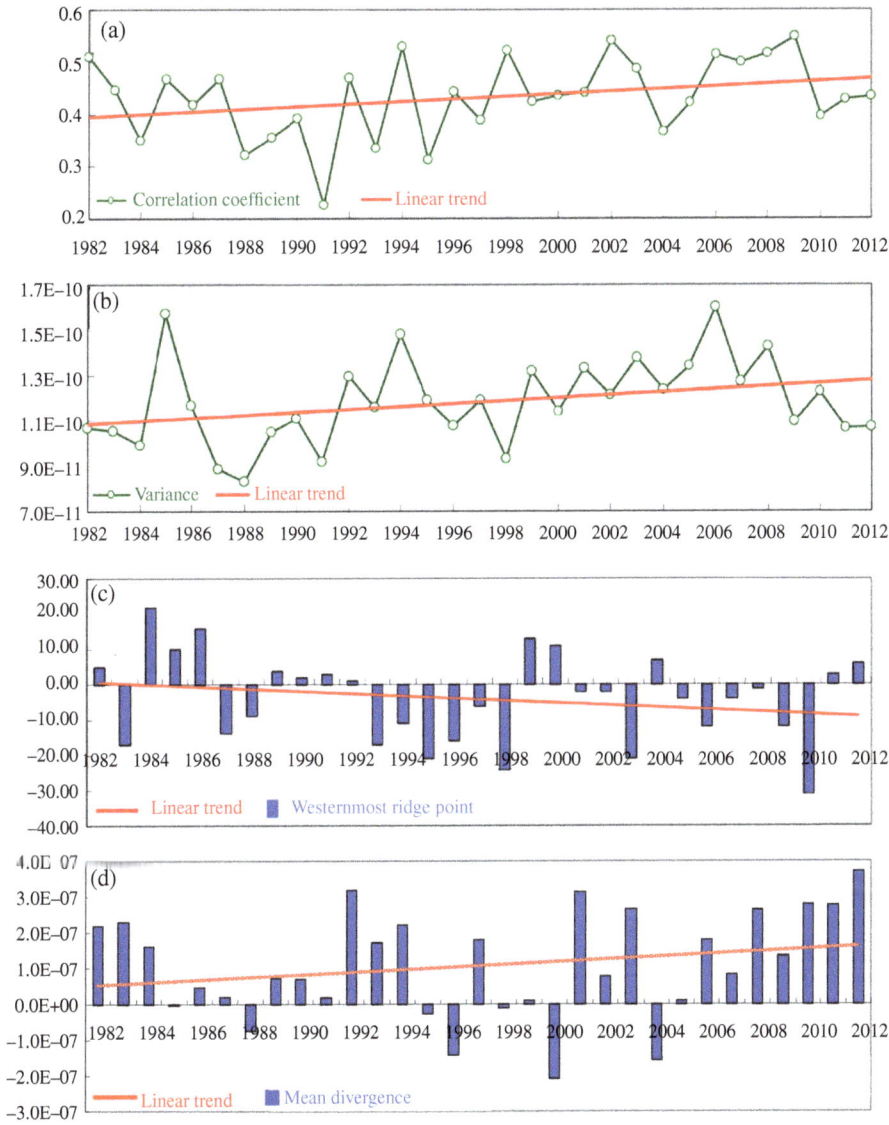

Figure 4. (a) Annual correlation between the averaged VOR at 850 hPa and the mean hourly precipitation over South China; (b) the annual variance of the vorticity at 850 hPa over South China (units: s^{-2}); (c) the anomaly of the westernmost ridge point of the western Pacific subtropical high; and (d) the annual mean divergence at 500 hPa (units: s^{-1}) over South China.

for providing the data. The authors also thank Editor Yan Yin and two anonymous reviewers for their valuable suggestions. This research was supported by the National Key Basic Research and Development Project of China (2012CB417201 and 2015CB453200) and the National Natural Science Foundation of China (41205027, 41375053 and 41575062).

where ζ is the vertical vorticity; $\mathbf{V}_h = u\mathbf{i} + v\mathbf{j}$ is the horizontal velocity vector, $\mathbf{i}, \mathbf{j}, \mathbf{k}$ stand for the unit vector points to the east, north and zenith respectively; $\nabla_h = \frac{\partial}{\partial x}\mathbf{i} + \frac{\partial}{\partial y}\mathbf{j}$ is the horizontal gradient operator; f is the Coriolis parameter; p is the pressure; $\omega = dp/dt$ and $\beta = \partial f/\partial y$.

Appendix A: The vorticity budget equation

The vorticity budget equation is following Kirk (2003):

$$\frac{\partial \zeta}{\partial t} = -\mathbf{V}_h \cdot \nabla_h \zeta - \omega\frac{\partial \zeta}{\partial p} + \mathbf{k} \cdot \left(\frac{\partial \mathbf{V}_h}{\partial p} \times \nabla_h \omega\right) - (\zeta + f)\nabla_h \cdot \mathbf{V}_h - \beta v$$

$$\text{HAV} \qquad \text{VAV} \qquad\qquad \text{TIL} \qquad\qquad \text{STR} \qquad\qquad \text{AF}$$

References

Black E. 2009. The impact of climate change on daily precipitation statistics in Jordan and Israel. *Atmospheric Science Letters* **10**: 192–200, doi: 10.1002/asl.233.

Buffoni L, Maugeri M, Nanni T. 1999. Precipitation in Italy from 1833 to 1996. *Theoretical and Applied Climatology* **63**: 33–40.

Chan JCL, Zhou W. 2005. PDO, ENSO and the early summer monsoon rainfall over south China. *Geophysical Research Letters* **32**: L08810, doi: 10.1029/2004GL022015.

Chen H, Zhou T, Yu R, Li J. 2009. Summer rain fall duration and its diurnal cycle over the US Great Plains. *International Journal of Climatology* **29**: 1515–1519.

Ding Y, Wang Z, Sun Y. 2008. Inter-decadal variation of the summer precipitation in East China and its association with decreasing Asian summer monsoon. Part I: observed evidences. *International Journal of Climatology* **28**: 1139–1161.

Fischer T, Gemmer M, Liu L, Su B. 2012. Change-points in climate extremes in the Zhujiang River Basin, South China, 1961–2007. *Climatic Change* **110**: 783–799, doi: 10.1007/s10584-011-0123-8.

Fu SM, Zhao SX, Sun JH, Li WL. 2010. One kind of vortex causing heavy rainfall during pre-rainy season in South China. *Chinese Journal of Atmospheric Sciences* **34**: 235–252, doi: 10.3878/j.issn.1006-9895.2010.02.01.

Grimm AM, Tedeschi RG. 2009. ENSO and extreme rainfall events in South America. *Journal of Climate* **22**: 1589–1609.

He C, Zhou TJ. 2015. Responses of the western North Pacific Subtropical High to global warming under RCP4.5 and RCP8.5 scenarios projected by 33 CMIP5 models: the dominance of tropical Indian Ocean – tropical western Pacific SST gradient. *Journal of Climate* **28**: 365–380, doi: 10.1175/JCLI-D-13-00494.1.

Hitchens NM, Brooks HE, Schumacher RS. 2013. Spatial and temporal characteristics of heavy hourly rainfall in the United States. *Monthly Weather Review* **141**: 4564–4575, doi: 10.1175/MWR-D-12-00297.1.

Horel JD. 1982. A rotated principal component analysis of the interannual variability of the Northern Hemisphere 500 mb height field. *Monthly Weather Review* **109**: 2080–2092.

Kaiser HF. 1958. The varimax criterion for analytic rotation in factor analysis. *Psychometrika* **23**: 187–200.

Karl TR, Knight RW. 1998. Secular trends of precipitation amount, frequency, and intensity in the United States. *Bulletin of the American Meteorological Society* **79**: 231–241.

Kendall MG. 1975. *Rank Correlation Methods*. Charles Grifin Company: London.

Kirk JR. 2003. Comparing the dynamical development of two mesoscale convective vortices. *Monthly Weather Review* **131**: 862–890.

Li YF, Leung LR. 2013. Potential impacts of the Arctic on interannual and interdecadal summer precipitation over China. *Journal of Climate* **26**: 899–917, doi: 10.1175/JCLI-D-12-00075.1.

Li DS, Sun JH, Fu SM, Wei J, Wang SG, Tian FY. 2015. Spatiotemporal characteristics of hourly precipitation over central eastern China during the warm season of 1982–2012. *International Journal of Climatology*, doi: 10.1002/joc.4543.

Liang N, Qian YF. 2009. Interdecadal change in extreme precipitation over South China and its mechanism. *Advances in Atmospheric Sciences* **26**: 109–118.

Lorenz EN. 1955. Available potential energy and the maintenance of the general circulation. *Tellus* **7**: 157–167.

Mann HB. 1945. Nonparametric tests against trend. *Econometrica* **13**: 245–259, doi: 10.2307/1907187.

Qian W, Lin X. 2005. Regional trends in recent precipitation indices in China. *Meteorology and Atmospheric Physics* **90**: 193–207.

Ramos AM, Trigo RM, Liberato MLR. 2014. A ranking of high-resolution daily precipitation extreme events for the Iberian Peninsula. *Atmospheric Science Letters* **15**: 328–334, doi: 10.1002/asl2.507.

Ren F, Wu G, Dong W, Wang X, Wang Y, Ai W, Li W. 2006. Changes in tropical cyclone precipitation over China. *Geophysical Research Letters* **33**: L20702, doi: 10.1029/2006GL027951.

Simmons A, Uppala S, Dee D, Kobayashi S. 2007. ERA-Interim: new ECMWF reanalysis products from 1989 on-wards. ECMWF Newsletter No. 110, ECMWF, Reading, UK, 25–35.

Stevenson SN, Schumacher RS. 2014. A 10-year survey of extreme rainfall events in the Central and Eastern United States using gridded multisensor precipitation analyses. *Monthly Weather Review* **142**: 3147–3162.

Tao SY. 1980. *Rainstorms in China*. Science Press: Beijing.

Tao SY, Ding YH. 1981. Observational evidence of the influence of the Qinghai–Xizang (Tibet) Plateau on the occurrence of heavy rain and severe convective storms in China. *Bulletin of the American Meteorological Society* **62**: 23–30.

Xia RD, Zhao SX. 2009. Diagnosis and modeling of meso-β-scale systems of heavy rainfall in warm sector ahead of front in South China (Middle part of Guangdong Province) in June 2005. *Chinese Journal of Atmospheric Sciences* **33**: 468–488, doi: 10.3878/j.issn.1006-9895.2009.03.06.

Xia RD, Zhao SX, Sun JH. 2006. A study of circumstances of meso-β-scale systems of strong heavy rainfall in warm sector ahead of fronts in South China. *Chinese Journal of Atmospheric Sciences* **30**: 988–1008, doi: 10.3878/j.issn.1006-9895.2006.05.26.

Yu RC, Li J. 2012. Hourly rainfall changes in response to surface air temperature over eastern contiguous China. *Journal of Climate* **25**: 6851–6861, doi: 10.1175/JCLI-D-11-00656.1.

Zhai PM, Zhang X, Wan H, Pan X. 2005. Trends in total precipitation and frequency of daily precipitation extremes over China. *Journal of Climate* **18**: 1096–1108.

Zhao YC, Wang YH. 2009. A review of studies on torrential rain during pre-summer flood season in South China since the 1980's. *Torrential Rain and Disasters* **28**: 193–202.

Zhao SX, Tao ZY, Sun JH, Bei NF. 2004. *Study on Mechanism of Formation and Development of Heavy Rainfalls on Meiyu Front in Yangtze River*. China Meteorological Press: Beijing.

Zhou TJ, Yu RC, Chen HM, Dai AG, Pan Y. 2008. Summer precipitation frequency, intensity, and diurnal cycle over China: a comparison of satellite data with rain gauge observations. *Journal of Climate* **21**: 3997–4010.

Permissions

List of Contributors

Ting-Chi Wu and Milija Zupanski
Cooperative Institute for Research in the Atmosphere, Colorado State University, Fort Collins, CO, USA

Kuo Zhou
College of Atmospheric Sciences, Chengdu University of Information Technology, China

Haiwen Liu
LASG, Institute of Atmospheric Physics, Chinese Academy of Sciences, Beijing, China
Department of Aviation Meteorology, Civil Aviation University of China, Tianjin, China

Liang Zhao and Yihua Lin
LASG, Institute of Atmospheric Physics, Chinese Academy of Sciences, Beijing, China

Yuxiang Zhu
CMA Training Center, China Meteorological Administration, Beijing, China

Fuying Zhang
College of Atmospheric Sciences, Nanjing University of Information Science and Technology, Nanjing, China

Ning Fu
Department of Aviation Meteorology, Civil Aviation University of China, Tianjin, China

Yulong Bai, Zhuanhua Zhang, Dehui Chen, Hongjun Bao and Lili Wang
College of Physics and Electrical Engineering, Northwest Normal University, Lanzhou, China
National Meteorological Centre, China Meteorological Administration, Beijing, P. R. China
School of Science, Nanchang Institute of Technology, China

Yanli Zhang
College of Geography and Environment Science, Northwest Normal University, Lanzhou, China

Unnikrishnan C.K., Saji Mohandas, Ashu Mamgain, E. N. Rajagopal and Gopal R. Iyengar
ESSO, MoES, National Centre for Medium Range Weather Forecasting, Noida, India

Biswadip Gharai and P. V. N. Rao
Atmospheric and Climate Sciences Group, Earth & Climate Science Area, National Remote Sensing Centre, ISRO, Hyderabad, India

Elisabeth Callen
Department of Geological and Atmospheric Sciences, Iowa State University, Ames, IA, USA

Donna F. Tucker
Department of Geography and Atmospheric Science, University of Kansas, Lawrence, KS, USA

Cyrille Meukaleuni, André Lenouo and David Monkam
Department of Physics, Faculty of Science, University of Douala, Cameroon

Hyacinth C. Nnamchi
Department of Geography, University of Nigeria, Nsukka, Nigeria

Fred Kucharski
Earth System Physics Section, The Abdus Salam International Centre for Theoretical Physics, Trieste, Italy
Center of Excellence for Climate Change Research/Department of Meteorology, King Abdulaziz University, Jeddah, Saudi Arabia

Noel S. Keenlyside
Geophysical Institute, University of Bergen, Norway
Bjerknes Centre for Climate Research, Bergen, Norway

Riccardo Farneti
Earth System Physics Section, The Abdus Salam International Centre for Theoretical Physics, Trieste, Italy

Edgar G. Pavia
Centro de Investigación Científica y de Educación Superior de Ensenada (CICESE), Ensenada, Mexico

Federico Graef
Centro de Investigación Científica y de Educación Superior de Ensenada (CICESE), Ensenada, Mexico
The Abdus Salam International Centre for Theoretical Physics (ICTP), Trieste, Italy

Ramón Fuentes-Franco
The Abdus Salam International Centre for Theoretical Physics (ICTP), Trieste, Italy

R. C. Ruscica and A. A. Sörensson
Centro de Investigaciones del Mar y la Atmósfera, Consejo Nacional de Investigaciones Científicas y Técnicas, Universidad de Buenos Aires, Argentina

C. G. Menéndez
Centro de Investigaciones del Mar y la Atmósfera, Consejo Nacional de Investigaciones Científicas y Técnicas, Universidad de Buenos Aires, Argentina
Departamento de Ciencias de la Atmósfera y los Océanos, FCEN, Universidad de Buenos Aires, Argentina

Linan Sun and Jiayao Wang
School of Resource and Environmental Sciences, Wuhan University, China
Zhengzhou Institute of Surveying and Mapping, China

Zuhan Liu, Xuecai Bao and Zhaoming Wu
Jiangxi Province Key Laboratory of Water Information Cooperative Sensing and Intelligent Processing, Nanchang, China
School of Information Engineering, Nanchang Institute of Technology, China

Bo Yu
College of Architecture and Environment, Sichuan University, Chengdu, China

Masakazu Taguchi
Department of Earth Science, Aichi University of Education, Kariya, Japan

Kazushi Takemura, Keiichi Ishioka and Shoichi Shige
Division of Earth and Planetary Sciences, Graduate School of Science, Kyoto University, Japan

Yanli Tang and Wenjie Dong
State Key Laboratory of Earth Surface Processes and Resource Ecology, Future Earth Research Institute, Beijing Normal University, China

Lijuan Li
State Key Laboratory of Numerical Modeling for Atmospheric Sciences and Geophysical Fluid Dynamics (LASG), Institute of Atmospheric Physics, Chinese Academy of Sciences, Beijing, China

Bin Wang
State Key Laboratory of Numerical Modeling for Atmospheric Sciences and Geophysical Fluid Dynamics (LASG), Institute of Atmospheric Physics, Chinese Academy of Sciences, Beijing, China
Ministry of Education Key Laboratory for Earth System Modeling, Center of Earth System Science (CESS), Tsinghua University, Beijing, China

Jian Wu and Caiyun Ling
Department of Atmospheric Science, Yunnan University, Kunming, China

Deming Zhao
Key Laboratory of Regional Climate-Environment for Temperate East Asia, Institute of Atmospheric Physics, Chinese Academy of Sciences, Beijing, China

Bin Zhao
Meteorological Bureau of Tengchong, China

Bo Wu and Tianjun Zhou
State Key Laboratory of Numerical Modeling for Atmospheric Sciences and Geophysical Fluid Dynamics, Institute of Atmospheric Physics, Chinese Academy of Sciences, Beijing, China
Joint Center for Global Change Studies (JCGCS), Beijing, China

Jianshe Lin
State Key Laboratory of Numerical Modeling for Atmospheric Sciences and Geophysical Fluid Dynamics, Institute of Atmospheric Physics, Chinese Academy of Sciences, Beijing, China
University of Chinese Academy of Sciences, Beijing, China

Yang Yang, Stuart Moore, Michael Uddstrom, Richard Turner and Trevor Carey-Smith
NIWA, Wellington, New Zealand

Lei Yin
Laboratory of Cloud-Precipitation Physics and Severe Storms (LACS), Institute of Atmospheric Physics, Chinese Academy of Sciences, Beijing, China
School of Earth Sciences, University of Chinese Academy of Sciences, Beijing, China

Fan Ping
Laboratory of Cloud-Precipitation Physics and Severe Storms (LACS), Institute of Atmospheric Physics, Chinese Academy of Sciences, Beijing, China
School of Geography and Remote Sensing, Nanjing University of Information Science and Technology, Nanjing, China

Jiahua Mao
School of Geography and Remote Sensing, Nanjing University of Information Science and Technology, Nanjing, China

Jin-Ho Yoon
School of Earth Sciences and Environmental Engineering, Gwangju Institute of Science and Technology, Gwangju, South Korea

Philip J. Rasch and Hailong Wang
Pacific Northwest National Laboratory, Richland, WA, USA

V. Vinoj
School of Earth, Ocean and Climate Sciences, Indian Institute of Technology Bhubaneswar, Odisha, India

Dilip Ganguly
Center for Atmospheric Sciences, Indian Institute of Technology Delhi, New Delhi, India

Sinclaire Zebaze and Clément Tchawoua
Department of Physics, Faculty of Science, University of Yaoundé I, Yaoundé, Cameroon

Amadou T. Gaye
Laboratoire de Physique de l'Atmosphère Siméon Fongang, ESP-Université Cheikh Anta Diop, Dakar, Sénégal

François M. Kamga
Department of Physics, Faculty of Science, University of Yaoundé I, Yaoundé, Cameroon
Faculté des Sciences et de Technologie, Université des Montagnes, Banganté, Cameroon

Xun Zou
School of Atmospheric Sciences, Nanjing University, China

Yiyuan Li and Jinxi Li
LASG, Institute of Atmospheric Physics, Chinese Academy of Sciences, Beijing, China

M. Zeri and G. Cunha-Zeri
Brazilian Center for Monitoring and Early Warnings of Natural Disasters (CEMADEN), São José dos Campos, Brazil

V. S. B. Carvalho
Instituto de Recursos Naturais, Universidade Federal de Itajubá, Brazil

J. F. Oliveira-Júnior and G. B. Lyra
Departamento de Ciências Ambientais, Instituto de Florestas, Universidade Federal Rural do Rio de Janeiro, Seropédica, Brazil

E. D. Freitas
Instituto de Astronomia, Geofísica e Ciências Atmosféricas, Universidade de São Paulo, Brazil

Haifeng Zhuo
State Key Laboratory of Numerical Modeling for Atmospheric Sciences and Geophysical Fluid Dynamics (LASG), Institute of Atmospheric Physics, Chinese Academy of Sciences, Beijing, China
University of Chinese Academy of Sciences, Beijing, China

Yimin Liu
State Key Laboratory of Numerical Modeling for Atmospheric Sciences and Geophysical Fluid Dynamics (LASG), Institute of Atmospheric Physics, Chinese Academy of Sciences, Beijing, China

Jiming Jin
College of Water Resources and Architectural Engineering, Northwest A&F University, Yangling, China
Department of Watershed Sciences and Plants, Soils and Climate, Utah State University, Logan, UT, USA

R. Manzanas
Meteorology Group, Instituto de Física de Cantabria, CSIC-Universidad de Cantabria, Santander, Spain

Martin Grabner
Department of Frequency Engineering, Czech Metrology Institute, Brno, Czech Republic

Pavel Pechac and Pavel Valtr
Department of Electromagnetic Field, Faculty of Electrical Engineering, Czech Technical University in Prague, Czech Republic

Michell Fontenelle Germano and Jefferson Inayan de Oliveira Souto
Federal University of Para (UFPa), Department of Meteorology, Belém, Brazil

Maria Isabel Vitorino and Júlia Clarinda Paiva Cohen
Federal University of Para (UFPa), Department of Meteorology, Belém, Brazil
Federal University of Para (UFPa), Graduate Program in Environmental Sciences (PPGCA), Belém, Brazil

Gabriel Brito Costa and Mayse Thais Correa Rebelo
Federal University of Western Para (UFOPa), Institute of Biodiversity and Forestry, Santarém, Brazil

Adriano Marlisom Leão de Sousa
Federal Rural University of Amazonia (UFRA), Socio-Environmental Institute, Belém, Brazil

Shenming Fu
Institute of Atmospheric Physics, Chinese Academy of Sciences, International Center for Climate and Environment Sciences, Beijing, China

Deshuai Li
College of Atmospheric Sciences, Lanzhou University, China

Jianhua Sun
Institute of Atmospheric Physics, Chinese Academy of Sciences, Laboratory of Cloud–Precipitation Physics and Severe Storms, Beijing, China

Dong Si and Fuyou Tian
China Meteorological Administration, National Climate Center, Beijing, China

Jian Ling
Institute of Atmospheric Physics, Chinese Academy of Sciences, LASG, Beijing, China

Index

www.ingramcontent.com/pod-product-compliance
Lightning Source LLC
Chambersburg PA
CBHW070155240326
41458CB00126B/4928